DIAGNOSIS OF ACTIVE SYSTEMS

T0214888

THE KLUWER INTERNATIONAL SERIES IN ENGINEERING AND COMPUTER SCIENCE

ANALOG CIRCUITS AND SIGNAL PROCESSING
Consulting Editor: Mohammed Ismail. *Ohio State University*

DIAGNOSIS OF ACTIVE SYSTEMS

Principles and Techniques

by

Gianfranco Lamperti

Dipartimento di Elettronica per l'Automazione,
Università degli Studi di Brescia, Brescia, Italy

and

Marina Zanella

Dipartimento di Elettronica per l'Automazione,
Università degli Studi di Brescia, Brescia, Italy

KLUWER ACADEMIC PUBLISHERS
DORDRECHT / BOSTON / LONDON

A C.I.P. Catalogue record for this book is available from the Library of Congress.

ISBN 978-90-481-7785-1

Published by Kluwer Academic Publishers,
P.O. Box 17, 3300 AA Dordrecht, The Netherlands.

Sold and distributed in North, Central and South America
by Kluwer Academic Publishers,
101 Philip Drive, Norwell, MA 02061, U.S.A.

In all other countries, sold and distributed
by Kluwer Academic Publishers,
P.O. Box 322, 3300 AH Dordrecht, The Netherlands.

Printed on acid-free paper

Dedicated to Alice.

Contents

List of Figures

List of Tables

Preface

This book is about model-based diagnosis of a class of discrete-event systems called active systems. Roughly, model-based diagnosis is the task of finding out the faulty components of a physical system based on the observed behavior and the system model. An active system is the abstraction of a physical artefact that is modeled as a network of communicating automata. For example, the protection apparatus of a power transmission network can be conveniently modeled as an active system, where breakers, protection devices, and lines are naturally described by finite state machines. The asynchronous occurrence of a short circuit on a line or a bus-bar causes the reaction of the protection devices, which aims to isolate the shorted line. This reaction can be faulty and several lines might be eventually isolated, rather than the shorted line only. The diagnostic problem to be solved is uncovering the faulty devices based on the visible part of the reaction. Once the diagnosis task has been accomplished, the produced results are exploited to fix the apparatus (and also to localize the short circuit, in this sample case).

Interestingly, the research presented in this book was triggered a decade ago by a project on short circuit localization, conducted by ENEL, the Italian electricity board, along with other industrial and academic European partners. Despite its link with power transmission networks, the evolution of the research has progressively been decoupled from its origin and the relevant principles and techniques have now a *raison d'être* that extends beyond any particular application domain.

Points of strength of the proposed diagnostic approach include the ability to cope with asynchronism both of unpredictable stimuli coming from the external world and of the state changes of the components of the considered artefact over time. Moreover, to make reasoning even more complex, input events of active systems are assumed to be unobservable (as they often are in the real world).

Several essential concepts proposed in the authors' work have a relevance on their own, independently of active systems. As a matter of fact, the motivations behind the modeling choices for behavioral models of system components and observation descriptions, particularly the classification of uncertainty requirements, are both independent of the context and the task in which they are exploited. In particular, logically and/or source uncertain observations could be considered in model-based diagnosis of static systems too. As to behavioral models, they fulfill very exacting modeling requirements that enable precedence constraints between faults, as well as intermittent and non-intermittent faults to be represented. Furthermore, a unified representation and management of the various causes triggering and affecting the system evolution is provided, these causes being either internal, such as faults, or external, such as exogenous actions (both spontaneous and controlled) and test/repair interventions.

The book is divided into four parts, preceded by an introduction. Concepts and techniques are introduced with the help of numerous examples, figures, and tables. When appropriate, the concepts are formalized in order to state significant theorems. Proofs of such theorems are confined to appendixes of relevant chapters. Besides, appendixes incorporate detailed algorithms expressed in pseudo-code.

Part I is devoted to background. Chapter 1 introduces the fundamentals of model-based diagnosis as originally conceived in the domain of static systems and shows its evolution towards the realm of dynamic (time-varying) systems. Chapter 2 compares several state-of-the-art approaches to model-based diagnosis of discrete-event systems.

Part II focuses on diagnosis of (asynchronous) active systems. Chapter 3 defines the class of considered systems and the basic elements of a diagnostic problem. Chapter 4 presents a monolithic approach to diagnosis of active systems, where the main problem is history reconstruction. Chapter 5 makes the diagnostic method modular, based on the recursive decomposition of the diagnostic problem into smaller subproblems, so as to support the diagnosis of large active systems.

Part III deals with an extended class of active systems that integrate both synchronous and asynchronous behavior, called polymorphic systems. Chapter 6 generalizes both monolithic and modular diagnoses to the extended class of polymorphic systems. Chapter 7 outlines rule-based diagnosis of polymorphic systems, which is supported by off-line knowledge compilation techniques. Chapter 8 introduces monitoring-based diagnosis, where diagnosis is required to be generated at each observable system event rather than at the end of the system reaction (a posteriori).

Part IV is devoted to advanced topics and a final sample application. Chapter 9 introduces the notion of an uncertain observation and defines diagnostic techniques for uncertain diagnostic problems. Chapter 10 defines the class of complex observations and relevant diagnostic methods. Chapter 11 (contributed by Roberto Garatti) extends the class of polymorphic systems by allowing uncertain events in component models. Chapter 12 copes with distributed diagnostic problems, where the behavior of the system is viewed by several (possibly overlapping) observers. Chapter 13 applies the diagnostic techniques to a sample application in the domain of power transmission networks.

Being a research monograph, this book is primarily intended for researchers in the field of Artificial Intelligence. However, since the subject of model-based diagnosis of DESs is shared with the community of Control Theory (as testified by the references), the book is also intended for researchers from this area. We assume that the reader has a basic familiarity with automata and programming concepts. Since the proposed techniques are possibly detailed by pseudo-coded algorithms, the book is also appropriate for engineers and software designers who are to develop tools for the supervision and monitoring of complex industrial systems. Finally, the book can be profitably adopted as an additional text-book for graduate courses (e.g. Artificial Intelligence or Algorithms and Data Structures) or continuing education initiatives.

Brescia, January 2003.

Gianfranco Lamperti
Marina Zanella

Acknowledgments

We are grateful to Roberto Garatti for his contribution to the book and for helping us debug the whole manuscript. We must acknowledge our debt to the people we met in several workshops and conferences, who gave us the opportunity to exchange ideas on the subject. We thank the anonymous reviewers, whose comments and suggestions greatly improved the content and presentation of the book. We express our affectionate gratitude to the staff of Kluwer Academic Publishers, in particular, Betty van Herk and Robbert van Berckelaer, for the professional handling of the project and for the unfailing courtesy.

Introduction

At the start of the new millennium many research directions of Artificial Intelligence are well defined and studied, while some others still require more work. For these areas, clear definitions and formalisms as well as efficient algorithms are needed to push the research edge forward and to lay the foundations for solving real world problems.

The research described in this book is focused on a task, diagnosis, whose automation, although a central topic in Artificial Intelligence and historically one of the first to be tackled, still needs a deeper understanding. As a matter of fact, up to the middle of the 1990s, diagnostic reasoning mainly focused on static and quasi-static systems.

In the last few years its applicability to the larger class of *dynamic systems*, that is, systems wherein the mapping from inputs to outputs changes over time, has been investigated. The present book is a contribution in this direction, in particular, its primary concern is to develop a model-based conceptual framework for fault diagnosis of a class of dynamic systems represented as untimed *discrete-event systems* (DESs).

Fault diagnosis is the task of finding out what is wrong with a physical system. Given the *observation* about how a physical system actually behaves, diagnostic reasoning has to find out what might be abnormal in that system and hence responsible for the observed behavior. Typically this means detecting any discrepancy between the normal (or intended) behavior of the considered system and isolating the causes of such discrepancies, called *faults*.

The expression *model-based diagnosis* cumulatively refers to all approaches to fault diagnosis that exploit an explicit model of the system to be diagnosed, where such a model allows one to make predictions about how the system should behave.

In order to reason about dynamic systems, it is more efficient, and often also reliable, to model some of their aspects as discrete. DESs [Cassandras and Lafortune, 1999] are dynamic systems with discrete inputs and outputs, whose behavior can be described in terms of discrete state changes. Leaving apart some trivial cases of physical systems that are commonly represented as discrete, such as digital circuits and telecommunication networks, most real-world dynamic systems can be viewed as DESs at some level of abstraction. In this scenario, researchers are in charge of finding out modeling primitives and reasoning mechanisms for performing various tasks based on DES models. The effort to model dynamic systems as discrete (for instance, [Lunze, 2000b] describes a method for generating discrete models of continuous variables systems) is largely compensated by the simplicity of reasoning.

In the most general case, the observed behavior of a DES is described by a sequence of input values and the corresponding sequence of output (or state) values, where each input or output value belongs to a finite domain of discrete values (or symbols). The model of a DES has to relate such sequences between each other and to a given initial state. While timed DES models [Cassandras and Lafortune, 1999, Lunze, 2000b] consider the time instants at which input and output symbols occur or their temporal distance, untimed models take into account just the order of input and output symbols.

In the literature, especially in recent years, Petri nets [Viswanadham and Johnson, 1988, Aghasaryan et al., 1998, Schullerus and Krebs, 2001] and automata [Sampath et al., 1995, Sampath et al., 1996, Sampath et al., 1998, Lunze, 2000b, Pencolé, 2000] have been adopted for representing the behavior of DESs for diagnostic purposes. The latter formalism is the favorite one in Artificial Intelligence. In particular, untimed nondeterministic communicating automata are used for representing the behavior of *active systems*, which constitute the class of DESs dealt with in this book.

An active system is a dynamic system modeled as a physically distributed DES whose inputs are not observable, for they are typically unpredictable events occurring in the external world. An active system can be viewed as a network of interconnected components, where each component remains inactive as long as no external event occurs. When excited by an external event, a component performs a state transition and, possibly, generates new events either directed to the external world or to other components that, in their turn, generate further events, and so on. Thus, an asynchronous sequence of state transitions in system components, namely a *reaction* of the active system, is generated. When the reaction finishes, the system goes in a *quiescent* state, where it remains until a new external event occurs, in which case a new reaction is started. Therefore, the behavior of an active system consists of a sequence of reactions separated by quiescent states.

A diagnostic problem inherent to an active system consists of the initial state and the observation (of the outputs only) of the system. The authors' perspective is that diagnosing an active system requires the reconstruction of what happened to it starting from the given initial state, based on the given observation and the (structural and behavioral) models of the considered system. This task is called *history reconstruction*. A reconstructed system history includes the unobserved input values coming from the external world and the state transitions such that the collected system observation is explained and provides important diagnostic information, since each transition is either normal or faulty. Therefore, the proposed technique lends itself to the notion of *explanatory diagnosis* [McIlraith, 1997], according to which diagnosis is the explanation of the behavior of the considered system, rather than the mere identification of a set of faulty components.

Another standpoint for coping with diagnosis of active systems is modularity, which consists in performing history reconstruction by adopting a problem-decomposition, solution-composition strategy, so as independent processing steps can be run in parallel, each of them dealing with a search space whose size is smaller with respect to that of the search space of a monolithic reconstruction.

Although reconstruction and modularity are underlying all the previous and current work on diagnosis of active systems, a variety of diagnostic approaches have been defined. This proliferation of approaches can be traced back to three main reasons:

(1) Extension of the modeling primitives;

(2) Duality of the performed task;

(3) Trade-off between on-line and off-line computation.

As stated above, a diagnostic problem inherent to an active system is characterized by an observation to account for. Therefore, the way a problem is solved depends on the nature of both system and observation models. The diagnostic method changes whenever modeling primitives are updated. The first reason for the proliferation of diagnostic approaches to active systems is that modeling primitives have been extended several times in our research. In fact, a primary aim of the research is generality, that is, the class of considered DESs has to be as large as possible: the size of such a class has progressively increased from the beginning of the research up to now, by introducing new modeling properties in the representation of physical systems. In this respect, active systems initially encompassed *asynchronous* DESs only (as opposed to *synchronous* ones), and this feature differentiated our work at its appearance from previous approaches to diagnosis of DESs in the literature. The chapters in Part II of the book refer to this initial notion of an active system.

Later, the class of considered systems was extended to those integrating both synchronous and asynchronous behavior. Such systems are called *polymorphic* in the present book, so as to distinguish them from the previous class. They are introduced in Part III.

Another major objective of the research is the ability to represent real world situations: in this respect, the modeling primitives relevant to both systems and observations have been progressively extended so as to mirror real world aspects, such as observations affected by loss or noise, uncertain temporal order between observed events, and so on. This is discussed in Part IV of the book.

Coming to the second reason for multiple diagnostic approaches for active systems, consider that distinct diagnostic methods can carry out the task of diagnosis taking different observations as input. In particular, *a posteriori diagnosis* assumes that the considered system has already been observed throughout a time interval, after which the system has come to a halt for its reaction has finished, and the gathered observation has to be processed for finding out possible prior faults. *Monitoring-based* (or *continuous*) *diagnosis*, instead, monitors the operation of the system and attempts simultaneously to detect and isolate faults. Within the research on model-based diagnosis of active systems, distinct methods for a posteriori diagnosis and monitoring-based diagnosis have been proposed, all of them adopting the same modeling primitives and the same reconstruction and modularity perspectives. A posteriori diagnosis of active systems is dealt with in Chapters 4 to 6, while monitoring-based diagnosis is addressed in Chapter 8.

Finally, the third spur that has induced the authors to make several proposals for model-based diagnosis of active systems is that behavior reconstruction, although accomplished in a modular way, may fail to meet temporal constraints on response times. In order to improve the efficiency of reasoning, the knowledge contained in the system model can be compiled once (and for all) off-line, that is, when no diagnostic process is ongoing, and then the result of this compilation can be exploited several times, whenever diagnostic outputs have to be produced, so as to reduce the effort of the needed computation. The diagnostic method embodying this strategy is presented in Chapter 7.

The variety of the above approaches devoted to model-based diagnosis of active systems has progressively emerged during the time span of the research. Such a research stemmed from the experience of the authors in the domain of fault diagnosis of power protection systems [Baroni et al., 1997a, Baroni et al., 1997b, Lamperti and Pogliano, 1997]: this is why this domain is considered in the last chapter of the book for providing an application example. Several published papers witness the evolution of the research.

The initial work on model-based diagnosis of DESs is presented in [Baroni et al., 1998, Baroni et al., 1999]. The emphasis is on the method for performing a posteriori diagnosis, called *on-line* diagnosis in these contributions, whose core is the reconstruction of the evolution of the system during the reaction on the basis of the observation gathered throughout the reaction itself. The main novelty of such a method is the ability to exploit just the behavioral models of system components, without needing the generation of any global behavioral model of the whole system. Another original feature is the class of considered DESs, namely (asynchronous) active systems.

In [Baroni et al., 2000] the concept of a *reconstruction plan* (renamed *reconstruction graph* in the present book) is formalized, that is, constraints are stated for a recursive decomposition of the reconstruction problem at hand into subproblems, and a technique is described for progressively composing the solutions inherent to subproblems into the solution of the whole given problem. The irrelevance of the global behavioral model of any (sub)system holds also when reconstruction is driven by a problem decomposition. The main advantages of this *modular* technique are that subproblems can be processed independently, possibly in parallel, and that it is not necessary to perform the reconstruction inherent to the whole system if attention is focused just on a subsystem. Besides, the aim of modularity is to reduce the computational difficulty of the reconstruction process. Distinct decompositions involve different computational efforts.

In [Lamperti and Zanella, 1999], the class of considered physical systems is extended by including also synchronous DESs as well as DESs that exhibit both synchronous and asynchronous behavior. The successive work [Lamperti and Zanella, 2000b] briefly introduces the concept of a (logically and temporally) uncertain observation and the relevant diagnostic method.

A further approach is proposed in [Lamperti and Zanella, 2000a], which, although based on the same modeling primitives and reasoning mechanisms, encompasses also *off-line* activities, so as to speed up the diagnostic process. Off-line activities largely consist in preprocessing the models of system components in order to automatically generate a set of *diagnostic rules*. Specifically, after the system has been decomposed into (not necessarily disjoint) subsystems (clusters), a set of diagnostic rules can be generated, possibly in parallel, for each cluster. Once the reaction has extinguished, the candidate diagnoses of a cluster whose set of diagnostic rules is available are determined on-line by checking each and every diagnostic rule against the actual cluster observation.

Furthermore, *off-line diagnosis* and *on-line diagnosis* can be integrated within a single diagnostic procedure, even though, in principle, each of them can be used on its own.

Still with the aim of reducing the complexity of on-line history reconstruction by performing the off-line compilation of the model of the considered system, according to a recent work [Garatti et al., 2002] a so-called *prospection knowledge* is generated off-line, which is inherent to the producer-consumer relationships between components as to exchanged events.

The diagnostic method for a posteriori diagnosis of asynchronous active systems is generalized in [Lamperti and Zanella, 2002] in order to support the new notions of an uncertain observation and of a complex observation.

In [Lamperti and Zanella, 2003] a software environment, called EDEN, which prototypes the approach to a posteriori diagnosis of active systems, is presented. The environment integrates a specification language, called SMILE, a model base, and a diagnostic engine. SMILE enables the user to create libraries of models and systems, which are permanently stored in the model base, wherein both final and intermediate results of the diagnostic sessions are hosted as well.

Work is in progress [Lamperti and Zanella, 2001] in order to solve an open problem, that is, how to automatically decompose the reconstruction problem in order to perform a modular reconstruction that is optimal according to some (efficiency) criteria. So far, our claim is that the proposed method for modular history reconstruction always produces the same result, given whichever (valid) decomposition.

This book does not only provide a systematic presentation of the previous theoretical work by the authors on model-based diagnosis of active systems, by making it smooth and organic what may seem fragmentary when reading a collection of papers, it also provides both a more thorough presentation and a scientific enhancement of each specific topic, and proposes novel issues as well.

I

BACKGROUND

Chapter 1

MODEL-BASED DIAGNOSIS

Abstract

Diagnosis has been a challenge for Artificial Intelligence from the middle 1970s. In the last two decades research has focused on model-based diagnosis, according to which a domain-independent reasoning process is capable of exploiting models representing the structure and behavior of the physical system to be diagnosed. The task of diagnosis is aimed at providing an explanation of the observation of a given system. Depending on the notion of an explanation that is adopted, two different characterizations of diagnosis can be given: consistency-based and abductive. Three diagnostic subtasks (and outputs) can be singled out: fault detection, fault isolation, and fault identification. The choice of models is crucial to obtain sound diagnoses as well as to reduce the computational complexity of the reasoning process. One of the major current (and future) research fields in the area is model-based diagnosis of dynamic (time-varying) systems: approaches that either rely on the simulation of the behavior of the system or not can be distinguished.

1.   Introduction

Diagnosis is one of the earliest task to which the application of Artificial Intelligence techniques was attempted and it is still an important research topic for the Artificial Intelligence community, not only for the variety and relevance of its potential applications but also because it is a challenge for several aspects of automated reasoning.

Historically, the first approach to diagnosis was *rule-based* (or *heuristic*): production rules, possibly organized according to decision trees, represented empirical knowledge about the associations between symptoms and faults in a given physical system to be diagnosed. Such a knowledge embodied rules of thumb elicited from human experts or retrieved from large data samples. The reasoning mechanism was capable of tracing back observed symptoms to their causes. MYCIN [Shortliffe, 1976], which was developed in the 1970s to assist physicians both in determining possible causes of bacteraemia and meningitis and in selecting an appropriate therapy, is perhaps the best known of the early rule-based systems. However, rule-based diagnosis is affected by a number of disadvantages:

- It can cope only with a bounded number of known symptoms and faults;

- Knowledge acquisition from human experts is difficult and time-consuming;

- Empirical associative knowledge is huge and hard to maintain;

- The available empirical knowledge is only inherent to the malfunctions that have already occurred, thus it is unlikely such a knowledge encompasses rare faults;

- Empirical knowledge does not exist for newly-born artifacts;

- Empirical knowledge is specific to the considered physical system and, in general, it cannot be reused for other systems;

- A large set of probabilities is possibly needed, as discussed below.

Rule-based diagnostic systems determine either all the possible sets of faults, each of which is a possible cause of the observed symptoms, or just the most likely set of faults, given the observed symptoms. In order to estimate the likelihood of any set of faults, a large set of probabilities is needed, $2^{n+m} - 1$ distinct probabilities in the worst case, where n is the number of faults and m the number of symptoms. The probability of a fault is a piece of information difficult to elicit and subjective in nature as it depends on the personal experience of each considered expert and on his/her recalled perception of such an experience.

Summing it up, the knowledge exploited by rule-based systems is *shallow* as it is subjective and/or derived from recorded experience, it does not take into account any piece of information about how a system was designed (in case it is an artifact) and how it is expected to behave.

By contrast, the chronologically successive approach of *model-based* diagnosis exploits *deep* knowledge, that is, objective knowledge about the structure and behavior of the considered system, which takes into account the physical principles governing the functioning of the system. Such a knowledge is easier to elicit than empirical knowledge as it can be drawn from design documents, textbooks, and first principles.

Therefore, such a knowledge is available also for newly conceived systems that have never been put into use before. Moreover, deep causal knowledge enables one to distinguish whether two symptoms are dependent or independent when a certain fault is assumed. Thus the number of probabilities to be considered in order to determine the most likely set of faults is reduced with respect to rule-based diagnosis.

Selected readings of the early years of model-based diagnosis can be found in [Hamscher et al., 1992].

The remaining of the chapter is organized as follows: Section 2 provides the basics of model-based diagnosis. In particular, Section 2.1 briefly presents the first fundamental contributions in the field, all addressing static systems; Section 2.2 classifies from a logical viewpoint the solutions to diagnostic problems into consistency-based and abduction-based; Section 2.3 identifies three diagnostic subtasks of model-based diagnosis of static systems; Section 2.4 introduces the principles of abstraction in modeling a physical system. Section 3 surveys some contributions inherent to the challenging task of model-based diagnosis of dynamic systems. Referenced contributions, both when dealing with static and dynamic systems, are from the Artificial Intelligence literature.

2.   Fundamentals of Model-Based Diagnosis

Model-based diagnosis brings to a crisp unambiguous definition of a *symptom*, according to which a symptom is a discrepancy between the observed behavior and the predicted behavior of a system. The *observed behavior* is the set of actual values of output variables, the *predicted behavior* is the set of the values of the same output variables computed based on the modeled normal behavior of the system. In both cases, values of input variables are assumed to be known.

A solution produced by a diagnostic session is a set of (*candidate*) *diagnoses*, each diagnosis being a set of faulty components, or a set of specific faults assigned to components, that *explains* the observation.

What is considered a component depends on the granularity level of the *structural model* of the system, which describes the topology of the system in terms of constituent parts (components) and interconnections among them. The structure of an artifact is available, at least theoretically, from its project, e.g. from CAD documents, wherein architectural modules are highlighted.

In general, however, the components to be considered by the diagnosis task are identified based on repair/replacement actions that can be carried out to fix the system. In fact, diagnosis is not a goal in itself, the ultimate goal being instead to increase the availability of the considered physical system, by repairing it in case it is affected by faults. Therefore, for supporting the task of diagnosis, it does not make sense to split a (macro)component C into several components at a finer granularity level if, in case any of these components contained in C is faulty, the whole unit C is replaced.

A fundamental assumption underlying most model-based diagnosis approaches is that the modeled structure of the system never changes, that is, the existing interconnections among components cannot be broken and no new interconnection can be added. When this is assumed, model-based diagnosis cannot find out any fault that changes the structure of the considered system, instead it can find out only faults that change the behavior of components.

What is considered a fault depends on the abstraction level of the *behavioral model* of each component of the considered system. The *normal behavior* of such a system, that is, how the artifact is intended to work according with the purpose it was designed and built for, is known from design documents. However, in model-based diagnosis such a normal behavior in general is not given in a global (system-level) description, instead it has to implicitly emerge from the composition of the explicit normal behaviors of its components.

Some of the basic principles of model-based diagnosis, as surveyed in [Mozetič, 1992], are the following (where the first two are a consequence of the "no function in structure" principle in the literature):

- The behavior of the system to be diagnosed is expressed by the behavioral models of its components;

- The behavioral model of a component is independent of any composite device the component may belong to;

- The interpreter of behavioral models is independent of the application domain of the considered system;

- The diagnostic task is independent of the specific representations adopted by structural and behavioral models.

Hence the architecture of a model-based diagnostic system is modular: it includes an interpreter of models, which is specific to the adopted modeling language, a diagnostic algorithm, and a library of models.

The above principles determine several advantages of model-based diagnosis over rule-based diagnosis, typically the reusability of behavioral knowledge, as the same model of a component can be used several times for diagnosing distinct systems the component belongs to. In principle, the same behavioral model of a component can be used also for accomplishing tasks other than diagnosis, provided that the module of the diagnostic algorithm in the architecture is replaced by a module performing another model-based task.

Another asset of model-based diagnosis is that the diagnostic reasoning mechanism is independent of the specific system to be diagnosed: the same diagnostic architecture can be used to diagnose distinct systems, even belonging to different application domains, provided that the models inherent to such systems are contained in the model library.

Whilst the component-oriented ontology sustained in the above principles is the most used in model-based diagnosis, a different conceptualization of domain knowledge, namely *causal networks*, can also be adopted (see e.g. [Console and Torasso, 1991a, Fattah and Provan, 1997]). Such networks, however, are not covered in this book.

2.1   Milestones

A glance to the *Artificial Intelligence* journal allows one to catch the major aspects of the research trends in diagnosis. Volume 24 highlights that in 1984 the favorite modeling style was that adopted by qualitative physics [de Kleer and Brown, 1984, Forbus, 1984, Kuipers, 1984], and digital circuits were the favorite application domain [de Kleer, 1984, Williams, 1984, Genesereth, 1984, Barrow, 1984]. A wide debate on the role of qualitative physics, as a modeling instrument for diagnostic applications, carried out in 1986, is hosted by volume 29, wherein [de Kleer and Brown, 1986] is faced by [Iwasaki and Simon, 1986a, Iwasaki and Simon, 1986b].

However, the manifesto of model-based diagnosis, which from 1987 has deeply influenced all research efforts on the topic, is the content of volume 32. A fundamental work that can be found there, Reiter's theory of diagnosis from first principles [Reiter, 1987], provides the first logical formalization of a diagnostic problem and of its solutions, as well as the relevant terminology adopted thereafter.

A system is a pair $(SD, COMPS)$, where SD is the (logical) description of the normal behavior of the system, and $COMPS$ is a set of constants representing its components. A diagnostic problem is a triple $(SD, COMPS, OBS)$, where OBS is the observed behavior of system $(SD, COMPS)$, i.e. the observed values of its output variables. OBS points out a misbehavior if there is a discrepancy between OBS and the predicted behavior, where, as already remarked, the predicted behavior is the set of values of the output variables of the system computed, based on SD, assuming that all of the components in $COMPS$ behave normally. Formally, there is a misbehavior if OBS is inconsistent with

SD, namely if

$$SD \cup OBS \models \bot .$$

A (candidate) diagnosis Δ is a minimal set of components in *COMPS*, $\Delta \subseteq COMPS$, such that, assuming that all of the components in Δ behave abnormally and all of the others behave normally, is consistent with $SD \cup OBS$, that is,

$$SD \cup OBS \cup \{Ab(C) \mid C \in \Delta\} \cup \{\neg Ab(C) \mid C \in (COMPS - \Delta)\} \not\models \bot,$$

where $Ab(C)$ is the predicate that represents that component C is abnormal. The requirement of minimality (principle of *parsimony*) imposed on each diagnosis means that no subset of a diagnosis is a diagnosis.

In the general case, the problem of determining the diagnoses corresponding to a given diagnostic problem is undecidable, as problem solving relies on a consistency check. However, there exist several practical cases wherein consistency is decidable and, therefore, diagnoses can be computed.

The resolution algorithm provided by Reiter needs a theorem prover capable of determining all the *conflicts* for the given diagnostic problem. The concept of a conflict, taken from [de Kleer, 1976], denotes any subset of components in *COMPS* such that the assumption the all of these components behave normally is inconsistent with $SD \cup OBS$ (that is, a component at least in a conflict is bound to be behaving abnormally). A *hitting set* for a collection of sets is defined as a set that contains at least one element from each set in the collection. All (minimal) diagnoses for a diagnostic problem can be computed as the minimal hitting sets for the collection of the minimal conflict sets of the problem.

Given a diagnostic problem, Reiter's algorithm may produce several competing diagnoses: a *measurement*, i.e. observing the value of a variable that was not observed before, is useful to discriminate between them. Another concept that is useful to the same purpose is *prediction*, meant as the predicted behavior of the system in correspondence of a diagnosis, when assuming that all of the components in the diagnosis behave abnormally and all of the others behave normally. Measurements and predictions, altogether, can lead to refute some diagnoses (or even all of them). However, if a diagnosis is refuted, there may be additional candidate diagnoses, each being a superset of the refuted diagnosis, that have not been computed by the algorithm. Thus, the diagnostic task performs a non-monotonic reasoning, as a candidate diagnosis may not survive a new measurement.

A domain-independent architecture, called GDE (General Diagnostic Engine), for accomplishing the diagnostic task, as defined by Reiter, is provided by de Kleer and Williams [de Kleer and Williams, 1987]. The same contribution faces the problem of how to choose the measurements to be performed in order to discriminate among competing diagnoses. Nowadays most model-based diagnostic systems can be regarded as some variant of GDE.

Given the probabilities of components to get faulty, and assumed that they get faulty independently from one another, GDE assigns an initial probability to each diagnoses it computes. Such probabilities are then progressively updated according to the Bayes law as new measurements are gathered. Fault probabilities are exploited also to select the measures to be performed, so as to minimize their number (and cost). The idea is to evaluate the discriminatory power of each hypothetical measurement with respect to the current competing diagnoses. Such an evaluation is based on Shannon's entropy [Shannon, 1948].

Unfortunately, fault probabilities may not be available or they may be unknown. The approach proposed in [de Kleer, 1990] copes with this lack of knowledge under some circumstances.

2.2   Characterizations

One of the main advantages of model-based diagnosis is that logical theories, in some variants of the first-order predicate calculus, have been used to characterize the solutions of a diagnostic problem in a formal unambiguous way. In this respect, all of the approaches in the literature define a diagnosis as an 'explanation' of the observed behavior. However, distinct approaches differ in the form and content of the system description and/or in the adopted notion of an explanation. In particular, based on two distinct notions of an explanation, two characterizations of diagnosis are given: consistency-based diagnosis and abductive diagnosis.

The diagnosis task as defined by Reiter is consistency-based, that is, a diagnosis is an explanation of the observation of the considered system in the sense that it is consistent with it, or, in other words, it does not contradict it. In the original consistency-based approach, *SD* consisted in the description of the normal behavior of the system only. Therefore, a (minimal) diagnosis does not give any information about how faulty components misbehave. Besides, each superset of a minimal diagnosis is still consistent with the observation and, then, is a consistency-based diagnosis itself. In fact, since only knowledge about how the components behave normally is exploited, no prediction can be made for the behavior of components that are hypothesized to be faulty and, therefore, no contradiction with the observation arises. That is why Reiter appeals to parsimony, which seems to be reasonable in practice.

Later the consistency-based approach was extended to possibly account for faulty behaviors of components too, which specify how the components work when certain faults occur. That is the case in SHERLOCK [de Kleer and Williams, 1989] and GDE+ [Struss and Dressler, 1989], both extensions of GDE.

However, the lack of knowledge on faulty behavioral modes does not result in any missing diagnosis, which is the major asset of consistency-based diagnosis. If the faulty behavioral models of a component are not specified, an undefined abnormal behavior is assumed (in the original Reiter's approach every component features an undefined abnormal behavior). On the other side, the use of an undefined faulty mode may lead to physically impossible diagnoses, which is a drawback of consistency-based diagnosis. In fact, the check for consistency with the observation may implicitly assign to a component any behavior other than the normal behavioral modes stated in *SD*. This, on the one hand, has the potential for finding any faults, including unforseen ones, on the other, it allows for the assignment of physically impossible behaviors to components.

If faulty behaviors are considered, a minimal consistency-based diagnosis is a set of assignments of abnormal behavioral modes to faulty components, where each of such components is either assigned a specific fault or it is generically assumed to exhibit a behavior that is not normal. In case faulty behaviors are considered, a superset of a minimal diagnosis is not bound to be a diagnosis itself, as shown in [de Kleer, 1990].

In the context of the basic consistency-based approach, given a system consisting of k components, the search space of the diagnostic process contains 2^k candidates. The size of the search space grows when the models of faulty behaviors are considered. A way to reduce the size of the search space is to compute only diagnoses with small cardinality. For instance, there are only k candidates for a *single fault* diagnosis, this being a diagnosis of cardinality one, i.e. consisting of exactly one faulty component.

A diagnosis containing several faulty components is a *multiple fault* diagnosis. Assuming that components get faulty one independently from the other, single fault diagnoses are more probable that multiple fault diagnoses.

Abductive approaches to diagnosis [Poole, 1989, Console et al., 1989] require that both normal and faulty behaviors of components be available and represented in *SD*. Unlike consistency-based diagnosis, an abductive diagnosis Δ cannot generically state that a faulty component exhibits a behavior which is not normal: it has to show the specific faulty behavior. In fact, an abductive diagnosis is a minimal set of assignments of specific behavioral modes to normal and faulty components that both is consistent with $SD \cup OBS$ and, together with SD, entails the observation, that is,

$$SD \cup \Delta \cup OBS \not\models \perp,$$
$$SD \cup \Delta \models OBS.$$

According to this definition, abductive diagnosis obeys to a stronger notion of an explanation with respect to consistency-based diagnosis. The price to pay is that all failure modes have to be known a priori. If fault models are not complete, abductive diagnosis may fail to generate diagnoses. Building models of faulty behaviors of components is usually much harder than describing the normal behavior only. However, there exist domains, such as, for instance, the medical one, wherein the converse holds. That is why abductive diagnosis has been explored mainly in medicine.

Since abductive diagnosis relies on a consistency check and on entailment, it is in general undecidable, like consistency-based diagnosis.

A generalized definition of diagnosis, which is abduction-based and, at the same time, exploits consistency constraints, is given in [Console and Torasso, 1991b]. Such a definition requires that the complete behavior of each component be specified. The generalized definition encompasses indeed a whole spectrum of definitions, ranging from consistency-based diagnosis, on the one end of the spectrum, to abductive diagnosis, on the other. In between, there are several intermediate approaches characterized by the subset of observations that is entailed, that, starting from the null set in consistency-based diagnosis, reaches the whole set in abductive diagnosis. Given a diagnostic problem, the task of abductive diagnosis results in a smaller set of candidate diagnoses than the task of consistency-based diagnosis. Therefore the former task is better than the latter in case the complete behavior of each component of the considered system is available.

If the complete behavioral description of a component is not available, one can introduce an *unknown fault mode*, which represents all the possible faulty ways the component may behave other than the ones that have been explicitly modeled. This idea is explored in GDE+ [Struss and Dressler, 1989] and in [Console et al., 1989, de Kleer and Williams, 1989, Friedrich and Lackinger, 1991] as well.

2.3 Subtasks

The classical theory of model-based diagnosis briefly described in the previous sections addresses a limited class of systems, called *static sys-*

tems, whose nominal behavior can be specified as a time invariant mapping from some input variables to some output variables. In model-based diagnosis of static systems, three diagnostic subtasks can be theoretically singled out:

- *Fault detection*, which consists in finding out *whether* a fault has occurred or not;

- *Fault isolation* (or *localization*), which ascertains *where* a fault has occurred;

- *Fault identification*, which determines *what* fault has occurred.

No diagnosis of static systems can be performed without fault detection, which is the only diagnostic subtask that interfaces directly the system being diagnosed. This subtask is typically performed by means of a discrepancy detector, which compares the measured outputs from the system with the predicted values from the model, according to same predefined metric.

If, instead, one or more discrepancies (symptoms) have been detected, fault isolation and/or fault identification is performed. The location of the fault depends on the structural description of the plant, that is, the constituent responsible of the fault is part of a process, a component, or simply a (set of) constraint(s).

For instance, in consistency-based diagnosis, in order to perform fault detection and fault localization, it is checked whether the observations inherent to the actual device behavior are consistent with the behavior predicted by the model of the correct device behavior. This is all can be done when fault models are not available. If, instead, they are available, in order to perform fault identification, the consistency of observations with models of faulty behavior is checked.

2.4   Modeling Issues

Modeling is defined by Poole [Poole, 1994] as "going from a problem P to a representation R_P of P in the formal language so that the use of the inference relation for the representation will yield a solution to the problem". Models are the heart of model-based diagnosis: they strongly influence the efficiency of the diagnostic process and the effectiveness of its results. Therefore, providing a good model is a crucial and central responsibility that involves several issues.

First, model-based diagnosis, while identifying faults, implicitly assumes that the models it exploits are dependable/reliable. However, models differ from the real systems they represent as every model depends on its own set of simplifying assumptions and approximations.

A model is good to a diagnostic purpose if such simplifications and approximations are acceptable in the case at hand. For example, the adopted models may feature too coarse a granularity for the current purpose. Such a modeling problem can be solved by providing more detailed models, which lead to a better approximation of the real system. This, however, inherently increases the computational complexity of the task, an effect that is particularly relevant when we scale up model-based diagnosis to large devices.

A means to cope with the complexity of model-based diagnosis is abstraction [Mozetič, 1992, Struss, 1992], which can be either structural or behavioral, depending on whether it is applied to structural or behavioral models. Abstraction is founded on the use of multiple models.

Specifically, *structural abstraction* composes several components at a level of abstraction into a single component at the next higher level. Many diagnostic systems that are based on structural abstraction use a predefined hierarchy of these models. For example, [Hamscher, 1991] can efficiently search a tree of models by focusing on components up in the hierarchy and expanding them into their sub-components only in case they have been isolated as faulty. Thus, while a system may consist of perhaps thousands of components at the lowest level, the diagnostic system only has to check a small number of them before reaching a diagnosis.

Behavioral abstraction, instead, consists in describing the behavior of the considered system at several different precision levels, such as, for instance, qualitative values at a lower level and quantitative at a higher one. Model-based diagnosis is first performed on the whole system at the higher abstraction level, which is computationally cheaper. Then, once a faulty component has been isolated, a more detailed behavior of such a component only is considered for a finer fault identification.

Since human experts, in making a diagnosis, exploit not only the knowledge of the structure and behavior of the considered system but also other epistemological kinds of knowledge, the use of two additional models, teleological and functional, to perform model-based diagnosis is suggested in [Chittaro et al., 1993]. The *teleological model* specifies the purpose of the system at hand by describing its goals and the relationships among them. The *functional model* links the teleological primitives with the behavioral ones, which are more detailed. The advantage is that the diagnostic process, starting from the teleological model, whose search space is smaller, and then passing through the functional model, can at the end focus attention on those parts only of the behavioral model that are responsible for the faulty behavior.

Research is ongoing in order to make explicit (as far as possible) the complete set of properties of any model, in order to support model adjustment, model switching or simply the combination of several models of the same artifact.

3.   Model-Based Diagnosis of Dynamic Systems

In the 1980s it was common to assume the system to be diagnosed as static and its observation was given at a single time point [Reiter, 1987, de Kleer and Williams, 1987]. A static behavioral model, possibly encompassing several behavioral modes, does not capture temporal aspects, it is just a set of constraints that restricts the set of states possible under each respective mode. When a system is observed, it is in a state and keeps on being in the same state as far as its outputs are measured, even if some additional measurements are accomplished later on in order to discriminate among candidate diagnoses.

Nowadays, research is focused on *dynamic systems*, that is, systems showing a complex behavior changing over time, and such systems are the subject of this book. A dynamic model of a component characterizes the evolution of a behavior over time, that is, it constrains not only the states of the component but also the relations between states across time.

In the domain of dynamic systems, the conceptual articulation and the complexity of the diagnosis task increases, since a temporal diagnosis has to account for both the values in the observation and their temporal location. In order to accommodate any distinct approach to temporal diagnosis within a lattice of definitions, the two aspects mentioned above, namely the presence of observed values and their temporal location, can be considered separately [Brusoni et al., 1998].

In principle, the approaches adopted for static systems could be applied also to dynamic ones. For example, consistency-based diagnosis also applies to dynamic systems, however it requires checking the consistency of device behaviors over time with the behaviors predicted by a dynamic model of the device. This amounts to:

- Tracking the actual behavior of the considered system over time;

- Simulating the modeled behavior;

- Comparing the actual outputs with the results of the simulation.

The simulation task and the comparison of behavior sets can often be computationally prohibitive, particularly when qualitative simulation yields ambiguous results or when fault identification requires simulation of many fault scenarios. In spite of this, the above idea underlies several approaches, sometimes called *simulation-based*, to model-based diagnosis of dynamic systems. In fact, a large group of approaches to diagnosis of dynamic systems are mainly based on modifications and extensions to Reiter's theory and tend to keep reasoning about temporal behavior separate from diagnostic reasoning.

In these attempts, attention is mainly focused on the problem of fault detection in the context of the dynamic behavior exhibited by the real system, whereas, once a fault has been detected, a variant of Reiter's diagnostic algorithm, normally not including significant temporal reasoning features, is applied. Therefore reasoning about temporal behavior in these approaches mainly concerns *system monitoring*, this being the activity of looking after the system and detecting symptoms over time, rather than true system diagnosis, that is, the activity of finding explanations for the symptoms. When the system is operating correctly there will be no discrepancy and model-based diagnosis will in effect be operating as a *condition monitor*.

In order to substantiate the above general claims, we consider some of these approaches and systems individually. The work of Pan [Pan, 1984] is among the earliest that address the problem of diagnosis of dynamic systems. It adopts a technique, called predictive analysis, which is capable to envision the possible behaviors of a mechanism in response to a given perturbation (namely, an hypothesized fault). Predictive analysis is used for verifying diagnostic hypotheses, which are supposed to be produced by a separate diagnostic engine. The diagnostic process is therefore carried out without any special diagnostic temporal reasoning oriented to dynamic systems.

Hamscher and Davis [Hamscher and Davis, 1984] consider the problem of the diagnosis of circuits with states and propose a representation that can be used to simulate circuit behavior over time (i.e. over a sequence of clock cycles). The authors remark that, in order to effectively diagnose a sequential circuit, it is necessary to examine its temporal evolution, since simply "hypothesizing a failure in a part means removing constraints in many time-slices. This in turn tends to leave large gaps in which it is impossible to deduce what actually happened". Indeed, this remark can be extended to the diagnosis of any dynamic system. In order to carry out diagnosis, Hamscher and Davis propose, however, to consider separately each time slice (i.e. each clock cycle), since for each slice the circuit can be seen as "a strictly combinational device", to which standard diagnostic methods can be applied. Also in this case, therefore, diagnosis is performed considering single time instants.

In a subsequent work, Hamscher [Hamscher, 1991] carries out a more detailed analysis about the diagnosis of complex sequential circuits, related to the development of a diagnostic system called XDE. Great attention is paid to the problem of representational abstraction, at the structural, temporal, and behavioral level. Again, models of component behaviors are used to make predictions about system output, whereas diagnosis in XDE is carried out according to the classical GDE scheme, extended to include fault models and to allow switching between different levels of abstraction.

An explicit separation between temporal and diagnostic reasoning is also adopted in [Console et al., 1992], by relying on the limiting assumption that the manifestations of a behavioral mode are instantaneous with respect to transitions between different modes, in other words, that observations concern only steady behavioral modes whereas there are no observations concerning transitions.

The work of Dvorak and Kuipers [Dvorak, 1992, Dvorak and Kuipers, 1992], called MIMIC, addresses continuous-variable dynamic systems, typical of industrial process plants whose operation is continuously monitored for abnormal behavior. The proposed approach includes the use of qualitative simulation to make a prediction of system behavior, which is then matched against the observations. If a discrepancy is detected, a fault hypothesis is generated via a precompiled decision tree, and qualitative simulation is used in turn to verify whether the hypothesis produces predictions consistent with observations or not. Also in this case, therefore, diagnosis is kept out of the temporal reasoning process. Moreover this approach depends on some important limiting assumptions: faults must occur one-at-a-time with respect to the sampling rate, they are assumed to be non-interacting, and only predefined combinations of faults are guaranteed to be correctly diagnosed.

A similar approach is adopted by Ng [Ng, 1991], that uses qualitative simulation to generate subsequent qualitative states of a dynamic system and then applies Reiter's classical diagnostic algorithm to generate fault hypotheses separately for each state: the iterated application to different states allows the progressive reduction of the set of hypotheses consistent with all states.

Along the same lines, Lackinger and Nejdl [Lackinger and Nejdl, 1991] introduce the DIAMON system for monitoring and diagnosis. Though this system includes hierarchical modeling and does not rely on fault models, its basic idea is similar to previously described approaches: it uses qualitative simulation to produce, step by step, predictions about the system, whereas classical diagnosis is applied just to a single state of the system when an inconsistency is detected.

Dressler and Freitag [Dressler and Freitag, 1994] also consider the problem of integrating monitoring and diagnosis. The novelty of their approach, however, is not in the diagnostic system, which is based on an assumption-based truth maintenance system (ATMS) for computing minimal conflicts, the same as GDE, but in a method for sharing predictions made at different time instants and logical contexts, thus saving computations.

Guckenbiehl and Schäfer-Richter, in the SIDIA system [Guckenbiehl and Schäfer-Richter, 1990], view the diagnosis of a dynamic system as the task of reconstructing the history of the system starting from some temporally located information about system attributes. They exploit an episode propagator to derive the values assumed by system variables over time from observations. The episode propagator is basically a forward-chaining system using production rules. The output of the propagator may evidence the presence of inconsistent episodes requiring diagnosis: then a classical consistency-based diagnosis scheme inspired by GDE is applied.

Friedrich and Lackinger [Friedrich and Lackinger, 1991] carry out a general analysis of the concept of temporal misbehavior and examine the distinction between *permanent faults* (i.e. faults having a permanent duration) and *transient faults* (i.e. temporally finite deviations from the normal behavior, also called *intermittent faults* , e.g. in [Raiman et al., 1991]). According to this analysis, they remark that "in order to distinguish between temporal properties of failures we have to model them explicitly. This is due to the fact that we can not draw conclusions about the temporal nature of failures by solely considering the temporal properties of their manifestations". Failures possibly affecting a component are therefore explicitly modeled and temporal diagnosis is then defined as a set of temporally located failure states, associated with components, such that they are consistent with observations. However, the behavior associated with each failure state is represented through a set of first order sentences, which are independent of time. Therefore this approach does not encompass failures whose consequences span over time, i.e. temporally evolving faulty behaviors. Moreover, similarly to the previously described approaches, diagnostic reasoning is carried out by verifying separately the consistency of the assumptions related to each relevant time instant.

A common feature of all the above approaches is that a simulation of the behavior of the dynamic system is carried out in order to compare, instant by instant, the simulated behavior with the observations. More recently, diagnostic tasks that disregard any temporal information, and, therefore, avoid simulation, have successfully been adopted also for some classes of dynamic systems.

According to the classification of these tasks provided in [Brusoni et al., 1998], *single-snapshot diagnosis* is performed on the data gathered by taking a single snapshot of the behavior of the device, while *symptom-collection diagnosis* is performed on a set of observations collected within a temporal window, without keeping track of their temporal locations and extents. In both cases, each candidate diagnosis explains the union of all the observed data, provided that they are globally consistent. Therefore these approaches avoid simulation and generate diagnostic candidates based on checking consistency of the model with observed states only (as opposed to observed behaviors, i.e. sequences of states). Consistency of observed states with the model is a necessary condition for the observed behavior to be consistent with the model. Struss [Struss, 1997] derives the assumptions under which it is also a sufficient condition. In particular, he presents theoretical foundations and practical considerations supporting the use of state constraints for consistency checking, while disregarding temporal behavior constraints, or using them for purposes other than simulation.

Struss proposes *state-based* diagnosis, by showing, both at a theoretical level and in a practical application, that it is possible to discriminate between different behavioral modes by considering only a single observation about the state of the system, without relating it to the system temporal behavior. This approach relies on the assumption that the behavioral mode of the system does not vary during the time interval considered for observation: in fact, it is assumed that the observation is sufficient to identify the actual behavioral mode (possibly a faulty one) of the current state. This limits the applicability of state-based diagnosis to a specific class of dynamic systems.

Other restrictions of the state-based approach concern the nature of the temporal constraints within behavioral models: it is assumed that only *CID-constraints* are present, namely constraints concerning continuity, integration and derivatives of single variables over time, whereas other kinds of constraints, for instance those binding the values of different variables, are not considered. The above assumption is related to a property called *homogeneity*, which states that temporal constraints are exactly the same in all the different behavioral models of the system. These assumptions are fairly restrictive: it is therefore possible to conclude that results about state-based diagnosis, though providing a substantial advancement for the diagnosis of a specific class of dynamic systems, should not be regarded as generally applicable.

In a contribution by Nejdl and Gamper [Nejdl and Gamper, 1994], diagnosis encompasses the reconstruction of a set of assumptions about the temporal arrangement of behavioral modes of the system. The reconstructed arrangement of behavioral modes has to entail all the collected observations, it has to be consistent with negative observations (obtained by the Closed World Assumption with respect to actual observations), and it has to fulfill constraints concerning the physically allowed behavioral evolutions of the system.

4.   Summary

- *Model-based diagnosis*: reasoning process, aimed at troubleshooting, that exploits models, typically structural and behavioral, of the physical system to be diagnosed.

- *Structural model*: topological description of a physical system, in terms of components and interconnections among them.

- *Behavioral model*: description that allows to make predictions about how the system should work; it can encompass also faulty operating modes.

- *Symptom*: discrepancy between the observed behavior (i.e. the actual values of the output variables) and the predicted normal behavior of a system.

- *Diagnosis*: set of behavioral modes assigned to components that explains the observation.

- *Consistency-based diagnosis*: diagnosis that is consistent with the observation, in the sense that the behavior that is predicted when such a diagnosis is assumed does not contradict the observation.

- *Abduction-based diagnosis*: diagnosis that entails the observation, in the sense that the behavior that can be predicted, when that diagnosis is assumed, entails the observation.

- *Abstraction*: use of multiple models, at different levels, in order to reduce the complexity of the diagnosis task.

- *Static systems*: systems that are described by means of time invariant mappings from input variables to output variables; they are addressed by the classical theory of model-based diagnosis.

- *Dynamic systems*: systems that are described by means of time-varying behavioral models; they are a challenging research topic.

- *Simulation-based approaches*: approaches to model-based diagnosis of dynamic systems that adopt the same paradigm used for diagnosis of static systems; that is, they simulate the normal behavior of the system, compare the result of simulation with the observation of the real system and, if there is a discrepancy, apply to it the same diagnostic algorithm conceived for static systems.

- *State-based diagnosis*: approach to model-based diagnosis of a limited class of dynamic systems that avoids simulation.

Chapter 2

DIAGNOSIS OF DISCRETE-EVENT SYSTEMS

Abstract

Model-based diagnosis of DESs encompasses many challenging aspects and it is of practical interest in many real contexts, therefore it has spurred research both in the Artificial Intelligence and Automatic Control communities. In this chapter a comparative analysis of several approaches in the literature is hinted, by pointing out both similarities and differences among them. The adopted comparison criteria are the specific subclass of DESs they address, the nature of the diagnosis task they support, the algorithms they propose, and the diagnostic output they produce. Later, five distinct state-of-the-art approaches other than the one that is the topic of the book, all based on untimed DES models, are briefly surveyed and their ways of working is illustrated based on a common example.

1. Introduction

The investigation on the task of diagnosis of DESs was prompted by some preliminary works within the Automatic Control area [Srinivasan and Jafari, 1993, Lin, 1994] and the Artificial Intelligence area [Cordier and Thiébaux, 1994]. Since it was soon discovered that, as stated in [Förstner and Lunze, 1999], "qualitative considerations have sufficient information to discriminate correct and faulty behaviors", from the middle '90s, the topic has been receiving an increasing interest from both communities (e.g. [Lamperti and Pogliano, 1997, Console et al., 2000a, Console et al., 2000b, Förstner and Lunze, 2000]).

The relationships between the two communities have been tight enough as far as diagnosis of DESs is concerned. In fact, a fundamental work [Sampath et al., 1995, Sampath et al., 1996, Sampath et al., 1998], proposing the so-called *diagnoser approach* (see Section 2) and com-

21

ing from the Control Engineering field, has been the basis not only for a subsequent contribution in the same field, the *decentralized protocol approach* [Debouk et al., 2000a, Debouk et al., 2000b] (see Section 5), but also for further research in the AI field [Rozé, 1997].

Still more, the *decentralized diagnoser approach* [Pencolé, 2000] (see Section 3) combines the diagnoser approach with the AI approach [Baroni et al., 1999, Baroni et al., 2000], called *active system approach* in the present chapter, to which this book is devoted.

Moreover, most approaches to diagnosis of DESs in the literature, whatever the area their authors come from, share many aspects, such as compositional modeling, modeling primitives, and ontology of time. In fact, in most approaches, systems are structurally modeled as distributed. The behavior of system components, being typically nondeterministic, owing to faults, parameter shifts, and also state quantization in case of discrete (qualitative) descriptions of analog systems, is represented by means of finite nondeterministic automata.

As to time abstraction, only few proposals, among which two [Chen and Provan, 1997, Rozé, 1997] based on the diagnoser approach, represent time constraints explicitly, while in all the others, including the diagnoser approach, the decentralized protocol approach, the active system approach and the decentralized diagnoser approach, time is regarded as a sequence of states and the only temporal relation is the one induced by their ordering.

However, distinct approaches differ in the specific subclass of DESs they address, in the nature of the diagnosis task they support, in the algorithms they propose, and in the diagnostic output they produce.

Basically, all contributions deal with synchronous systems, while the active system approach copes with asynchronous ones and, as will be introduced in Part III of this book, also with *polymorphic* systems, that is, systems integrating both synchronous and asynchronous behavior.

As to the nature of the task, diagnosis is either carried out within a monitoring environment, while the system is operating (possibly in a faulty way), or *a posteriori*, while the system is either out of work or idle. A method that performs *monitoring-based* diagnosis can be adopted also for a posteriori diagnosis while the vice versa does not hold.

The former is the case with the diagnoser approach and its temporal extensions [Chen and Provan, 1997, Rozé, 1997], and the *quantized system approach* [Förstner and Lunze, 1999] (see Section 7). The latter is the case with the *process algebra approach* [Console et al., 2000a, Console et al., 2000b] (see Section 6).

The active system approach originally performed diagnosis just *a posteriori*; a method for carrying out diagnosis during monitoring has recently been introduced, as detailed in Chapter 8. Analogously, the decentralized diagnoser approach performs *a posteriori* diagnosis while its extension, the incremental *decentralized diagnoser approach* [Pencolé et al., 2001] tackles (a kind of) monitoring.

As to diagnostic algorithms, the current shared prospect [Cordier and Largouët, 2001] is that, in the general case, the specific faults of a dynamic system cannot be inferred without first finding out what has happened to the system itself.

In this respect, in spite of slightly different terminologies, such as *histories* in the active system approach, *paths* in the process algebra approach, *situation histories* or *narratives* [Barral et al., 2000], and *trajectories* [Kurien and Nayak, 2000, Cordier and Largouët, 2001], all the distinct approaches describe the evolution of a DES as a sequence interleaving states and transitions since, as already remarked, the favorite behavioral models of DESs in the literature are automata.

Since finding out the system evolutions is a computationally expensive and, therefore, inefficient process (see, for instance, [Rozé, 1997] about the computational difficulties of the diagnoser approach [Sampath et al., 1995, Sampath et al., 1996], or the worst case computational complexity analysis in [Baroni et al., 1999], or the discussion in [Kurien and Nayak, 2000]), most of the approaches exploit a trade-off between *off-line* and *on-line* computation.

By off-line computation we mean preprocessing, that is computation performed when no diagnostic processing is ongoing: it generates compiled knowledge once for all, and such a knowledge can be exploited several times on-line.

By on-line processing we mean the specific diagnostic computation carried out in order to solve each given diagnostic problem. Such a computation, unlike the off-line one, is often submitted to time constraints and, therefore, efficiency is required.

Based on the method for tracking the evolutions of the system that explain a given observation, two broad categories of approaches to diagnosis of DESs can be basically singled out:

- Those that first generate (a concise/partial model of) all possible evolutions and then retrieve only the evolutions that explain the observation;

- Those that generate in one shot the evolutions explaining the observation.

The first category includes some works from both the AI area [Laborie and Krivine, 1997, Cordier and Largouët, 2001] and the Automatic Control area, among which the diagnoser approach, the decentralized protocol approach and the quantized system approach.

The diagnoser approach proposes algorithms for the off-line creation of a data structure to be efficiently exploited for diagnosis during monitoring. Like most other approaches, it can focus attention just on the whole system at hand, or on any structurally independent part of it [Rozé, 1997], and it requires the (automated) generation of a global behavioral model of the system to be diagnosed.

Embodied in the second category are some approaches of the AI area, such as [Kurien and Nayak, 2000], the active system approach and the decentralized diagnoser approach.

The decentralized diagnoser approach [Pencolé, 2000] draws off-line a local diagnoser for each component. Such a diagnoser is an automaton whose states and (observable) transitions are labeled with compiled knowledge about unobservable paths and interacting components, respectively. Each local diagnoser is employed on-line for both a more efficient reconstruction of all the possible evolutions of the relevant component that comply with the observation and a more efficient merging of the histories of distinct components into global system histories.

While the active system approach originally performed only *on line* computation, a recent extension [Garatti et al., 2002] compiles the models of the system off-line, thus generating a so-called *prospection knowledge* to be exploited for a more efficient on-line history construction.

The active system approach is capable of drawing the diagnoses of any arbitrary subsystem and it processes only the models of the involved components, without generating any global behavioral model.

Coming to diagnostic outputs, the active system approach computes diagnoses at several abstraction levels, as will be described in Chapter 4, whereas the output of the diagnoser approach is just a set of faults, the same as in the decentralized protocol approach and in [Rozé, 1997].

A survey of a few recent proposals in the literature, specifically focused on model-based diagnosis of DESs, is provided in the subsequent sections. The presentation is interleaved by examples, which, besides increasing the concreteness of the discussion, are also a means of comparing the different approaches. However, the presentation is self-contained even if examples are disregarded.

2.   Diagnoser Approach

According to the diagnoser approach [Sampath et al., 1995, Sampath et al., 1996, Sampath et al., 1998], each system consists of several distinct physical components and is equipped with a set of sensors. The complete (i.e. both normal and abnormal) behavior of each component is represented by means of a nondeterministic finite automaton: each transition, whose label is an event, is either observable, typically a command issued by the system supervisor, or unobservable (silent), typically a failure event.

Failures are classified into disjoint classes corresponding to different failure types: the diagnosis task is required to univocally identify not the failure event itself but only the type of failure. The global sensor map asserts the correspondence between each composition of the states of all components and the set of discrete outputs of all sensors.

As already remarked in the previous section, the diagnoser approach distinguishes on-line activities, which are carried out while monitoring the system at hand, and off-line ones, which are accomplished beforehand. Diagnosis is performed on-line whereas the verification of the diagnosability properties of the system and the automatic generation of the data structure, called *diagnoser*, that will be exploited on-line are accomplished off-line.

The diagnoser is derived from the behavioral model of the whole system, hereafter called system model. A systematic procedure is presented in [Sampath et al., 1996] for generating the system model based on the behavioral models of the components of the system and on the sensor map. Such a procedure basically consists in performing the synchronous composition of the component models (see, e.g., [Lafortune and Chen, 1991]) for the definition of the standard operation of synchronous composition on state machines), thus capturing the interaction among components, and then in redefining the observability of the obtained transitions, possibly introducing additional states, so as to reflect the discrete outputs of the available sensors.

In the system model, which is a nondeterministic automaton, unobservable events are either failures or events that cause changes in the system state which are not recorded by sensors. The system model is both assumed not to incorporate unobservable cycles and to generate a live language. These two assumptions together guarantee that a failure event is, sooner or later, followed by an observable event.

Example 2.1. The top of Figure 2.1 displays the behavioral models (observable events are in bold) of the two components A (left) and B (right) of the system we consider in this example.

On the bottom, the synchronous composition of the two automata is shown, wherein each state is the composition of the two corresponding component states in the lattice. So, for instance, state 1 is the composition (A_1, B_1).

Suppose that the system is endowed with a single sensor, which practically distinguishes state A_1 from state A_2. The set of the discrete

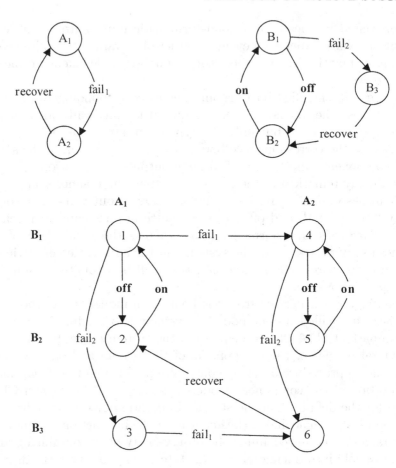

Figure 2.1. Component models for Example 2.1 and their synchronous composition.

outputs of the sensor is $\{norm, ab\}$. Using the same formalism as in [Sampath et al., 1996], the global sensor map is the following:

$$h(A_1, \bullet) = norm,$$
$$h(A_2, \bullet) = ab,$$

where h is the function providing the output of the sensor and the bullet denote an indifference condition, that is, the value of h does not depend on the state of component B.

Figure 2.2 portrays the construction of the system model, based on both the result of the synchronous composition above (bottom of Figure 2.1) and the sensor map. Shaded nodes represent the states added to the synchronous composition in order to account for the discrete values given by the sensor: in this respect, the observable events are the

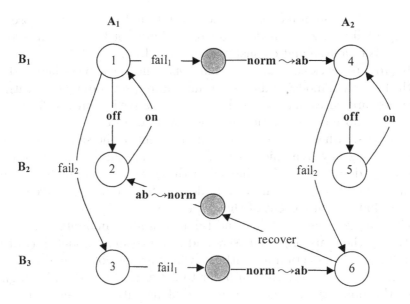

Figure 2.2. System model for Example 2.1.

changes of the sensor value, that is, the change from *norm* to *ab* and vice versa, indicated by *norm* $\rightsquiggle$ *ab* and *ab* $\rightsquiggle$ *norm*, respectively. Numbered states are the same as in the synchronous composition. ◇

2.1 Diagnosability

Two related notions of diagnosability are introduced in [Sampath et al., 1995], the first being stronger than the second, which is called I-diagnosability and is better detailed in [Sampath et al., 1996]. Such notions are defined with respect to the projection of the state transitions of the considered system (in other words, the set of observable events) and the partition of failure events (that is, the types of failures that have to be detected). Besides, in order to define I-diagnosability, one or more observable *indicator* events are associated to every class of failure events.

Roughly, a system is *diagnosable* or *I-diagnosable* if, given any occurrence of whichever failure event, it is possible to detect such a failure within a finite number of events (i.e. a finite number of transitions of the system model) following it or one of the indicator events associated with such a kind of failure, respectively.

The check for necessary and sufficient conditions for diagnosability requires accounting for both the system model and the diagnoser in case every dynamic evolution of the system cannot include several failures of the same type. If this condition does not hold, the check requires to process both the system model and another automaton, say a diagnoser for multiple failures, which is derived from it (and not from the diagnoser) by means of a systematic construction procedure.

In order to check necessary and sufficient conditions for I-diagnosability, both the system model and another automaton, say a diagnoser with indicators, are necessary where the latter is derived from the former by means of another systematic construction procedure.

Note that, when more than one failure event of the same failure type occurs along any dynamic trace of the system, the definition of diagnosability does not require that all of these events be detected, it only requires to conclude within a finite delay after the first occurrence of a failure that a failure event of that type occurred.

Checking for diagnosability and I-diagnosability amounts to cycle detection in the (various) diagnosers and in the system model: any of the standard cycle detection algorithms (which are of polynomial complexity) may be used. However, in order to prove that a system is diagnosable, the global system model is needed along with the diagnoser or one of its variations, which, in the general case, are still larger than it.

The diagnoser and its variations are drawn from the global system model. In the worst case, the state space of the diagnoser is exponential in the state space of the system model [Sampath et al., 1995]. As underlined in [Rozé, 1997, Pencolé et al., 2001], building the system model, although via a well known operation of synchronization (which, however, is not enough in general since also the sensor map has to be considered), is unrealistic due to its intractable size for large systems. Hence, also the diagnoser and its variations cannot be produced. Thus, in general, the diagnosability of a system is an intractable problem (as well as its diagnosis by means of the diagnoser approach).

Example 2.2. The (already defined) set of observable events of the system of Example 2.1 is

$$\{on,\ off,\ norm \rightsquigarrow ab,\ ab \rightsquigarrow norm\}$$

and all of such events are displayed in bold in the system model (Figure 2.2). The set of all failure events of the system is $\{fail_1, fail_2\}$ and now we partition it into two distinct sets, that is, we assume that each of the two failure events has to be detected in a distinct way.

Let the initial state of the system be state 1. Given this projection and this partition of failure events, the system is diagnosable. Any proof of this claim based on the method embodied by the diagnoser approach is skipped as it would require the generation of the diagnoser for multiple failures since there are infinite traces including multiple instances of failures of the same type (all traces cycling either through states $1, 3, 6, 2$ or through $1, 4, 6, 2$ include multiple instances of both $fail_1$ and $fail_2$).

However, thanks to the small size and simplicity of the considered system, we can see that the system is diagnosable based on the definition of diagnosability: in fact, starting from state 1, the first occurrence of failure $fail_1$ is detected in 1 transition following it ($norm \rightsquigarrow ab$), the first occurrence of failure $fail_2$ is detected in 4 transitions at most, i.e. $fail_1$, $norm \rightsquigarrow ab$, $recover$, $ab \rightsquigarrow norm$ (otherwise in 2, i.e. $recover$, $ab \rightsquigarrow norm$), further occurrences of the two failures need not be detected. Therefore the definition of diagnosable system is fulfilled.

It goes without saying that the system is also I-diagnosable since diagnosability implies I-diagnosability: it is easy to see, for instance, that, if $norm \rightsquigarrow ab$ and $ab \rightsquigarrow norm$ are chosen as indicator events for $fail_1$ and $fail_2$, respectively, then either fault can be detected with no delay (i.e. 0 transitions) after the corresponding indicator event. $\diamond$

In [Sampath et al., 1998] a generalized definition of diagnosability and a procedure to check for the diagnosability of systems generating nonlive observable languages are presented. The definition is the same as that for live languages except that the diagnosability condition is required to hold for terminating traces as well.

2.2 Diagnoser

The diagnoser, besides being necessary off-line in order to establish the diagnosability properties of a DES whose evolutions over time cannot be affected by several failures at the same time, is needed on-line for diagnosis. However, as already remarked in the previous section, the construction of the diagnoser is intractable for large systems.

The generation of the diagnoser, which is a deterministic, completely observable automaton, starts from a known initial state, which is assumed to be normal. This means that the initial state of every evolution to be considered is known in advance.

The purpose of the diagnoser while monitoring the system is to infer about past occurrences of failures, based on observed events. Monitoring the DES amounts to following on-line a path in the diagnoser, progressing along a transition each time an event is observed.

If a sequence of events has already been observed, then, following the (only) path marked by this sequence of events, a node of the diagnoser is reached which contains the estimate of all the possible current states of the DES, where each state is associated with a set of faults.

Each state of the DES contained in the node is reachable in the system model by means of a sequence of transitions producing the given observation, while its corresponding set of faults contains all the faults occurred along that sequence, that is, the set of faults is a candidate diagnosis.

Thus, each node of the diagnoser contains all the candidate diagnoses relevant to the sequence of events observed so far. A node of the diagnoser detects a fault if all of its candidate diagnoses include that fault.

Example 2.3. Figure 2.3 displays the diagnoser of the diagnosable system of Example 2.1, assuming, as already done, state 1 as the initial state of the system.

Label N stands for *normal* (no fault), while F_1 and F_2 are shorthands for faults $fail_1$ and $fail_2$, respectively. The initial node of the diagnoser is $1N$, where 1 is the given initial state of the system and N corresponds to the assumption that the system is initially free of faults.

Suppose that, while monitoring the system, *norm* $\leadsto$ *ab* is the first event that is observed. By following the (only) transition exiting from the initial node that is labeled by this observable event, node $4F_1\ 6F_1F_2$ is reached, this meaning that the current state of the system (according to the model of Figure 2.1) can be either 4 of 6, in the former case the system being affected by fault $fail_1$, and in the latter also by fault $fail_2$. Thus, after *norm* $\leadsto$ *ab* has been observed, fault $fail_1$ is certain whereas fault $fail_2$ is uncertain. This uncertainty, however, is removed once the next event is observed. In fact, if event *off* is observed, the system is affected by fault $fail_1$, otherwise (that is, if event *ab* $\leadsto$ *norm* is observed) it is affected by both faults.

This is another facet of diagnosability: a system is diagnosable if any uncertainty about the diagnosis (i.e. the set of faults affecting the system) is solved in a finite number of transitions of the system model (and, consequently, in a finite number of observed events during monitoring). $\diamond$

If the same state in the system model can be reached by means of two or more distinct paths, producing the same observation but each characterized by a distinct set of faults, then the set of faults associated with such a state within the (only) node of the diagnoser reachable through that observation is the intersection of all such sets. Besides an A is added for denoting ambiguity. While constructing the diagnoser, this ambiguity is propagated to the successor states of the ambiguous state in the next nodes as far as no failure events are encountered. Once a fault has been encountered, the set of faults is updated and the A is dropped.

A theorem, proven in [Sampath et al., 1995], states that, if the DES is diagnosable or I-diagnosable, the diagnoser is capable of detecting any failure event in a bounded number of events after its occurrence or the

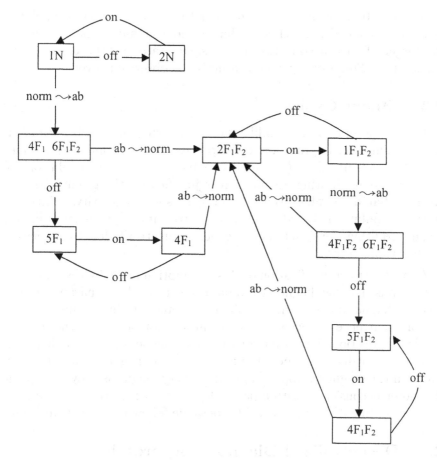

Figure 2.3. Diagnoser for Example 2.3.

occurrence of one of its indicator events, respectively. Besides, a bound on the detection delay corresponding to each failure type is given based on the diagnoser only. Thus the diagnoser is proved to be a structure capable of providing the needed diagnoses of diagnosable systems.

Nothing is added about the diagnosis of systems that are neither diagnosable nor I-diagnosable. What can be evinced is that the diagnoser can be exploited for them as well and, in general, each time an event is observed, a set of candidate diagnoses is output where such a set is not complete in case a node including an A (ambiguity) has been encountered.

Diagnosis of systems generating nonlive observable languages [Sampath et al., 1998] is performed based on the diagnoser derived from a system model which has been modified with respect to the real one so as to generate a live language. In order to extend the system model, a transition is added for each state having no successors: the source and the target of such a transition is the same state and the transition is labeled by a *Stop* event, which is a fictitious observable event.

2.3   Silent Cycles

The diagnoser approach, unlike the active system approach, the decentralized diagnoser approach, and the process algebra approach, assumes that there are no silent (i.e. unobservable) cycles in the evolutions of a system. This limiting assumption is justified on the grounds that it ensures that observable events occur with some regularity. Since detection of failures is based on observable transitions of the system, this means requiring the system not to generate arbitrarily long sequences of unobservable events.

Another strong justification of this assumption made by the diagnoser approach is that the diagnoser is a data structure which guarantees that, given a diagnosable system, a fault be detected within a bounded number of occurrences of events. This claim cannot be true if the number of silent transitions interleaving two observable ones is unbounded. The assumption at hand has possibly influenced the diagnostic method proposed in the diagnoser approach, in that every diagnosis provided by the diagnoser is consistent with a path whose latest transition is an observable one, while it does not consider possible following silent transitions.

3.   Decentralized Diagnoser Approach

The decentralized diagnoser approach by Pencolé [Pencolé, 2000] is an attempt to combine two techniques for the diagnosis of DESs: the diagnoser approach and the (on-line) active system approach. The 'decentralized' attribute means that the method, similarly to the active system approach and unlike the diagnoser approach, does not need generating the global behavioral model of the system to be diagnosed.

The method features compositional modeling, the same as its two inspiring approaches, and basically uses the same modeling primitives, that is, an automaton for describing the complete behavior of each component.

Pencolé embraces two peculiar features of the active system approach that differentiate it from the diagnoser approach. In fact, he adopts the definition of diagnosis of the active system approach, that is, a diagnosis is a history, while the diagnostic outputs of the diagnoser approach are sets of faults, and his approach, the same as the active system approach, performs *a posteriori* diagnosis whereas the diagnoser approach produces diagnosis while monitoring the system.

Example 2.4. In its formal description, the decentralized diagnoser approach is more restrictive than necessary. In particular, it assumes that the only events coming from the external world that affect the system behavior are failure events and the only events exchanged between the components of the system represent failure propagations.

These restrictions do not exist in the active system approach, which inspired the system modeling of the decentralized diagnoser approach, and, indeed, they could be removed from the latter as well.

If these restrictions hold, it is impossible to represent a huge class of DESs, such as, for instance, the system of Example 2.1, which we have already used for discussing the diagnoser approach, since:

- The two transition marked by *on* and *off* are neither triggered by exogenous failure events nor by internal events sent by a component to another;

- The event exchanged between the two components (*recover*) is not generated by a faulty transition.

However, the diagnostic method proposed by the decentralized diagnoser approach, which is deeply rooted in that of the active system approach and is meant to produce the same result, i.e. the sequences of transitions consistent with the given observation, do work also for systems for which the above restrictions do not hold.

Hence, we will force the modeling primitives of the decentralized diagnoser approach just what suffices to deal with the system of Example 2.1, so as the different approaches can be compared on a common ground.

The models of the two components of the system of Example 2.1 are depicted in Figure 2.4 according to the modeling primitives of the decentralized diagnoser approach. Bold arrows represent observable transitions. Each transition is labeled by its triggering event, placed on the left of the bar, and by its output event(s), placed on the right.

Bold events are observable. Homonymous non-null events in distinct component models represent the same event, which is generated as output by a component and received as input by another. In the example there is just one such event, *recover*. The null event, denoted by ϵ, can be the triggering event of a transition: this is the artifice we have adopted for representing transitions that are neither triggered by exogenous failure events nor by internal events, thus forcing the primitives of the decentralized diagnoser approach. ◇

$\Sigma_{obs} = \{$ on, off$\}$

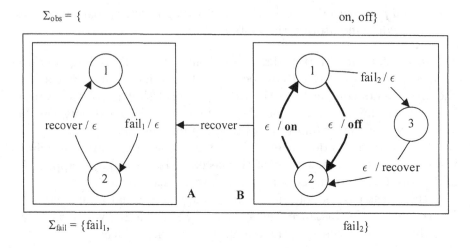

$\Sigma_{fail} = \{fail_1,$ fail$_2\}$

Figure 2.4. Component models for Example 2.4.

In defining a diagnostic problem, the decentralized diagnoser approach supports a notion of an observation which is restricted with respect to that of the active system approach (see Chapter 9), although being more articulated than the single totally temporally ordered sequence of observed events of the diagnoser approach. Such an observation consists of several totally temporally ordered sequences of precise observed events, each pertaining to a single component. Besides, the sender component of each observed event is assumed to be known, while this assumption has been relaxed in the active system approach.

Moreover, the decentralized diagnoser approach tackles synchronous systems only, the same as the diagnoser approach, while the active system approach addresses asynchronous systems and also polymorphic systems (see Part III of this book). However, unlike the diagnoser approach, the decentralized diagnoser approach does not account for any explicit sensor map.

3.1   Global Diagnosis

The aim of the decentralized diagnoser approach is to improve the efficiency of the active system approach while obtaining the same results, that is, all the possible evolutions of the considered physical system over time, starting from a known initial state and complying with a given observation. Each of such evolutions is a history and, altogether, histories make up the solution, called *global diagnosis*, where the 'global' attribute highlights that such a solution is inherent to the whole system.

Example 2.5. Figure 2.5 displays a global diagnosis, which is the solution of a diagnostic problem inherent to the system of Example 2.4.

According to such a problem, the initial state of each system component is 1 and the observation that has to be explained is null for component A and *on* for component B.

The resulting global diagnosis, denoted as $\Delta((1,1)(\epsilon, on))$, is an automaton. Each node of this automaton contains two pairs, the first being the composition of the states of the two components of the system, and the latter the observations inherent to the two components as registered along any path leading to the current node starting from the initial node.

Therefore, the initial node contains the pairs $((1,1)(\epsilon, \epsilon))$, denoting the given initial states of the two components and a null observation for both components.

Each arrow represents a component transition that is applicable given the system state of the source node, thus transforming it into the system state of the target node. If the applied transition is observable, the observation is updated accordingly.

So, for instance, the transition from state 2 to state 1 of component B, marked by $\epsilon|\mathbf{on}$, is applicable to node $((2,2)(\epsilon, \epsilon))$ since the node represents a situation wherein component B is in state 2.

The target node contains the new state of component B, that is, 1, and the new observation of component B, which is the concatenation of the previous observation (ϵ) of component B, with the event produced by the considered transition (on), thus obtaining the sequence ϵ *on*, which becomes *on* for short. Then the target node is $((2,1)(\epsilon, on))$.

The global diagnosis represents all the possible evolutions of the system starting from the given initial state and explaining the given observation. Each of such evolutions is a sequence of transitions leading from the initial node to a final node, where each final node is denoted by a double ellipse. A node is final if it is characterized by the same observation as that given by the diagnostic problem, that is (ϵ, on) in our example. Therefore, state $((2,1)(\epsilon, on))$ is final.   ◇

3.2 Local Diagnoser

To achieve the purpose of improving the efficiency of the active system approach, a so-called *local diagnoser* is drawn off-line from each component behavioral model. Such a diagnoser resembles that of the diagnoser approach, as it is an automaton wherein each transition is observable and, therefore, marked by an observable label.

However, a local diagnoser differs from the traditional diagnoser of the diagnoser approach in states, as each state contains compiled knowledge about unobservable paths (possibly including unobservable cycles).

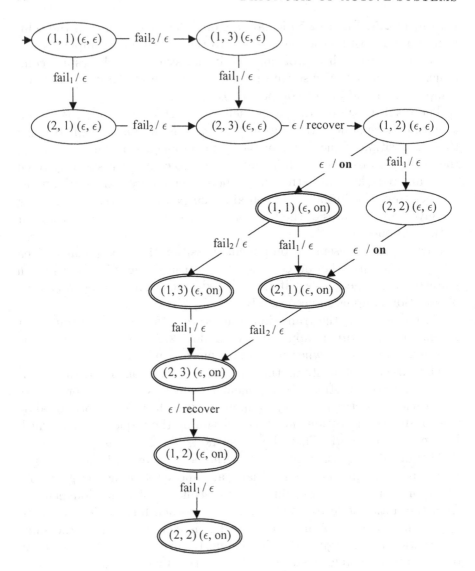

Figure 2.5. Global diagnosis $\Delta((1,1,)(\epsilon, on))$ for Example 2.5.

Most importantly, although not highlighted in [Pencolé, 2000], while a traditional diagnoser is deterministic, a local diagnoser may be nondeterministic, that is, there may be several transitions leaving the same state of the local diagnoser which are marked by the same observable event.

Example 2.6. Figure 2.6 displays the local diagnoser inherent to component B of the system of Example 2.4. Such a diagnoser is built off-line from the component model of B, which is displayed in Figure 2.4.

Each state of the local diagnoser, depicted as a box, contains a pair: a component state that, in the component model, is the target of an observable transition, and all the unobservable subgraphs that, in the component model, are reachable starting from the state.

The first element of the pair is denoted by a component state with an unrooted entering arrow: let's say that it is the initial state of the box. Thus the number of boxes in the local diagnoser amounts to the number of states that are the target of observable transitions in the relevant component model. In the case of component B, there are two such states, 1 and 2. So, the box $x_{\Delta_B}(1)$ contains state 1 and the subgraph reachable from it by means of unobservable transitions. The box $x_{\Delta_B}(2)$, instead, does not contain any unobservable subgraph reachable from state 2: in fact, there aren't any in the component model of B.

The transitions between the boxes are all and only the observable transitions of the component model (in the local diagnoser there may be several instances of the same component transition, if needed): the actual source state of each of such transitions is a piece of implicit compiled knowledge (it is a state contained in the source box), while their target is the initial state of the target box.

The diagnoser of component A consists of a single box, called $y(1)$, which contains exactly the component model of A and wherein the initial state is 1. ◇

3.3 Local Diagnosis

Each local diagnoser is exploited for reconstructing all the possible evolutions of the relevant component, called *local histories*, that comply with the component observation. Altogether local histories make up a *local diagnosis*.

First, the sequences of observable transitions that produce the given sequence of observed events are determined. This search is more efficient than it would be if performed within the component model since it avoids performing a depth-first search (DFS) within unobservable transitions, however it has still to make a DFS within a nondeterministic graph, that is, within the observable transitions of the local diagnoser.

Later, the unobservable paths that connect each pair of observable transitions belonging to such sequences are determined (likely by means of an efficient indexing mechanism), thus obtaining the local histories.

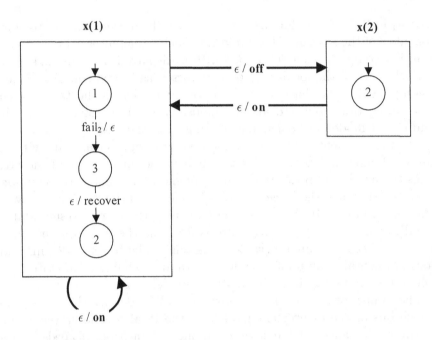

Figure 2.6.   Local diagnoser of component B for Example 2.6.

Example 2.7. The local diagnoser inherent to component B of the system of Example 2.4, displayed in Figure 2.6, is built off-line just once and then it can be exploited several times on-line for computing local diagnoses.

A local diagnosis inherent to a component is an automaton containing all the sequences of transitions explaining a local sequence of observed events.

Consider for instance the problem of determining the local diagnosis inherent to component B, given 1 as the initial state of the component and *on* as the (only) observed event. Such diagnostic problem is denoted as $\Delta_B(1, on)$.

Based on the local diagnoser of B, we obtain quite easily the graph, wherein each node is a box of the diagnoser, that explains the given observation: such a box graph is displayed in the top of Figure 2.7.

The initial box is univocally identified by the initial state: state 1 identifies box $x_{\Delta_B}(1)$. There is only one observable transition exiting from such a box that generates the given observed event, transition $\epsilon|\textbf{on}$, and the target box is once again $x_{\Delta_B}(1)$, which is the final node of the search, as denoted by the double edge.

Now, the obtained box graph is converted into the local diagnosis by simply translating each box into the relevant subgraph. For a box which

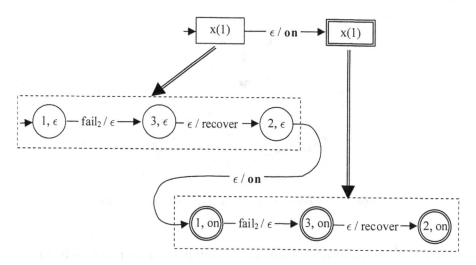

Figure 2.7. Local diagnosis $\Delta_B(1, on)$ for Example 2.7.

is not final in the box graph, the relevant subgraph is the one contained in the corresponding box of the diagnoser, rooted in the initial state of such a box and ending in the state from which the observable transition is exiting (state 2 in the case we are considering).

For a final box of the box graph, the relevant subgraph is the whole subgraph contained in the corresponding box of the diagnoser, wherein all states can be final states. This way we obtain the local diagnosis displayed on the bottom of Figure 2.7, wherein each sequence of transitions starting from the (given) initial state and ending in a final state is a possible evolution of the component that explains the given observation.

The construction of the local diagnosis $\Delta_A(1, \epsilon)$ is displayed in Figure 2.8. ◇

3.4 Merging

Once the local histories of all the components have been reconstructed, they have to be merged so as to produce the global histories.

Example 2.8. The global diagnosis of Figure 2.5 can be obtained by applying the classical composition operation with a synchronization on the internal events to the two local diagnosis of Figures 2.7 and 2.8, that is, formally

$$\Delta((1, 1)(\epsilon, on)) = \Delta_B(1, on) \odot \Delta_A(1, \epsilon).$$

◇

In the general case, the above composition is done by exploiting a *reconstruction graph*, which, as in the active system approach (see Section 6 of

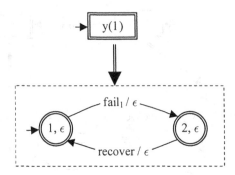

Figure 2.8. Local diagnosis $\Delta_A(1, \epsilon)$ for Example 2.7.

Chapter 5), is a multilevel hierarchy specifying how to recursively merge the reconstructed histories inherent to subsystems so as to obtain the histories of the whole system.

However, while in the active system approach the reconstruction graph is built before any reconstruction is carried out, in the decentralized diagnoser approach it is built on the fly after the reconstruction of the local histories has been completed, so as to exploit the information about component interactions implicit in the local histories.

The rationale is to merge beforehand the sets of histories which are deeply interconnected, that is, that frequently exchange events with each other, and to perform the combination of independent histories only thereafter.

The information about the components a history interacts with is annotated in the observable transitions of the history itself. This annotation is simply copied from the corresponding transition in the local diagnoser while the history is reconstructed, since this is also compiled knowledge included in the local diagnoser.

The local diagnoser approach is appealing in the exploitation of interaction information for the progressive composition of histories. In fact, composing deeply interacting histories first is a good heuristics for reducing the size of the reconstruction space, as possible incompatible hypotheses are rapidly eliminated.

Generating the reconstruction graph in advance, as done in the active system approach, can benefit from the information as to potential interactions, which are a superset of the actual interactions claimed by the reconstructed histories.

In the active system approach, such information is mapped to the structural model of the system. However, in the active system approach the generation of a reconstruction graph can account for the repetitiveness of structural patterns, maybe accompanied by a repetitiveness of the observations, which is not considered by the decentralized diagnoser approach. The different kinds of heuristics exploited by the two approaches, indeed, are not mutually exclusive but, rather, complementary, and could cooperate within a unifying processing frame.

In the decentralized diagnoser approach, the local histories of all components have to be reconstructed before any composition is accomplished. This leads to a notion of a reconstruction graph less general than the active system approach, wherein a reconstruction graph can include also the composition of histories which are not specialized by any observation. This composition, which is not supported by the decentralized diagnoser approach, is very useful when the reconstructed histories are generated once and then exploited several times for being specialized by different known observations.

4. Incremental Decentralized Diagnosis Approach

The *incremental decentralized diagnosis approach* [Pencolé et al., 2001] represents the incremental version of the decentralized diagnoser approach. The technique, unlike the decentralized diagnoser approach, which performs a posteriori diagnosis, performs a kind of generalized monitoring that produces the histories of the system by taking into account a packet of observed events at each step (not necessarily a single observed event as usually done in monitoring).

The technique is incremental in that there is a current set of already computed diagnoses at every moment and such a set is updated at each step based on the newly acquired packet of observed events.

A packet is inherent to a temporal window and consists of several temporally ordered sequences of observed events, each pertaining to a single component.

Under the assumption that the sequence of observable events received from each component is exactly the sequence of observable events emitted by the same component in the considered temporal window, the technique is the same as the one adopted by the decentralized diagnoser approach for a posteriori diagnosis, that is, local histories are determined based on the local diagnosers and the local observations, then they are merged into the global diagnosis.

The only difference is that, at each time step, the construction of local histories starts from several initial states instead of one, where such states are the component states contained in the final states of the current global histories.

Such a technique does not work if the above assumption is relaxed, that is, if the sequence of observable events emitted by a component within a temporal window may be randomly split into successive observation packets.

Example 2.9. In order to understand why the technique is ineffectual when the above assumption does not hold, consider, for instance, two observable transitions t_A and t_B of two distinct components A and B of a system S, which generate the observable events a and b, respectively.
Suppose that

- Transition t_A produces as output an event that triggers t_B;

- System S has actually performed the 'chain' of transitions $t_A\ t_B$, thus emitting the sequence of observable events $a\ b$;

- Such a sequence is split into two successive observation packages, a and b, respectively.

Now we apply the technique (which is valid when the assumption holds) to the problems inherent to the two packages, for which the assumption does not hold.

The local diagnosers of components A and B are now exploited to determine the local histories relevant to the observation package a.

Transition t_B does not belong to any history of B as such a transition is observable while no observable event coming from B has been received.

Transition t_A, instead, belongs to a history (at least) of component A, since a is the observable event it generates. However, when the synchronization for determining the global diagnosis is performed, any history including t_A is discarded since it cannot be synchronized with any history of component B. So, after observation package a has been processed, no global history includes t_A.

Analogously, when the next observation package (b) is considered, no local history of component A includes t_A since such a transition is observable while no observable event coming from A has been received.

This prevents the global diagnosis from including any history containing t_A and, therefore, the chain $t_A\ t_B$, which is the actual evolution of the system, is missed. $\diamond$

The incremental technique was extended in order to compute diagnoses even when the simplifying assumption about the observation does not hold. According to the extended technique, a larger set of global candidate diagnoses is produced, which summarizes both trajectories that explain the events observed in the considered window and some trajectories that explain also a set of hypothetical unreceived events supposed to have been emitted but not yet received by the observer. This set of global diagnoses is obtained by synchronizing properly extended local diagnoses, determined by means of the local diagnosers of the decentralized diagnoser approach.

The authors of the incremental decentralized diagnosis approach state they "propose to complete the observations by guessing what is lacking". To this purpose, they argue that, if the hypothesis is made that there is a bounded number of local observations at the same time in the communication channel associated with each component, then the sequence of hypothetical unreceived events of each component is finite.

Each local diagnosis based on the relevant local observation inherent to a temporal window has to explain the whole set of observation sequences whose prefix is what has actually been observed in the temporal window.

The main concern of the technique is how to hypothesize unreceived observable events so as not to miss any history of the considered system.

Possible wrong hypotheses made during the computation of the successive extended diagnoses are removed with the help of the diagnosis of the last window, when no more observable events are expected.

5. Decentralized Protocol Approach

The work by Debouk, Lafortune, and Teneketzis [Debouk et al., 2000a, Debouk et al., 2000b], which proposes the *decentralized protocol* approach, is focused on monitoring, the same way as the diagnoser approach from which it stems.

Specifically, it addresses a decentralized processing architecture consisting of several separate local sites (actually two local sites are considered in the quoted papers), each communicating information to a coordinator that is responsible for diagnosing the failure occurred in a given system.

The modeling abstraction is the same as in the diagnoser approach, therefore, as already remarked, it accounts for synchronous systems.

In addition, a system can be considered by the decentralized protocol approach if there is a set of (categories of) observable events that makes the system diagnosable according to the definition of diagnosability given in the diagnoser approach. This set is decomposed into several (not necessarily disjoint) subsets, so that the system is not diagnosable with respect to any of them. Each subset contains the categories of events which are observable by a distinct local site.

Example 2.10. Consider the system described in Example 2.1, whose system model is provided in Figure 2.2. As discussed in Example 2.2, given state 1 as the initial state, such a system is diagnosable if the events

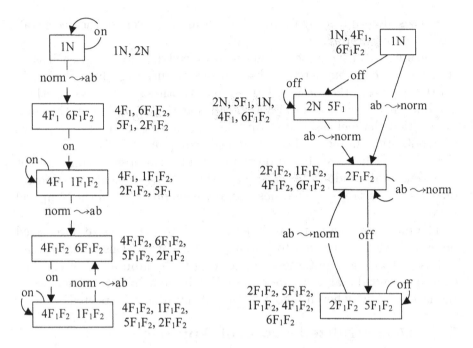

Figure 2.9. Diagnosers for the two local sites of Example 2.10.

in bold in Figure 2.2 are observable and failure events are partitioned into sets $\{fail_1\}$ and $\{fail_2\}$.

Now, assume that $\{on, norm \leadsto ab\}$ and $\{off, ab \leadsto norm\}$ are the sets of observable events at the first and second site, respectively. The diagnosers at the two sites are shown in Figure 2.9. Each node in the diagnosers is sided by a shaded piece of information whose meaning will be clarified in Example 2.11.

As required by the approach, the system is not diagnosable with respect to either set of locally observable events, as it can be intuitively ascertained as follows.

In the diagnoser on the left, once node $4F_1\ 1F_1F_2$ has been reached, according to which fault $fail_1$ is certain and fault $fail_2$ is uncertain, if an indefinite sequence of *on* observable events is received, it is impossible to discriminate whether $fail_2$ has occurred or not. Therefore, if the current state of the system is 1, which means that $fail_2$ has occurred, it is not guaranteed that such a failure event is detected within a finite delay, which contradicts the definition of diagnosability.

Likewise, in the diagnoser on the right, node $2N\ 5F_1$ is uncertain as regards fault $fail_1$ and such uncertainty is not removed in a finite delay owing to transition *off* cycling on the node.   ◇

A local site monitors the whole system at hand based on its set of observable events, by exploiting the global diagnoser corresponding to this set, where such a diagnoser is built following the diagnoser approach.

The sites are allowed to report to the coordinator a concise processed version of their results, according to a communication protocol. A protocol consists of the diagnostic rules and communication rules used at the local sites, and of the decision rules used by the coordinator to infer the occurrence of failures.

The challenge is to enable the coordinator, which is assumed not to be provided with any model of the system, to 'perform as well' as the (centralized) diagnoser corresponding to the whole set of observable events, that is, the set with respect to which the system is diagnosable.

The centralized diagnoser is capable of detecting the occurrence of every fault in a diagnosable system within a finite delay (that is, within a finite number of transitions in the centralized system model). Then, in the intentions of the authors of the decentralized protocol approach, the decentralized architecture should be able to do the same, even if, maybe, with different delay values.

The challenge, which has been faced both in case the information transmitted by the local sites are received by the coordinator in the order they are sent and in case they are received with a bounded temporal error, has been won even if not in the general case but rather under a set of assumptions as to the diagnosers exploited by the local sites.

5.1 Protocol

According to a protocol described in [Debouk et al., 2000a, Debouk et al., 2000b], where it is denoted as *Protocol 2*, it is assumed that each site knows the set of events observable by the other site. This is important as, since there may exist events that are observable by both sites, whenever one of such events is observed, each site sends a message to the coordinator including the information that its local diagnostic conclusion has to be considered together with that of the other site.

In such a case, the coordinator will compute the estimate of the current state and faults of the system as the intersection of the estimates performed by both sites, that is, the intersection of the contents of their current diagnoser nodes. In fact, each time a site observes an event, it sends to the coordinator the content of the node of the relevant local diagnoser and, in addition, the *unobservable reach* corresponding to such a node.

The unobservable reach of a node of a local diagnoser is the set of system states, each accompanied by the relevant failure labels, that are reachable by means of a sequence of transitions starting from any system state belonging to the considered node, where all such transitions are not observable by the local site and the final transition of the sequence is observable by the other site.

Intuitively, since a node of the diagnoser contains a (possibly non-univocal) estimate of the state of the system and the set of faults affecting the system exactly on the time an observable event is received, the unobservable reach represents an estimate of all the states the system can be in after such an event has been observed and the next observable event has not been received yet.

In this respect, the unobservable reach consists of the pairs (*state, faults*) of the current node and of all the pairs reachable starting from them by means of unobservable transitions.

The constraint that the final transition has to be observable by the other site is due to the fact that, this way, the unobservable reach represents any possible situation of the system when an event is observed by the other site (j) but not by the current one (i).

If an event is observed by site j and not by site i, the actual situation of the system has to belong to both the current node of the diagnoser of site j and the unobservable reach of site i.

Therefore, the coordinator will compute the estimate of the current state and faults of the system as the intersection of the content of the current node of the diagnoser of site j and of the content of the current unobservable reach of site i.

Example 2.11. The unobservable reach of each node of the local diagnosers in Figure 2.9 is shaded on the side of the node itself.

Assume that the system to be diagnosed emits the sequence of observable events

$$norm \rightsquigarrow ab \; off \; on \; off \; on \; off.$$

The projections of such a sequence on sites 1 and 2 are

$$norm \rightsquigarrow ab \; on \; on,$$

and

$$off \; off \; off,$$

respectively. The former leads diagnoser 1 to a F_2-uncertain state, from which only the observation of a $norm \rightsquigarrow ab$ event can provide an exit.

The latter leads diagnoser 2 to a F_1-uncertain state, where such an uncertainty can be removed only if an $ab \rightsquigarrow norm$ event is observed.

The application of Protocol 2 to the given trace is depicted in Table 2.1, where:

1 *Observed event* is the observable event emitted by the system; such an event is observed by either Site 1 or Site 2 (since in this example

Table 2.1. Application of Protocol 2 to Example 2.11.

Observed event	$\mathcal{N}_1$	UR_1	$\mathcal{N}_2$	UR_2	Diagnostic result
	$1N$	$1N, 2N$	$1N$	$1N, 4F_1, 6F_1F_2$	$1N$
$norm \rightsquigarrow ab$	$4F_1, 6F_1F_2$	$1N, 2N,$ $5F_1, 2F_1F_2$	"	"	$\{\mathcal{N}_1\} \cap \{UR_2\} =$ $4F_1, 6F_1F_2$
off	"	"	$2N, 5F_1$	$2N, 5F_1, 1N,$ $4F_1, 6F_1F_2$	$\{\mathcal{N}_2\} \cap \{UR_1\} =$ $5F_1$
on	$4F_1, 1F_1F_2$	$4F_1, 1F_1F_2,$ $2F_1F_2, 5F_1$	"	"	$\{\mathcal{N}_1\} \cap \{UR_2\} =$ $4F_1$
off	"	"	$2N, 5F_1$	$2N, 5F_1, 1N,$ $4F_1, 6F_1F_2$	$\{\mathcal{N}_2\} \cap \{UR_1\} =$ $5F_1$
on	$4F_1, 1F_1F_2$	$4F_1, 1F_1F_2,$ $2F_1F_2, 5F_1$	"	"	$\{\mathcal{N}_1\} \cap \{UR_2\} =$ $4F_1$
off	"	"	$2N, 5F_1$	$2N, 5F_1, 1N,$ $4F_1, 6F_1F_2$	$\{\mathcal{N}_2\} \cap \{UR_1\} =$ $5F_1$

the domains of observable events of the two sites are disjoint); its initial value is void;

2 $\mathcal{N}_{1(2)}$ is the current node of diagnoser 1(2), as determined by Site 1(2); its initial value is $1N$ since 1 is the initial state of the considered system, as assigned by the diagnostic problem, and, the same as in the diagnoser approach, it is assumed that the system is initially normal;

3 $UR_{1(2)}$ is the unobservable reach of the current node of diagnoser 1(2), as determined by Site 1(2); it is initialized with the unobservable reach of the initial node of diagnoser 1(2), i.e. the unobservable reach of $1N$;

4 *Diagnostic result* is the diagnostic conclusion drawn by the coordinator, where such a conclusion is inherent to the whole system (the diagnostic rule applied by Protocol 2 at each step to determine the conclusion is reported between braces); its initial value is $1N$ for the reasons explained at point 2 above.

Fields 2 and 3 above are the only pieces of information available to the coordinator, their content being received from the local sites. In the table, every time any of such fields is updated, the new content (which, incidentally, may equal the previous one) is reported in the relevant slot while, if it is not updated, a symbol " is used.

In this example, the result provided step by step by the decentralized diagnoser approach is exactly the same as that produced by the diagnoser approach. In fact, given the first column as input, the diagnoser approach produces the last column as output, as can be proven by following the path indicated by the first column on the diagnoser displayed in Figure 2.3.   ◇

5.2   Remarks

In the decentralized protocol approach, each site takes into account the whole system and then needs a diagnoser of the whole system. The need for several global diagnosers significantly reduces the advantages of compositional modeling and makes the approach not applicable in practice to large systems, since the diagnoser generation runs into significant computational difficulties, as pointed out in Section 2.1 of this chapter.

The decentralized protocol approach does not aim to reduce the computational complexity of the task. Rather, it addresses the problem of limiting the amount of data exchanged among the workstations that are in charge of monitoring and diagnosis.

In the decentralized protocol approach what is actually decentralized is the observation and the observer: each observer, at its own site, makes a diagnosis based on its partial observation.

The notion of decentralized observation, which has been adopted also by the active system approach (see Chapter 12) has a relevance on its own, regardless of the processing architecture, either centralized or decentralized, which may be adopted for accomplishing the diagnostic task. In fact, it is realistic that distinct types of (physical) events are sent to the observer(s) by means of distinct communication channels.

However, the decentralized protocol approach assumes that the temporal order of the events received by the observer on each channel is total, while this simplifying assumption is removed in the active system approach.

6.   Process Algebra Approach

In order to carry out diagnosis, Console, Picardi and Ribaudo [Console et al., 2000a, Console et al., 2000b] model distributed DESs by means of a process algebra, PEPA [Hillston, 1996], which is usually adopted to the purpose of evaluating the performances of concurrent processes. Thus, we refer to it as the process algebra approach.

In the process algebra approach, PEPA terms are algebraic descriptions provided by the user to represent

- Any component behavior;

- Any aggregation of several components;

- Any observation;

- The synchronization of a system description with an observation.

The untimed semantics of a PEPA term is amenable to a representation as a nondeterministic automaton, called Labeled Transition System (LTS), which can be automatically computed using the PEPA Workbench [Gilmore and Hillston, 1994], a software tool that supports modeling and analysis with PEPA.

The LTSs of the four aforementioned items respectively represent

- The behavioral model of the relevant component;

- The synchronous composition of the behavioral models of the aggregated components;

- A graph depicting the whole system observation;

- The global histories of the considered system, compliant with the given observation.

Example 2.12. The system of Example 2.1 can be described in PEPA terms as follows.

Behavior of components

$$A ::= fail_1.recover.A$$
$$B ::= off.on.B + fail_2.recover.on.B$$

The first equation describes the behavior of component A, the second that of component B. Intuitively we can see that, in the equations, A and B represent the initial states of the two components, corresponding to A_1 and B_1 in the component models of Figure 2.1, respectively.

A is defined as $fail_1.recover.A$ by means of operator $::=$, since, after the sequence of actions $fail_1$ and $recover$ has been performed, the component is in state A again. According to this formalism, state A_2 of Figure 2.1 is $recover.A$.

In general, the same behavior can be described by several alternative equations. However, the way equations are written dictates which the initial states of components are.

The *choice* operator $+$ in the second equation captures the possibility of selecting among alternatives. In this case, there are two alternative behaviors, both starting from B and leading back to B. States B_2 and B_3 of Figure 2.1 corresponds to $on.B$ and $recover.on.B$, respectively.

Structure of the device

$$SD ::= A \bowtie_{\{recover\}} B$$

In this equation the behaviors of components A and B are composed by means of the *cooperation operator* $\bowtie$, below which the *cooperation set* $\{recover\}$ is reported. This means that such behaviors proceed asynchronously as far as actions not contained in the cooperation set are concerned, while they cooperate for any action in such a set. The fact that action *recover* is shared guarantees that it is executed synchronously by the two components (homonymous actions in distinct equations are otherwise executed asynchronously). The LTS corresponding to SD is actually the synchronous composition obtained in the diagnoser approach and displayed in the lower part of Figure 2.1.

Diagnosis *Diag* is characterized as the synchronization of the system with the observation.

Observation

$$Obs ::= on$$

In order to represent a diagnostic problem inherent to system SD, we have first to express the observation the problem refers to. According to the above observation, only *on* has been observed.

Diagnosis

$$Diag ::= SD \bowtie_{\{on,off\}} Obs/H$$

where

$$H ::= \{fail_1, fail_2, recover\}.$$

The cooperation set for determining the LTS of *Diag* is the set of observable actions. We assume, as in Figure 2.1, that only actions *on* and *off* are observable. The hiding operator / cuts off the actions in set H from the current analysis, that is, they are deemed as invisible in the construction of the relevant LTS. Therefore, from the point of view of our diagnostic problem, H is the set of non-observable actions.

It could be objected that the information carried by H is redundant, as such a set is the complement of the cooperation set. This, however, is not a general property of the hiding set inherent to a cooperation operation, instead, it is a consequence of the specific (diagnostic) semantics we are describing.

In the general case, the cooperation set can be any subset of the complement of the hiding set, that is, the synchronization can be performed on any subset of visible actions.

The LTS corresponding to *Diag* is actually the global diagnosis obtained in the decentralized diagnoser approach for the same problem and displayed in Figure 2.5. ◇

6.1 Observation

PEPA syntax, although quite expressive, is very cumbersome as regards observations: it allows the representation of sequences of observed events and alternative observations, the latter by possibly factorizing subsequences. Thus, representing by means of a PEPA term an uncertain or complex observation, as will be defined in Chapters 9 and 10, basically amounts to listing all the relevant alternatives, which is exactly what the active system approach wants to avoid. In the active system approach the description of the observations taken as input by the diagnostic process is very concise as it specifies just the logical and temporal constraints between observed events, and techniques are given for processing these constraints directly, without generating all the relevant observation alternatives.

Unlike the other approaches, the PEPA behavioral descriptions of components make no assumption on which events can be observed. The decision as to which events are to be regarded as observable is made just when the behavioral descriptions are composed with the description of the observation.

The authors of the process algebra approach underline that this separation between the system to be diagnosed and the observation enables a flexible testing of the diagnosability properties of the system.

However, in our opinion, a system behavioral description that is completely independent from the communication means between the system and the observer(s) makes it difficult to represent all the possible alterations affecting the observation.

In fact, if such alterations are not represented in the system description, they have to be represented in the observation. But effects in the observation are the result of the composition of multiple primitive effects: those manifesting themselves within the system, those affecting the communication channels, and those concerning the observer. Therefore, they are very hard to compute.

Think, for instance, how difficult it would be to represent within the observation the fact that the observable events transmitted on a particular channel may be randomly lost. In other words, the diagnosability of a dynamic system cannot be referred just to a system and a set of observable labels but, rather, it depends also on the communication channels delivering the observations to the observer(s) and on the observer(s)' sensitivity.

6.2   Remarks

In the papers introducing the process algebra approach, the whole system description is synchronized with the whole observation, without performing any problem decomposition. By composing the whole system with the whole observation, the generation of the LTS of the whole system is forced, which is quite huge. The process algebra approach would benefit, instead, from a multilevel decomposition of the synchronization process, analogous to that performed by the active system approach and the decentralized diagnoser approach.

In our view, the major asset and, at the same time, the major drawback of the process algebra approach is relying on existing tools. In fact, on the one hand, this brings the advantage of exploiting past experiences as well as a very expressive formalism and powerful interpretation tools. On the other hand, the formalism is not specific for describing diagnostic problems, and therefore it is scarcely intuitive from the user point of view, and very verbose. Moreover, in the process algebra approach just a subset of all the PEPA expressive features are used: useless features are likely to cause an interpretation overhead.

7.   Quantized System Approach

In their *discrete-event quantized system* approach Förstner and Lunze [Förstner and Lunze, 1999] propose an algorithm for the construction of a set of qualitative models of a quantized continuous-variable system. Such models are then exploited for accomplishing a consistency-based diagnosis of the considered system.

A quantized system is a quantitative system whose input and output signals are measured through quantizers, which are devices generating an event each time a signal changes its qualitative value. A qualitative value represents an interval of continuous values. A quantized state represents a partition of the continuous-variable state space of the considered system. The quoted paper is focused on modeling autonomous (i.e. without inputs) quantized systems.

Even if the algorithm for generating the model is out of the scope of this discussion, the adopted models are of interest here for their similarities with those of the approaches presented in the previous sections: they are in fact nondeterministic automata wherein the occurrence of an event causes a state transition. However, the approach, unlike the other ones presented in this chapter, does not feature compositional modeling. In fact, each automaton, as generated by the aforementioned algorithm, describes the observable behavior of the overall system in presence of a specific fault. There is an automaton also for representing the normal behavior, that is, the behavior corresponding to the null fault.

The behavioral models, since global, may be huge, and every state transition, unlike the transitions of the models of the other approaches, is observable. In fact, silent transitions having the effect of generating internal events are useless in the discrete-event quantized system approach, whose concern is modeling the global system behavior and not the internal interactions among components.

7.1 Diagnosis of Quantized Systems

In the discrete-event quantized system approach diagnosis is performed while monitoring the considered physical system. Two major differences of the discrete-event quantized system approach with respect to the others are the following.

First, the initial state of the system before the beginning of the diagnostic process is uncertain, that is, it may range over a set of states. That is why the modeled system behavior, either faultless or in presence of a specific fault, is nondeterministic in general.

Second, the assumption is made that a fault occurs before (or exactly at the time) the diagnostic process begins and is present throughout such a process.

The diagnostic method is carried out on-line iteratively, once for each observed event, assuming as the initial state(s) of the current step the final state(s) reached at the previous step. At each step the diagnostic process checks whether the current event is consistent with the evolutions described by the given system models. If the current event does not comply with a model, such a model is discarded, that is, it is not considered anymore during the diagnostic process. This means that, based on the observed behavior, the faults corresponding to the discarded models are refuted. At the end of the diagnostic process, each (single or multiple) fault corresponding to a model that has not been discarded is a candidate diagnosis since it justifies the observed behavior.

The approach is consistency-based as it produces as output a set of candidate diagnoses that is complete but not sound. This, however, does not depend on the diagnostic method but, rather, on the considered models, as generated by the algorithm proposed in [Förstner and Lunze, 1999]: such modes represent all physically possible behaviors as well as spurious behaviors that can not occur in the quantized system, which is a general property of qualitative modeling.

In fact, since the behavior of the quantized system does not possess in general the Markov property, every model that possesses the Markov property, like (stochastic) automata, can only be an approximate representation of the quantized system. Hence the effort to determine a model of the system which is complete, in other words, which encompasses all the physically possible behaviors, while, at the same time, representing a minimal set of spurious behaviors. Different (complete) models of the same system yield different diagnostic results.

In the discrete-event quantized system approach attention can be focused on subsystems. This can be achieved by considering only the system models corresponding to possible faults affecting such subsystems. However, in any case, such models are global and the observation of the whole system has to be taken into account.

By contrast, in the active system approach, decentralized diagnoser approach, incremental decentralized diagnosis approach, and process algebra approach, focusing attention on a specific subsystem means taking into account only the models of its own components and its own observation.

The method, as described above, is untimed in nature as all the previous ones. However, as suggested in [Lunze, 1999], the approach can be improved by using temporal information. In fact, if the output of the diagnostic method consists of several candidate diagnoses, they can be discriminated based on the temporal distances of the events taken into account.

7.2   Chronicles and Further Work

The rationale behind the discrete-event quantized system approach is the same as for chronicle-based approaches [Laborie and Krivine, 1997], namely checking on-line the observed behavior against a signature of the modeled behavior. In the former approach the signature is a set of automata, in the latter it is a set of chronicles. However, as proposed by Laborie and Krivine in their *automatic chronicle generation* approach [Laborie and Krivine, 1997], chronicles can indeed be derived from automata as compiled knowledge.

In particular, in the automatic chronicle generation approach, a (timed) nondeterministic automaton describing the complete behavior of the considered class of systems is exhaustively simulated to generate a set of traces. Afterwards, from the set of traces a set of chronicles, which covers all the possible behaviors and optimizes some quality criterion, is drawn. During on-line monitoring chronicle recognition is performed.

Therefore the quantized system approach is less efficient than the automatic chronicle generation approach, as, owing to state quantization, the considered automata are nondeterministic. The automatic chronicle generation approach, instead, is very efficient on-line but requires a great amount of processing off-line for the exhaustive simulation of the model, which is not guided by any observation, and for the optimization of the set of chronicles.

The discrete-event quantized system approach is one side of the research coordinated by Lunze on diagnosis of quantized systems [Lunze, 2000b, Lunze, 2000a], the other being a discrete-time quantized system approach, wherein the behavioral model is a semi-Markov process. Each transition in the automaton modeling the system behavior represents the possible evolution of the system within a constant sampling time. Such a research addresses also the stochastic properties of the quantized system. In particular, a probability can be assigned to each state transition and thus each resulting diagnosis has an associated probability. Finally, extensions of the discrete-time stochastic method to time-varying faults are discussed in [Schiller et al., 2000].

8. Summary

- *Diagnoser approach*: untimed model-based approach that performs monitoring-based diagnosis of distributed synchronous DESs. The complete behavior of each system component is described by a non-deterministic automaton, and all component models are subdued beforehand to a synchronous composition so as to obtain the global behavioral model of the system, from which a completely observable deterministic automaton, called the diagnoser, is drawn. The diagnoser is used throughout the monitoring process and provides at each step the estimates of the current system state and, for each estimated state, the corresponding candidate diagnosis, this being a set of specific faults, where faults are assumed to be persistent. The major drawback of the approach is that it relies on the global system model, whose generation is infeasible in practice for real size systems owing to its computational cost.

- *Decentralized diagnoser approach*: untimed model-based approach that performs a posteriori diagnosis of distributed synchronous DESs and that is deeply rooted in the active system approach, while trying to improve its efficiency by exploiting off-line compilation of models, as in the diagnoser approach. Each system component, which is described by a nondeterministic automaton, is processed beforehand so as to obtain a so-called local diagnoser, which is a completely observable automaton hiding unobservable component transitions in its macro-states. The local diagnoser is exploited in order to produce a local diagnosis, this being the set of all the possible dynamic evolutions of the component complying with the given observation; then all local diagnoses are merged in a modular way to produce the global diagnosis, where the order of merging steps is decided on the fly, based on component interactions implicit in local diagnoses, so as to reduce the size of the spaces resulting from merging. Each fault may be either transient or persistent.

- *Incremental decentralized diagnosis approach*: version of the decentralized diagnoser approach for performing monitoring-based diagnosis, where at each monitoring step a packet of observed events is considered; the difficulties of reasoning when the observed events are randomly split into successive observation packets are highlighted.

- *Decentralized protocol approach*: version of the diagnoser approach running on a distributed architecture, where each local site gathers its own observation of the whole system and performs monitoring-based diagnosis by exploiting a diagnoser of the whole system; the diagnostic results produced at each step by the local sites are sent to a coordinator site that is in charge of determining the current candidate diagnoses based on a simple processing protocol, without needing any global model or diagnoser of the system. Each distinct local site is able to observe a set of events, which possibly overlaps those of other sites. The major drawback of the approach is the same as for the diagnoser approach.

- *Process algebra approach*: untimed model-based approach to diagnosis of distributed DESs that models the behavior of system components, the structure of the system, the observation, and the diagnostic problem by means of a process algebra (PEPA). The behavioral model of the whole system is drawn and then synchronized with the description of the observation. The description of the structure of the system specifies the events on whose occurrences the (otherwise asynchronous) behavior of components have to synchronize. Synchronizing the global system model with the observation means deriving all and only the system evolutions generating such an observation. Each fault may be either transient or persistent. The major drawbacks of the approach are the cumbersome descriptions (of observations, in particular) and the absence of any modular processing.

- *Quantized system approach*: untimed approach to monitoring-based diagnosis of continuous-variable systems modeled as DESs. The complete behavior of a system is described by a set of completely observable nondeterministic automata, one corresponding to the normal behavior of the overall system and each of the others corresponding to a faulty behavior of the overall system owing to a distinct fault. Faults are assumed to be persistent and present since the beginning of the monitoring process (or even before it). At each step an input is given and an event is observed; such an event is checked against every automaton, after which all automata that are not compliant with it are discarded. The major drawback of the approach is the lack of compositional modeling, that is, the behavior of the system has to be described at system level and not on a component basis.

II

DIAGNOSIS OF ACTIVE SYSTEMS

Chapter 3

ACTIVE SYSTEMS

Abstract

An active system is a network of *components*. Each component is endowed with input and output *terminals*, which are connected with terminals of other components of the system through proper connection *links*. Interactions among components are represented by *events* produced and received on terminals through connection links. System components may also produce observable outputs. These are modeled as events on a specific terminal, which are assumed to be collected in some way by an external *observer*. The behavioral model of a component is described by a communicating finite automaton, which incorporates both nominal and faulty behavior. In the transition function, the current state and the input event at a given time instant give rise to the production of some output events and to an instantaneous state transition. The behavior of system components in the presence of a relevant input event coming from the external world is the *reaction* of the system. The observable outputs produced by the system during its reaction represent the *observation* of the system. The system observation is always associated with a relevant observer, that is, the mode in which the observable outputs can be actually seen from the external world. In fact, only a subset of the set of observable outputs are actually visible by an observer. Furthermore, the total ordering of the observed events can be lost and transformed into partial ordering, thereby making the system observation temporally underconstrained. A *diagnostic problem* is the association among an observer, a system observation, and the state of the system prior to the reaction.

1. Introduction

Chapter 2 has introduced several approaches to model-based diagnosis in the realm of DESs. This chapter is concerned with the definition of a specific class of DESs called active systems. The notion of an active system is given in terms of formal properties rather than concrete, physical characteristics. Depending on the context and the nature of the task to be performed (simulation, supervision, diagnosis, etc.), a physical system can be modeled in different ways. For example, the protection apparatus of a power transmission network (see Chapter 13) can be either modeled as a continuous system or a DES.

In fact, a physical system is not continuous or discrete *per se*, although practical reasons may suggest a particular modeling choice.

However, choosing between continuous and discrete modeling is only the first problem, as several abstraction levels can be considered in either case. Roughly, the more detailed the model, the more complicated the reasoning task of the diagnostic technique, which, in the worst case, is bound to become overwhelming in space and time complexity.

The model of the system is likely to influence its diagnosability too. Intuitively, the more observable the system, the more accurate the diagnosis (the smaller the set of candidates). Ideally, the set of candidate diagnoses should be a singleton, specifically, a single set of faulty components. At any rate, the set of candidates should be sound and complete with respect to the given diagnostic problem, that is, all the constraints imposed by the model and the system observation should be accounted for by the diagnostic task.

Unlike other approaches in the literature where DESs are synchronous, in this chapter active systems are assumed to be asynchronous[1]. This means that events generated by components are stored within connection links before being (asynchronously) consumed.

Thus, the modeling ontology involves two classes of elements: components and links. Topologically, an active system is a network of components that are connected to each other by means of links. The behavior of each component is specified by a finite automaton whose transitions between states are triggered by event occurrences. Such events may either come form the external world or from neighboring components through links. The component model is assumed to be complete, thereby including both normal and faulty behavior.

Unlike components, which exhibit active behavior, links are passive elements that incorporate the events exchanged between components.

Active systems are compositionally modeled in terms of the topology (the way components are connected to one another through links) and the behavior of each component (the way the component react to events)

---

[1] A generalized class of DESs that integrate synchronous and asynchronous behavior is introduced in Chapter 6.

and of each link[2]. Therefore, the actual model of an active system is implicitly defined by its topology, and component and link models. Such a model is in turn a finite automaton, where a state encompasses component and link states[3] while events are component transitions.

The diagnostic task is carried out based not only on the model of the system but also on the system observation and on the mode such an observation is perceived. In its simplest form, a system observation is the sequence of observable events generated by the system. The qualification of an event as observable refers both to the model of the system and to the observer of it. There are events that are (virtually) observable and other that are unobservable. However, a virtually observable event may be unobservable by the specific observer relevant to the diagnostic problem. Besides, the total temporal ordering of observable events may be lost, so that the observation perceived by the observer is temporally underconstrained.

In the remainder of the chapter, Sections 2, 3, and 4 introduce the notions of a component, a link, and a system, respectively. The concept of a system reaction (the way a system reacts to an external event) is presented in Section 5. Sections 6, 7, and 8 deals with observers, observations, and diagnostic problems, respectively. The introduced concepts are summarized in Section 9.

2.   Component

Components are the basic elements of active systems. Each component is characterized by a structure, called the *topological model*, and a mode in which it reacts to incoming events, called the *behavioral model*.

Topologically, a component is described by a set of input terminals and a set of output terminals, which can be thought of as pins through which events are fetched and generated, respectively.

Both nominal and faulty behaviors of a component are uniformly described by a single finite automaton, that is, by a set of states and a transition function. A state transition is triggered by an input event ready on one input terminal. During the transition, before reaching the new state, a (possibly empty) set of output events are generated on output terminals.

The specification of the topological model and the behavioral model of a component C can be abstracted from the specific component and unified in a single notion, namely the *component model*, as follows.

---

[2]Links may differ in the number of storable events and in the mode they behave when the buffer of events is full.
[3]A link state is simply the queue of events stored in it.

A component model M is a communicating automaton,

$$M = (\mathbb{S}, \mathbb{E}_{in}, \mathbb{I}, \mathbb{E}_{out}, \mathbb{O}, \mathbb{T})$$

where:

$\mathbb{S}$ is the set of states;

$\mathbb{E}_{in}$ is the set of inputs;

$\mathbb{I}$ is the set of input terminals;

$\mathbb{E}_{out}$ is the set of outputs;

$\mathbb{O}$ is the set of output terminals;

$\mathbb{T}$ is the (nondeterministic) transition function:

$$\mathbb{T} : \mathbb{S} \times \mathbb{E}_{in} \times \mathbb{I} \times 2^{\mathbb{E}_{out} \times \mathbb{O}} \mapsto 2^{\mathbb{S}}.$$

An input (output) event is a pair (e, ϑ), where e is an input (output) and ϑ an input (output) terminal. A *component* is the instantiation of a component model.

A transition T from state S_1 to state S_2 is triggered by an input event $\alpha = (e, I)$ at an input terminal I and generates the (possibly empty) set of output events $\beta = \{(e_1, O_1), \ldots, (e_n, O_n)\}$ at output terminals $O_1, \ldots, O_n$, respectively. This is denoted by

$$T = S_1 \xrightarrow{\alpha|\beta} S_2.$$

When α is *consumed*, S_2 is reached and the set of output events are generated at the relevant output terminals. An output terminal O may appear at most once in β.

Three *virtual terminals* are implicitly defined in a model:

$In \in \mathbb{I}$, the *standard input*;

$Out \in \mathbb{O}$, the *standard output*;

$Flt \in \mathbb{O}$, the *fault terminal*.

An output event (e, Out) is a *message*. If T is labeled with a message, T is *observable*, denoted by $Observable(T)$, otherwise T is *silent*.

Among events, only messages are visible outside the system. An event (e, Flt) is a *faulty event* and the relevant transition T, a *faulty transition*, denoted by $Faulty(T)$.

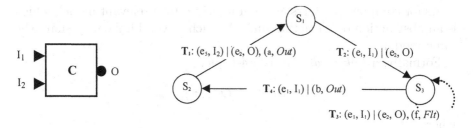

Figure 3.1. Topological model (left) and behavioral model (right) of component C.

Example 3.1. Shown on Figure 3.1 are the topological model (left) and the behavioral model (right) of a component C. The topology of C consists of two input terminals, I_1 and I_2, denoted by triangles, and one output terminal, O, denoted by a circle. Note that virtual terminals are not depicted, as they are implicitly assumed. The behavioral model is an automaton incorporating three states, $S_1 \cdots S_3$, and four transitions, $T_1 \cdots T_4$, where:

$T_1 = S_2 \xrightarrow{(e_5, I_2) \,|\, (e_2, O), (a, Out)} S_1$ is triggered by the input e_5 at terminal I_2 and generates the output e_2 at terminal O, along with the observable output a. As such, T_1 is observable.

$T_2 = S_1 \xrightarrow{(e_1, I_1) \,|\, (e_2, O)} S_3$ is triggered by the input e_1 at terminal I_1 and, as T_1, generates the output e_2 at terminal O, without, however, any further observable event. Therefore T_2 is silent.

$T_3 = S_3 \xrightarrow{(e_1, I_1) \,|\, (e_2, O), (f, Flt)} S_3$ is triggered by the input e_1 at terminal I_1 and, similarly to T_1 and T_2, generates the output e_2 at terminal O, along with the fault event (f, Flt). Thus, since the set of output events do not include an observable output, T_3 is a faulty-silent transition (faulty transition are depicted in dotted arrows). Note that no state change occurs.

$T_4 = S_3 \xrightarrow{(e_1, I_1) \,|\, (b, Out)} S_2$ is triggered by the input e_1 at terminal I_1 and generates the observable output b. So, T_4 is observable. $\diamond$

3. Link

In active systems, components are connected with each other by means of links. Specifically, a link connects an output terminal O of a component with an input terminal I of another component. This way, when an output is generated at O, it is queued within the link and, if it is the head of the queue, it becomes a triggering event for the component relevant to terminal I.

As for components, links are characterized by a relevant model, which is an abstraction of the actual link. As such, several links may share the same model.

Formally, a *link model* M_L is a 4-tuple,

$$M_L = (I, O, \chi, \pi),$$

where:

I is the *input terminal*;

O is the *output terminal*;

χ is the *capacity*;

π is the *saturation policy*.

A *link* is an instantiation of a link model, that is, a directed communication channel between two different components C_1 and C_2, where an output terminal of C_1 and an input terminal of C_2 coincide with the input and output terminal of L, respectively.

Specifically, if O_1 and I_2 are respectively an output terminal of C_1 and an input terminal of C_2, a link L connecting O_1 to I_2, denoted

$$L = (C_1.O_1 \Rightarrow C_2.I_2),$$

is a connection through which each output e generated at O_1 at time t is available at I_2 at time $t' > t$ (no assumptions are made about the duration of the interval $[t, t']$).

Within L, e is said to be a *queued event* of L. The capacity χ of L is the maximum number of queued events in L. The sequence of queued events of L at a given time is denoted by $\|L\|$. The cardinality of $\|L\|$ is denoted by $|L|$. If $|L|$ equals χ, L is *saturated*.

The first consumable event in $\|L\|$ is the *ready event*. When a ready event is consumed, it is dequeued from $\|L\|$. Events are consumed one at a time.

When L is saturated, the semantics for the triggering of a transition T of a component C that generates a new output event (e, O) on L is dictated by the saturation policy π of L, which can be either:

LOSE: e is lost;

OVERRIDE: e replaces the last event in $\|L\|$;

WAIT: the transition T cannot be triggered until L becomes unsaturated, that is, until at least one event in $\|L\|$ is consumed.

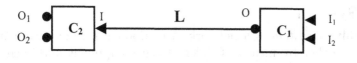

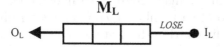

Figure 3.2. Link model $M_L = (I_L, O_L, 3, LOSE)$ (bottom), and instance L of M_L (top).

Let

$Head(L)$ denote the first consumable element in $\|L\|$;

$Tail(L)$ denote the queue of events in $\|L\|$ following the first event;

$App(L, e)$ denote the queue of events obtained by appending e to $\|L\|$;

$Repl(L, e)$ denote the queue of events obtained by replacing the last event in $\|L\|$ with e.

The *Ins* function is defined as follows:

$$Ins(L, e) = \begin{cases} App(L, e) & \text{if } |L| < \chi(L) \\ \|L\| & \text{if } |L| = \chi(L), \pi = LOSE \\ Repl(L, e) & \text{if } |L| = \chi(L), \pi = OVERRIDE. \end{cases}$$

Example 3.2. Shown on the bottom of Figure 3.2 is the pictorial representation of a link model $M_L = (I_L, O_L, 3, LOSE)$. An instantiation L of M_L is displayed on top of the same figure. L is a link connecting the output terminal O of component C_1 with the input terminal I of C_2, thereby $I_L = O$ and $O_L = I$. When a new event is generated by C_1 on terminal O and L is not saturated (the length of the queued events is less than 3), the event is appended to the queue. For example, if the current queue of events in L is $\langle e_1, e_2 \rangle$ (L is not saturated), the operation $Ins(L, e_3)$ will be equivalent to $App(L, e_3)$, thereby generating the new queue $\langle e_1, e_2, e_3 \rangle$, where $Head(L) = e_1$ and $Tail(L) = \langle e_2, e_3 \rangle$. At this point, a further attempt to insert a new event e into L will result in the loss of e. Instead, if the ready event e_1 is consumed, L will be no longer saturated, resulting in the tail of the old queue, namely $\langle e_2, e_3 \rangle$, where e_2 becomes the new ready event. ◇

4.   System

Roughly, a *system* is a network of components which are connected with each other by means of links. Each component and each link in the system is characterized by a relevant model. Several components, as well as several links, may share the same model, respectively. A system may incorporate a number of component terminals that are not connected with any link within the system. These are called the *dangling terminals* of the system.

Accordingly, a system Θ is a triple $(\mathbb{C}, \mathbb{L}, \mathbb{D})$, where

$\mathbb{C}$ is the set of components;

$\mathbb{L}$ is the set of links among terminals of components in $\mathbb{C}$;

$\mathbb{D}$ is the set of dangling terminals.

Component terminals cannot be overloaded, that is, at most one link is connected with each of the component terminals in Θ. Consequently, there is a functional dependency from a terminal x of a component C to the relevant link L. This makes it is possible to unambiguously write $Link(x)$ to denote the (possibly virtual) link relevant to terminal x. Furthermore, by definition, if $\alpha = (e, x)$ is an event, $Link(\alpha) = Link(x)$.

The set of dangling terminals is the union of two disjoint sets,

$$\mathbb{D} = \mathbb{D}_{\text{on}} \cup \mathbb{D}_{\text{off}}$$

where

$\mathbb{D}_{\text{on}}$ is the set of *on-terminals*;

$\mathbb{D}_{\text{off}}$ is the set of *off-terminals*.

If $\mathbb{D}_{\text{on}} \neq \emptyset$, the system Θ is *open*, otherwise Θ is *closed*. If Θ is closed, no events are available at input terminals in $\mathbb{D}_{\text{off}}$, while events generated at output terminals in $\mathbb{D}_{\text{off}}$ are lost.

Instead, if Θ is open, we assume that the dangling terminals in $\mathbb{D}_{\text{on}}$ are connected with links outside Θ, in other words, Θ is supposed to be incorporated within another (virtually unknown) system.

Therefore, events generated at output terminals in $\mathbb{D}_{\text{on}}$ are buffered within the relevant (unknown) link external to Θ. Similarly, the component corresponding to an input terminal ϑ in $\mathbb{D}_{\text{on}}$ is assumed to be sensitive to events ready at ϑ. That is, state transitions may be triggered by events buffered in the (external) link connected with ϑ.

Example 3.3. Shown on the top of Figure 3.3 is the topology of a system $\Psi = (\mathbb{C}, \mathbb{L}, \mathbb{D})$, where

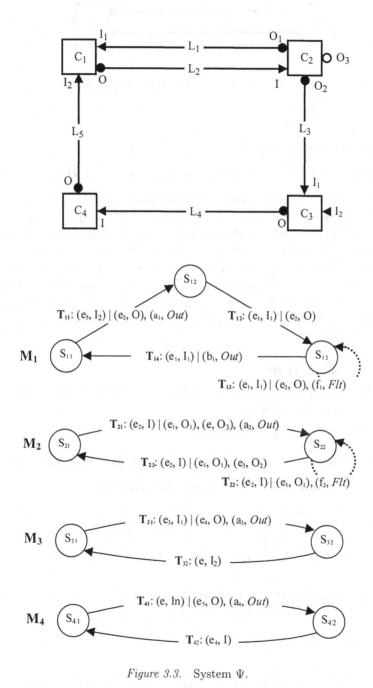

Figure 3.3. System Ψ.

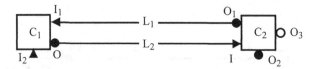

Figure 3.4.   Subsystem $\Psi' \subseteq \Psi$ (see Example 3.4).

$\mathbb{C} = \{C_1, C_2, C_3, C_4\}$;

$\mathbb{L} = \{L_1, L_2, L_3, L_4, L_5\}$;

$\mathbb{D} = \{O_3(C_2), I_2(C_3)\}$, where $\mathbb{D}_{\mathrm{on}} = \{I_2\}$ and $\mathbb{D}_{\mathrm{off}} = \{O_3\}$.

Since $\mathbb{D}_{\mathrm{on}} \neq \emptyset$, Ψ is open, specifically, component C_3 is sensitive to events ready at input terminal I_2, which is assumed to be connected with a link external to Ψ. Instead, no link is supposed to be connected with output terminal O_3 (dangling terminals in $\mathbb{D}_{\mathrm{off}}$ are denoted in white).

Displayed on the bottom of the system topology are the behavioral models $M_1, \ldots, M_4$ of components $C_1, \ldots, C_4$, respectively. The only faulty transitions are T_{13} and T_{22}, that is, only C_1 and C_2 are possibly misbehaving. We assume that all of the link in $\mathbb{L}$ share the same model $(I, O, 2, LOSE)$. $\diamond$

4.1   Subsystem

Let $\Theta = (\mathbb{C}, \mathbb{L}, \mathbb{D})$ be a system. A *subsystem* Θ' of Θ, denoted $\Theta' \subseteq \Theta$, is a system $(\mathbb{C}', \mathbb{L}', \mathbb{D}')$ where:

$\mathbb{C}' \subseteq \mathbb{C}$ is the set of components;

$\mathbb{L}' \subseteq \mathbb{L}$ is the set of links in $\mathbb{L}$ among terminals of components in $\mathbb{C}'$;

$\mathbb{D}' \subseteq \mathbb{D}$ is the set of dangling terminals such that:

$$\mathbb{D}'_{\mathrm{on}} = \{\vartheta \mid \vartheta \in \mathbb{D}_{\mathrm{on}}, \vartheta \text{ of } C, C \in \mathbb{C}'\} \cup$$
$$\{\vartheta \mid \vartheta \notin \mathbb{D}_{\mathrm{on}}, \vartheta \text{ of } C, C \in \mathbb{C}', Link(\vartheta) \notin \mathbb{L}'\}$$
$$\mathbb{D}'_{\mathrm{off}} = \{\vartheta \mid \vartheta \in \mathbb{D}_{\mathrm{off}}, \vartheta \text{ of } C, C \in \mathbb{C}'\}.$$

Example 3.4. With reference to system Ψ in Example 3.3, a subsystem $\Psi' = (\mathbb{C}', \mathbb{L}', \mathbb{D}') \subseteq \Psi$ is displayed in Figure 3.4, where:

$\mathbb{C}' = \{C_1, C_2\}$,
$\mathbb{L}' = \{L_1, L_2\}$,
$\mathbb{D}' = \mathbb{D}'_{\mathrm{on}} \cup \mathbb{D}'_{\mathrm{off}}$, where $\mathbb{D}'_{\mathrm{on}} = \{I_2(C_1), O_2(C_2)\}$, $\mathbb{D}'_{\mathrm{off}} = \{O_3(C_2)\}$.

$\diamond$

5. Reaction

A system Θ, while operating, is either *quiescent* or *reacting*. A quiescent system has no buffered events within its links and it neither changes its state nor produces any output event.

Let $\mathbb{L}(\Theta) = \{L_1, \ldots, L_m\}$. The *queue* of system Θ at a certain instant time is the record of sequences of events relevant to each link in $\mathbb{L}$:

$$Queue(\Theta) = \|\mathbb{L}(\Theta)\| = (\|L_1\|, \ldots, \|L_m\|).$$

If Q is a queue of a system Θ incorporating a link L, then

$$Q[L] = \|L\|.$$

The set of ready events in Q is denoted by $Ready(Q)$. Accordingly, system Θ is quiescent when $Ready(Q) = \emptyset$, in other words, when all links in Θ are empty.

A quiescent system becomes reacting upon the arrival of an event on either the standard input (In) or a dangling input on-terminal of a component.

The *state* of a system Θ, where

$$\mathbb{C}(\Theta) = \{C_1, C_2, \ldots, C_n\},$$
$$\mathbb{L}(\Theta) = \{L_1, L_2, \ldots, L_m\},$$

is a pair (σ, Q), where:

$$\sigma \in \mathbb{S}(C_1) \times \mathbb{S}(C_2) \times \cdots \times \mathbb{S}(C_n),$$
$$Q = Queue(\Theta).$$

5.1 History

Before the reaction starts, every component is waiting in a state; afterwards, during the reaction, it undergoes a sequence of transitions, namely a *history* of the component, which connects the initial state with a final state.

The notion of a history can be straightforwardly extended to a system. The history of a system is a sequence of transitions, each one relevant to a component included in the system, which connects the initial (quiescent) state of the system (that is, the composition of the initial states of its components, while all links are assumed to be empty) with a final state, which, on its turn, is required to be quiescent.

Example 3.5. Considering component C_1 in system Ψ outlined in Figure 3.3, assuming the initial state S_{11}, a possible history of C_1 is:

$$h(C_1) = \langle T_{11}, T_{12}, T_{14} \rangle.$$

Table 3.1.   State transitions relevant to history $h(\Theta)$ of Equation (3.1).

Ψ	C_1	C_2	C_3	C_4	L_1	L_2	L_3	L_4	L_5
Ψ_0	S_{11}	S_{21}	S_{31}	S_{41}	$\emptyset$	$\emptyset$	$\emptyset$	$\emptyset$	$\emptyset$
Ψ_1	S_{11}	S_{21}	S_{31}	S_{42}	$\emptyset$	$\emptyset$	$\emptyset$	$\emptyset$	$\langle e_5 \rangle$
Ψ_2	S_{12}	S_{21}	S_{31}	S_{42}	$\emptyset$	$\langle e_2 \rangle$	$\emptyset$	$\emptyset$	$\emptyset$
Ψ_3	S_{12}	S_{22}	S_{31}	S_{42}	$\langle e_1 \rangle$	$\emptyset$	$\emptyset$	$\emptyset$	$\emptyset$
Ψ_4	S_{13}	S_{22}	S_{31}	S_{42}	$\emptyset$	$\langle e_2 \rangle$	$\emptyset$	$\emptyset$	$\emptyset$
Ψ_5	S_{13}	S_{21}	S_{31}	S_{42}	$\langle e_1 \rangle$	$\emptyset$	$\langle e_3 \rangle$	$\emptyset$	$\emptyset$
Ψ_6	S_{11}	S_{21}	S_{31}	S_{42}	$\emptyset$	$\emptyset$	$\langle e_3 \rangle$	$\emptyset$	$\emptyset$
Ψ_7	S_{11}	S_{21}	S_{32}	S_{42}	$\emptyset$	$\emptyset$	$\emptyset$	$\langle e_4 \rangle$	$\emptyset$
Ψ_8	S_{11}	S_{21}	S_{32}	S_{41}	$\emptyset$	$\emptyset$	$\emptyset$	$\emptyset$	$\emptyset$

Supposing the initial (quiescent) state[4] $\Psi_0 = (S_{11}, S_{21}, S_{31}, S_{41})$ for Ψ, a possible history of Ψ is:

$$h(\Psi) = \langle T_{41}, T_{11}, T_{21}, T_{12}, T_{23}, T_{14}, T_{31}, T_{42} \rangle. \qquad (3.1)$$

The sequence of state transitions for system Ψ is shown in Table 3.1, where for each state Ψ_i of Ψ reached by the i-th transition of $h(\Psi)$, $i \in [1..8]$, the relevant states of components and sequences of queued events are outlined. $\diamond$

5.2   Signature

The totally temporally ordered sequence of messages generated by the system during a reaction, while performing the whole sequence of transitions in its history h, is the *reaction signature* of the system, denoted by $Sign(h)$.

Each message in the signature is endowed with a logical content, a temporal content, and a source.

The *logical content* is the (unique) value bringing the information of the message; such a value belongs to a finite domain of discrete values, named *labels*, associated with the system;

The *temporal content* is the (unique) position of the message within the sequence of messages, which is ordered according to the message emission order;

---

[4] When specifying the initial state of a system Θ, we assume by default that Θ be quiescent, that is, $Queue(\Theta) = (\emptyset \cdots \emptyset)$.

The *source* of a message is the (unique) component that generated such a message.

Example 3.6. If system Ψ during a reaction followed history $h(\Psi)$ of Equation 3.1, then its signature would be the following:

$$Sign(h(\Psi)) = \langle (a_4, C_4), (a_1, C_1), (a_2, C_2), (b_1, C_1), (a_3, C_3) \rangle. \qquad (3.2)$$

Incidentally, since the sets of messages of the components in Ψ are disjoint, the information on the source is redundant. Generally speaking, however, the same message might be shared by several components (typically, when such components are characterized by the same model), thereby making it necessary the association of the message with the relevant source. $\qquad \diamond$

6. Observer

The diagnostic process exploits the observable events generated by a system during the reaction to an external event. Virtually, such information is represented by the signature of the reaction, as defined in Section 5.2.

However, such a scenario is more ideal than realistic in practice. That is, what is actually observed is not in general the reaction signature but, rather, a transformation of it. The discrepancy between the system reaction and what is actually available to the external observer is caused by two main reasons:

(1) Not all of the system components are actually visible by the external observer;

(2) The total ordering among observable events may get lost during the transmission.

The first point means that the fact that an output is generated at terminal *Out* is only a necessary condition for its being visible from the external observer. In other words, the external observer might happen to be incapable to perceive that message. Such an insensitivity can be either:

Real: for practical reasons, the messages of a component cannot be sent to the observer, or

Virtual: for formal reasons, the diagnostic process makes some limitations on the set of exploitable messages, as discussed in Chapter 5.

Point (2) is related to the mode in which messages are transmitted from the system to the observer. Specifically, since we assume that messages are not timestamped when generated (that is, no time information is associated with messages), the total temporal ordering imposed by the reaction signature might be relaxed by the communication channel (from the system to the observer) into a set of sequences, each one corresponding to the messages relevant to a different subset of the system components.

In order to cope with these requirements, we introduce the notion of an observer, which allows the diagnostic process to exploit the system observation appropriately.

Let $\Theta = (\mathbb{C}, \mathbb{L}, \mathbb{D})$ be a system. An *observer* Ω of Θ is a pair

$$\Omega(\Theta) = (\mathbb{V}, \mathbb{P})$$

where

$\mathbb{V}$ is a subset of $\mathbb{C}$ corresponding to the components whose messages are visible to the observer, called the *visible set*;

$\mathbb{P}$ is a partition of $\mathbb{C}$, called the *ordering partition*, where each part defines a domain of temporally ordered messages relevant to the corresponding components.

As such, $\mathbb{V}$ and $\mathbb{P}$ substantiate requirements (1) and (2) above, respectively. On the one hand, $\mathbb{V}$ implicitly represents the domain of messages which are actually observable by the observer Ω, thereby assuming that messages relevant to components in $\mathbb{C} - \mathbb{V}$ are not visible to Ω. On the other, $\mathbb{P}$ implicitly defines the mode in which the total temporal relationships among messages in the reaction signature are relaxed into a set of sequences of messages, where each sequence corresponds to the messages relevant to the components within a part in $\mathbb{P}$.

If $\mathbb{V} = \emptyset$, the observer $\Omega(\Theta)$ is said to be *blind*[5], denoted *Blind*$(\Omega(\Theta))$.

Example 3.7. A possible observer of the system Ψ displayed in Figure 3.3 is $\Omega(\Psi) = (\mathbb{V}, \mathbb{P})$, where

$$\mathbb{V} = \{C_1, C_2, C_3\},$$
$$\mathbb{P} = \{\{C_1, C_2\}, \{C_3, C_4\}\}.$$

According to $\mathbb{V}$, the only invisible messages to $\Omega(\Psi)$ are those generated by components in $\mathbb{C} - \mathbb{V}$, namely C_4. Under this assumption and the ordering partition $\mathbb{P}$, the reaction signature outlined in Equation (3.2) will be relaxed into the two sequences $\langle a_1, a_2, b_1 \rangle$ and $\langle a_3 \rangle$. Such sequences are what is actually observed by $\Omega(\Psi)$.   ◇

---

[5]The notion of a blind observer, which sounds paradoxical, is introduced for formal reasons, as clarified in Section 8.

6.1 Observer Restriction

The *restriction* of an observer $\Omega(\Theta) = (\mathbb{V}, \mathbb{P})$, $\mathbb{P} = \{\mathbf{C}_1, \ldots, \mathbf{C}_n\}$, on a subsystem $\Theta' \subseteq \Theta$, denoted by $\Omega_{\langle \Theta' \rangle}(\Theta)$, is an observer $\Omega(\Theta') = (\mathbb{V}', \mathbb{P}')$ such that:

$$\mathbb{V}' = \mathbb{V} \cap \mathbb{C}(\Theta')$$
$$\mathbb{P}' = \{\mathbf{C}'_i \mid \mathbf{C}'_i = \mathbf{C}_i \cap \mathbb{C}(\Theta'), i \in [1 \mathinner{.\,.} n], \mathbf{C}'_i \neq \emptyset\}.$$

Example 3.8. Considering Example 3.7, the restriction $\Omega_{\langle \Psi' \rangle} = (\mathbb{V}', \mathbb{P}')$, where $\mathbb{C}(\Psi') = \{C_2, C_3, C_4\}$, is such that

$$\mathbb{V}' = \{C_1, C_2, C_3\} \cap \mathbb{C}(\Theta') = \{C_2, C_3\}$$
$$\mathbb{P}' = \{\{C_1, C_2\} \cap \mathbb{C}(\Theta'), \{C_3, C_4\} \cap \mathbb{C}'\} = \{\{C_2\}, \{C_3, C_4\}\}.$$

Instead, the restriction $\Omega_{\langle \Psi'' \rangle} = (\mathbb{V}'', \mathbb{P}'')$, where $\mathbb{C}(\Psi'') = \{C_3, C_4\}$, is such that

$$\mathbb{V}'' = \{C_1, C_2, C_3\} \cap \mathbb{C}(\Psi'') = \{C_3\},$$
$$\mathbb{P}'' = \{\{C_3, C_4\} \cap \mathbb{C}(\Psi'')\} = \{\{C_3, C_4\}\}.$$

$\diamond$

7. Observation

The *observation* of a system consists of what was observed by an observer during a system reaction. Such an observation is taken as input by the diagnosis task in order to find out whether the system evolved normally or not.

Let Θ be a system and $\Omega(\Theta) = (\mathbb{V}, \mathbb{P})$ an observer of Θ, where $\mathbb{P} = \{\mathbf{C}_1, \ldots, \mathbf{C}_n\}$. An *observation item* $obs(\mathbf{C})$ of a set of components $\mathbf{C} \in \mathbb{P}$ is a temporal, totally-ordered sequence of messages relevant to visible components in $\mathbf{C}$:

$$obs(\mathbf{C}) = \langle m_1, \ldots, m_p \rangle.$$

An *observation* $OBS(\Theta)$ of Θ is the record of the $n \geq 1$ observation items $obs(\mathbf{C}_i)$, $i \in [1 \mathinner{.\,.} n]$:

$$OBS(\Theta) = (obs(\mathbf{C}_1), \ldots, obs(\mathbf{C}_n)).$$

If $OBS(\Theta)$ is composed of a single item only, it is called a *linear observation*, otherwise it is called an *itemized observation*.

7.1 Unknown Set

The *unknown set* of an observation item $obs(\mathbf{C}_i)$ is defined as follows:

$$Ukn(obs(\mathbf{C}_i)) = \mathbf{C}_i - \mathbb{V}.$$

The *visible set* of $obs(\mathbf{C}_i)$ is defined as follows:

$$Vis(obs(\mathbf{C}_i)) = \mathbf{C}_i \cap \mathbb{V}.$$

Similarly, the unknown set of a system observation $OBS(\Theta)$ is defined as follows:

$$Ukn(OBS(\Theta)) = \bigcup_{i=1}^{n} Ukn(obs(\mathbf{C}_i)) = \bigcup_{i=1}^{n} (\mathbf{C}_i - \mathbb{V}) = \mathbb{C} - \mathbb{V}. \qquad (3.3)$$

The visible set of $OBS(\Theta)$ is the visible set of the relevant observer:

$$Vis(OBS(\Theta)) = \mathbb{V}.$$

7.2   Observation Index

An *index* K of $OBS(\Theta)$ is a record of n integer variables,

$$K = (k_1, \ldots, k_n),$$

which is isomorphic to $(\mathbf{C}_1, \ldots, \mathbf{C}_n)$, such that

$$\forall i \in [1 .. n] \ (k_i(K) \in [0 .. |obs(\mathbf{C}_i)|]).$$

K is said to be *complete*, denoted by $Complete(K)$, when

$$\forall i \in [1 .. n] \ (k_i(K) = |obs(\mathbf{C}_i)|).$$

Intuitively, each element k_i of the index denotes a prefix of the corresponding observation item $obs(\mathbf{C}_i)$. In particular, when $k_i = 0$, the prefix is the empty string, while when $k_i = |obs(\mathbf{C}_i)|$, that is, when k_i equals the length of $obs(\mathbf{C}_i)$, the prefix coincides with the entire observation item.

Example 3.9. With reference to the observer $\Omega(\Psi)$ of Example 3.7, denoting $\mathbf{C}_1 = \{C_1, C_2\}$ and $\mathbf{C}_2 = \{C_3, C_4\}$, an observation of Ψ is the following:

$$OBS(\Psi) = (obs(\mathbf{C}_1) = \langle a_1, a_2, b_1 \rangle, obs(\mathbf{C}_2) = \langle a_3 \rangle).$$

Considering the definition of unknown set and observation restriction, we have

$$Ukn(obs(\mathbf{C}_1)) = \mathbf{C}_1 - \mathbb{V} = \{C_1, C_2\} - \{C_1, C_2, C_3\} = \emptyset,$$
$$Ukn(obs(\mathbf{C}_2)) = \mathbf{C}_2 - \mathbb{V} = \{C_3, C_4\} - \{C_1, C_2, C_3\} = \{C_4\},$$
$$Ukn(OBS(\Psi)) = \mathbb{C} - \mathbb{V} = \{C_1, C_2, C_3, C_4\} - \{C_1, C_2, C_3\} = \{C_4\}.$$

A possible index value for $OBS(\Psi)$ is $(2, 1)$, which denotes the record of prefixes $(\langle a_1, a_2 \rangle, \langle a_3 \rangle)$. Such an index is complete in correspondence of the value $(3, 1)$.   ◇

7.3 Observation Restriction

The *restriction* of an observation item $obs(\mathbf{C})$ on a set of components $\mathbf{C}' \subseteq \mathbf{C}$, denoted by $obs_{\langle \mathbf{C}' \rangle}(\mathbf{C})$, is the subsequence of messages in $obs(\mathbf{C})$ relevant to visible components in $\mathbf{C}'$.

Let $OBS(\Theta) = (obs(\mathbf{C}_1), \ldots, obs(\mathbf{C}_n))$ be an observation relevant to an observer $\Omega(\Theta)$. The restriction of $OBS(\Theta)$ on a subsystem $\Theta' \subseteq \Theta$ is an observation $OBS(\Theta')$ relevant to an observer

$$\Omega(\Theta') = \Omega_{\langle \Theta' \rangle}(\Theta) = (\mathbb{V}', \mathbb{P}')$$

and defined as follows:

$$OBS(\Theta') = (obs'(\mathbf{C}_i) \mid obs'(\mathbf{C}_i) = obs_{\langle \mathbb{C}(\Theta') \cap \mathbb{V}' \rangle}(\mathbf{C}_i), i \in [1 \mathinner{\ldotp\ldotp} n'], \mathbf{C}_i \in \mathbb{P}'),$$

where $n' \leq n$.

Example 3.10. Consider the observation $OBS(\Psi)$ of Example 3.9. Let Ψ' and Ψ'' be the subsystems of Ψ such that $\mathbb{C}(\Psi') = \{C_2, C_3, C_4\}$ and $\mathbb{C}(\Psi'') = \{C_3, C_4\}$. The following restrictions hold:

$$OBS_{\langle \Psi' \rangle}(\Psi) = (\langle a_2 \rangle, \langle a_3 \rangle) \text{ relevant to observer } \Omega(\Psi'),$$
$$OBS_{\langle \Psi'' \rangle}(\Psi) = (\langle a_3 \rangle) \text{ relevant to observer } \Omega(\Psi'').$$

where (see Example 3.8)

$$\Omega(\Psi') = \Omega_{\langle \Psi' \rangle}(\Psi) = (\mathbb{V}', \mathbb{P}'), \mathbb{V}' = \{C_2, C_3, C_4\}, \mathbb{P}' = \{\{C_2\}, \{C_3, C_4\}\},$$
$$\Omega(\Psi'') = \Omega_{\langle \Psi'' \rangle}(\Psi) = (\mathbb{V}'', \mathbb{P}''), \mathbb{V}'' = \{C_3\}, \mathbb{P}'' = \{\{C_3, C_4\}\}.$$

$\diamond$

7.4 Observation Extension

The *extension* of $OBS(\Theta)$, denoted by $\| OBS(\Theta) \|$, is the finite set of sequences of messages defined as follows:

$$\| OBS(\Theta) \| = \begin{cases} \{\emptyset\} & \text{if } \forall \, obs(\mathbf{C}) \in OBS(\Theta) \, (obs(\mathbf{C}) = \emptyset) \\ \{Q_1, \ldots, Q_p\} & \text{otherwise} \end{cases}$$

where each *instance* $Q = \langle m_1, \ldots, m_q \rangle \in \| OBS(\Theta) \|$, where

$$q = \sum_{i=1}^{n} |obs(\mathbf{C}_i)|,$$

is such that $\forall i \in [1 \mathinner{\ldotp\ldotp} n]$, the restriction of Q_i on $\mathbf{C}_i$ equals $obs(\mathbf{C}_i)$.

Example 3.11. Considering the observation $OBS(\Psi) = (\langle a_1, a_2, b_1 \rangle, \langle a_3 \rangle)$ of Example 3.9, we have

$$\|OBS(\Psi)\| = \{Q_1, Q_2, Q_3, Q_4\}$$

where

$$Q_1 = \langle a_1, a_2, b_1, a_3 \rangle,$$
$$Q_2 = \langle a_1, a_2, a_3, b_1 \rangle,$$
$$Q_3 = \langle a_1, a_3, a_2, b_1 \rangle,$$
$$Q_4 = \langle a_3, a_1, a_2, b_1 \rangle.$$

$\diamond$

8.   Diagnostic Problem

Given a system reaction, the diagnostic task can be accomplished based on some information pertinent to the mode in which such a reaction has been observed. This involves the knowledge of both the system observation and the relevant observer. Additionally, the initial state of the system is assumed to be known[6]. All these information constitute a diagnostic problem, that is, the input of the diagnostic process.

More precisely, let Θ be a system, $\Omega(\Theta)$ an observer of Θ, $OBS(\Theta)$ an observation made by $\Omega(\Theta)$, and Θ_0 the state of Θ before the reaction started, namely, the initial state of Θ.

The triple

$$\wp(\Theta) = (\Omega(\Theta), OBS(\Theta), \Theta_0)$$

is a *diagnostic problem* for Θ. When the observer $\Omega(\Theta)$ is blind (that is, the visible set $\mathbb{V}$ is empty), $\wp(\Theta)$ is said to be a *blind diagnostic problem*, and $OBS(\Theta)$ is denoted by *null*.

Example 3.12. A diagnostic problem for the system Ψ depicted in Figure 3.3 is $\wp(\Psi) = (\Omega(\Psi), OBS(\Psi), \Psi_0)$, where

$$\Omega(\Psi) = (\{C_1, C_2, C_3\}, \{\{C_1, C_2\}, \{C_3, C_4\}\}),$$
$$OBS(\Psi) = (\langle a_1, a_2, b_1 \rangle, \langle a_3 \rangle),$$
$$\Psi_0 = (S_{11}, S_{21}, S_{31}, S_{41}).$$

$\diamond$

---

[6]The knowledge on the initial state of the system is not essential to the diagnostic process. However, it allows the latter to reduce considerably the size of the search space, as shown in Chapter 6.

8.1 Diagnostic Problem Restriction

The *restriction* of $\wp(\Theta)$ on a subsystem $\Theta' \subseteq \Theta$, denoted by

$$\wp_{\langle\Theta'\rangle}(\Theta)$$

is a diagnostic problem $\wp(\Theta') = (\Omega(\Theta'), OBS(\Theta'), \Theta'_0)$ such that:

$$\Omega(\Theta') = \Omega_{\langle\Theta'\rangle}(\Theta),$$

$$OBS(\Theta') = OBS_{\langle\Theta'\rangle}(\Theta),$$

Θ'_0 is the projection of Θ_0 on the components in Θ'.

Example 3.13. With reference to Example 3.12, the restriction of the diagnostic problem $\wp(\Psi)$ on the subsystem $\Psi' \subset \Psi$ such that $\mathbb{C}(\Psi') = \{C_3, C_4\}$, is the diagnostic problem $\wp(\Psi') = (\Omega(\Psi'), OBS(\Psi'), \Psi'_0)$ where

$$\Omega(\Psi') = \Omega_{\langle\Psi'\rangle}(\Psi') = (\{C_3\}, \{\{C_3, C_4\}\}),$$
$$OBS(\Psi') = OBS_{\langle\Psi'\rangle}(\Psi') = (\langle a_3\rangle),$$
$$\Psi'_0 = (S_{31}, S_{41}).$$

 $\diamondsuit$

9. Summary

- *Component*: the basic element of an active system. It is described by a topological model and a behavioral model. The former accounts for both input and output terminals. The latter expresses the (either nominal or faulty) mode in which the component reacts to events.

- *Link*: the means of connecting components to each other. Events exchanged among components are temporally buffered within links before being consumed. A link connects an output terminal of a component with the input terminal of another component.

- *System*: a network of components which are connected to each other through links. It can be either open or closed, depending on whether it embodies input dangling terminals that are connected to other components outside the system or not, respectively.

- *History*: the sequence of transitions performed by a system as a consequence of the occurrence of an event external to the system. It moves the system between two quiescent states, thereby generating a sequence of observable output events, the system signature.

- *Observer*: the mode in which a system signature is actually seen by the external world. It accounts for both the subset of visible components (visible set) and the rules on (partial) temporal ordering of the observed events (ordering partition).

- *Observation*: a set of observation items isomorphic to the ordering partition of the relevant observer, each item being a sequence of observable events (messages) generated during the reaction of the system.

- *Diagnostic Problem*: an association among an observer, a relevant system observation, and the state of the system before the reaction. A diagnostic problem is blind when the visible set of the associated observer is empty (in such a case, the observation is *null*).

Chapter 4

MONOLITHIC DIAGNOSIS

Abstract

Solving a diagnostic problem means generating the set of candidate diagnoses associated with it. However, the generation of such candidates requires the diagnostic engine to make up the behavior of the system based on the information provided by the diagnostic problem, specifically, the observer, the system observation, and the state of the system prior to the reaction. The reconstructed behavior of the system is intensionally (concisely) represented by an automaton called *active space*. Each path from the root of an active space to a final node is a *history* of the system, that is, a reaction consistent with the diagnostic problem. A history is either nominal or faulty, depending on whether it incorporates faulty transitions or not. In a simple view, a diagnosis associated with a history is the (possibly empty) set of faulty components relevant to the transitions within the history. Given a history h, three types of diagnoses are envisaged, namely, *shallow* diagnosis, *deep* diagnosis, and *dynamic* diagnosis. A shallow diagnosis of h is the set of faulty components involved in h. A deep diagnosis of h is the set of faults relevant to h, each fault being the association of a fault event with a component. Finally, a dynamic diagnosis of h is the sequence of (possibly duplicated) faults relevant to h. Even though, owing to cycles, an active space may contain an unbounded number of histories, the number of shallow/deep diagnoses is always finite, insofar as the number of possible faults is bounded. The diagnostic information associated with a diagnostic problem can be accommodated within a *diagnostic hierarchy*, which involves the set of shallow diagnoses, the set of deep diagnoses, the set of dynamic diagnoses (implicitly represented by a *diagnostic graph*), and the set of histories.

1. Introduction

This chapter focuses on the central task of diagnosis of active systems based on given diagnostic problems. The ingredients of a diagnostic problem are the system observation, the relevant observer, and the initial state of the system. Solving a diagnostic problem amounts to generating the set of candidate diagnoses that are consistent with the diagnostic problem. Such candidates correspond to possible reactions of the system, each reaction being a system history rooted in the initial state of the system. In order for a history to be consistent with the diagnostic problem, the set of observable events generated by the history must comply with the observation. Specifically, all observable events visible by the observer are required to be actually part of the observation and the temporal constraints of such observable events within the history must comply with the (possibly relaxed) temporal constraints imposed by the observation. Thus, the central problem of the diagnostic task is the reconstruction of the system behavior.

Owing to possible unobservable cycles within component models, the number of possible histories may be unbounded. In fact, a new history must be considered at each new iteration of an unobservable cycle.

One may argue at this point that reconstructing the system behavior is impossible unless some limiting assumptions are formulated on the system model and/or the diagnostic problem. Roughly, this argument can be phrased as follows:

(1) Solving a diagnostic problem means finding out the set of candidate diagnoses that comply with the diagnostic problem (in particular, with the system observation).

(2) A candidate diagnosis is the set of faulty components relevant to a system history consistent with the diagnostic problem.

(3) Therefore, generating the set of candidate diagnoses amounts to generating the whole set of system histories consistent with the diagnostic problem.

(4) Since the number of system histories consistent with the diagnostic problem is possibly infinite, the (unbounded) set of system histories cannot be generated.

(5) Consequently, generally speaking, the set of candidate diagnoses is expected to be, in the best of cases, sound but incomplete.

The implicit assumption of this claim is that the reconstruction generates the system histories by enumeration.

Instead, in contrast with this assumption, the (possible unbounded) set of system histories can be concisely represented by a finite graph where each path from the root to a final node is a history that comply with the diagnostic problem. Such a graph, called active space, is a finite automaton where nodes are system states and edges are marked by component transitions. That is, the active space relevant to a diagnostic problem is the intensional representation of the whole set of system histories that are consistent with the given diagnostic problem[1].

However, the reconstruction of the system behavior (or more precisely, of all possible system reactions) is only one side of the problem. Once generated the active space, we need to enumerate all possible candidate diagnoses[2] based on the whole set of possible histories implicitly encompassed by the active space. Since the number of system histories is possibly unbounded, a new argument may be formulated as follows:

(1) The (possibly unbounded) set of system histories relevant to a diagnostic problem can be intensionally represented by a finite automaton (active space).

(2) However, candidate diagnoses can be determined by considering all possible histories within the active space: the (possibly empty) set of faulty components involved in the history is a candidate diagnosis.

(3) Since the number of histories may be unbounded, we should consider an infinite number of histories in order to be sure about the completeness of the set of candidate diagnoses.

(4) Consequently, generally speaking, the set of candidate diagnoses is expected to be, in the best of cases, sound but incomplete.

In this case, the third claim above is misleading, as the complete set of candidate diagnoses can be generated without any enumeration of the system histories. Consider a history (a path within the active space) that encompasses an unobservable cycle. As a matter of fact, the relevant set of faulty components does not depend on the number of times such a cycle is iterated, as duplicated faulty components are removed from the candidate diagnosis. Therefore, the set of candidate diagnoses can be determined by considering only a limited number of histories within the active space, thereby allowing for a sound and complete diagnostic result.

---

[1]Equivalently, an active space is a projection (restriction) of the system model on the histories relevant to the diagnostic problem
[2]Even if the set of possible system histories is unbounded, the set of possible diagnoses is always bounded by the powerset of system components.

A further problem to be faced is the nature of the diagnosis. So far, we have assumed that a diagnosis is a set of faulty components. This is called a shallow diagnosis. However, such a definition is rather coarse-grained and does not completely exploit the models of components and the dynamic nature of the system. As outlined in Chapter 3, a faulty transition in a component model is characterized by a relevant faulty event. Thus, the same component may be faulty in different ways and the knowledge of the mode in which the component is misbehaving might be of interest to the diagnostic task[3]. Accordingly, a finer-grained notion of a diagnosis is the set of faults relevant to a system history, where each fault is an association between a component and a relevant faulty event. This is called a deep diagnosis[4].

We can make a step further by considering a notion of diagnosis that accounts for the complete sequence of faults generated by a history. Instead of removing duplicated faults from the diagnosis, we maintain not only such duplicates but also the temporal ordering in which they have been generated. This is called a dynamic diagnosis. Generally speaking, a dynamic diagnosis is bound to include an infinite number of faults, specifically, when such faults are relevant to a cycle within the history. However, as for histories, we can make an intensional representation of all such dynamic diagnoses by means of a finite graph. Such a graph maintains all the dynamic information on how the system has possibly misbehaved during the reaction.

Even if the reconstruction of the system behavior is a precondition for the generation of candidate diagnoses, the way in which the active space is made up is not unique. This chapter focuses on 'monolithic' diagnosis, meaning that the active space is generated by a single step. However, as outlined in Chapter 5, such a generation may be accomplished by means of several steps, where each step focuses either on the reconstruction of the behavior of a subpart of the system (yielding a partial active space) or on the merging of a set of partial active spaces. Whatever the reconstruction strategy, the final active space is the same.

This chapter is organized as follows. Section 2 copes with the reconstruction technique and formally defines the notion of an active space. Section 3 introduces the ontology of diagnosis notions and combines them into a so called diagnostic hierarchy. A summary is given in Section 4. Appendix A provides a pseudo-coded implementation of the

---

[3]For example, a tripped breaker may be faulty by either not opening or closing.
[4]A shallow diagnosis can be obtained from a deep diagnosis by projecting the latter on relevant components.

reconstruction algorithm, while Appendix B contains the proofs of relevant theorems.

2. Behavior Reconstruction

The reconstruction of the system reaction is a central task for the diagnosis of active systems. Given a diagnostic problem

$$\wp(\Theta) = (\Omega(\Theta), OBS(\Theta), \Theta_0),$$

the goal of the reconstruction is to find out the whole set of histories of Θ that explain the observation $OBS(\Theta)$, viewed by the observer $\Omega(\Theta)$, and starting at the initial state Θ_0. The number of possible histories may be very large and, possibly, infinite (when circular patterns of unobservable transitions occur). This is not a problem, since all such histories can be represented by a finite graph, called *active space*. An active space is a finite automaton where nodes represent possible system states, while edges denote component transitions. Thus, each path from the initial state of the automaton (corresponding to the initial state Θ_0 of the system) to a final state (where the system is quiescent) is a possible history of the system.

In order to generate the corresponding active space, the reconstruction technique performs a simulation of the reactive behavior of Θ, starting at Θ_0, by applying the *triggerable* transitions at the current state of Θ based on the behavioral models of the involved components. In order for a transition to be triggerable, the following conditions must be true:

(1) The transition is among those leaving one of the component states in the current node of the active space;

(2) If the input event of the transition comes from a link L of the system, the relevant input must be the first event (ready) in the queue of events within L;

(3) If the transition generates a message, the latter must be consistent with the ordering relationship among messages in $OBS(\Theta)$;

(4) If the transition generates an event directed to a neighboring component, it must be consistent with the management policy of the relevant link.

If no transition is triggerable, backtracking is required. In order to maintain information on the current state of the reconstruction process, each node N of the active space is required to store three different pieces of information:

$$N = (\sigma, K, Q)$$

where

σ is the record of states of system components;

K is the *index* of the observation, that is, a record of integers denoting the messages already generated up to the current node;

Q is a queue of Θ, that is, the set of queues of messages within the internal links of Θ.

A node of the active space is *final* if and only if:

(1) The index K is *complete*, that is, all of the messages in $OBS(\Theta)$ have been generated;

(2) Q is empty.

When the reconstruction process relevant to a diagnostic problem $\wp(\Theta)$ is based on a linear observation (K is simply an integer), it can be described by Algorithm 4.1, where nodes and edges generated during the search are stored in variables $\aleph$ and $\mathcal{E}$, respectively.

Algorithm 4.1. (*Reconstruction from linear observation*)

1. $\aleph := \{N_0\}$ (N_0 is unmarked); $\mathcal{E} := \emptyset$;

2. Repeat Steps 3 through 5 until all nodes in $\aleph$ are marked;

3. Get an unmarked node $N = (\sigma, K, Q)$ in $\aleph$;

4. For each $i \in [1 .. p]$, for each transition T within the model of component C_i, if T is triggerable, that is, if its triggering event is available within the link and T is consistent with both $OBS(\Theta)$ and the link policy (when T generates output events on non-virtual terminals), do the following steps:

   (a) Create a copy $N' = (\sigma', K', Q')$ of N;

   (b) Set $\sigma'[i]$ to the state reached by T;

   (c) If T is observable (a message is generated), then increment K' by one;

   (d) If the triggering event E of T is relevant to an internal link L_j, then remove E from $Q'[j]$;

   (e) Insert the internal output events of T into the relevant queues in Q';

   (f) If $N' \notin \aleph$ then insert N' into $\aleph$; (N' is unmarked)

   (g) Insert edge $N \xrightarrow{T} N'$ into $\mathcal{E}$;

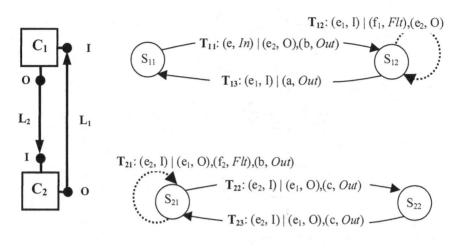

Figure 4.1. System Φ.

5. Mark N;

6. Remove from $\aleph$ all the nodes and from $\mathcal{E}$ all the edges that are not on a path from the initial state N_0 to a final state in $\aleph$. $\diamond$

Example 4.1. With reference to the system Φ shown in Figure 4.1, for which we assume both links L_1 and L_2 with capacity $\chi = 1$ and saturation policy $\pi = OVERRIDE$, consider the following diagnostic problem:

$$\wp(\Phi) = (\Omega(\Phi), OBS(\Phi), \Phi_0) \tag{4.1}$$

where

$$\Omega(\Phi) = (\mathbb{V}, \mathbb{P}), \mathbb{V} = \mathbb{C}(\Phi) = \{C_1, C_2\}, \mathbb{P} = \{\mathbb{C}(\Phi)\},$$
$$OBS(\Phi) = \langle b, b, c, a, b, c, a \rangle,$$
$$\Phi_0 = (S_{11}, S_{21}).$$

Shown in Figure 4.2 is the active space for our example. The initial state is denoted by a circle entered by an isolated arrow (top of the figure), while the (only) final state is denoted by a double circle. Edges are labeled with the identifier of a component transition possibly followed by the relevant message (within brackets). As usual, faulty transitions (T_{12} and T_{21}) are depicted as dotted arrows. The dashed part of the graph represents the part of the search space that is inconsistent with the diagnostic problem. For example, at node $(S_{12}, S_{22}, 3, \langle \rangle, \langle e_2 \rangle)$ no transition can be triggered, as transition T_{23} (which is in principle triggerable by the stored event e_2) would generate message c, which is inconsistent

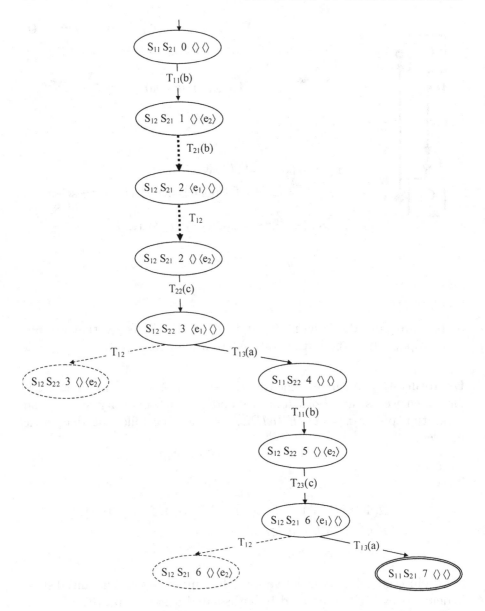

Figure 4.2.   Active space for the diagnostic problem (4.1).

with the current index value (the next (fourth) message in $OBS(\Phi)$ is a). According to the graph, only one history is consistent with the given diagnostic problem, namely

$$h(\Phi) = \langle T_{11}, T_{21}, T_{12}, T_{22}, T_{13}, T_{11}, T_{23}, T_{13}\rangle.$$

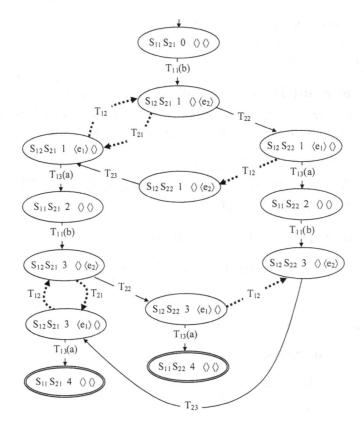

Figure 4.3. Active space relevant to the diagnostic problem (4.2).

◇

Example 4.2. Consider a variation of Example 4.1, where we have the following diagnostic problem:

$$\wp'(\Phi) = (\Omega'(\Phi), OBS'(\Phi), \Phi_0) \tag{4.2}$$

where

$$\Omega'(\Phi) = (\mathbb{V}', \mathbb{P}), \mathbb{V}' = \{C_1\},$$
$$OBS'(\Phi) = (\langle b, a, b, a \rangle).$$

What makes this example different from the previous one is that the observer can see the messages of component C_1 only.

The resulting active space is displayed in Figure 4.3. Notice that, in contrast with the previous result depicted in Figure 4.2, the number of possible histories is unbounded. This is due to the circular paths involving the silent transitions T_{12} and T_{21}. However, every history within the active space is consistent with the system observation (transitions of component C_2 are not marked by the relevant message as the latter is invisible to the observer).   ◇

2.1   Formal Definition of Active Space

We can now formalize the notion of an active space as follows. Let $\wp(\Theta) = (\Omega(\Theta), OBS(\Theta), \Theta_0)$ be a diagnostic problem, where

$$\mathbb{C}(\Theta) = \{C_1, \ldots, C_p\},$$
$$\forall i \in [1 .. p] \text{ the model of } C_i \text{ is } (\mathbb{S}_i, \mathbb{E}_{\text{in}i}, \mathbb{I}_i, \mathbb{E}_{\text{out}i}, \mathbb{O}_i, \mathbb{T}_i),$$
$$\mathbb{P}(\Omega(\Theta)) = \{\mathbf{C}_1, \ldots, \mathbf{C}_n\}.$$

The *active space* of $\wp(\Theta)$ is an automaton

$$Act(\wp(\Theta)) = (\mathbb{S}, \mathbb{E}, \mathbb{T}, S_0, \mathbb{S}_{\text{f}})$$

where

$\mathbb{S}$ is the set of states;

$\mathbb{E}$ is the set of events;

$\mathbb{T}$ is the transition function, $\mathbb{T} : \mathbb{S} \times \mathbb{E} \mapsto \mathbb{S}$;

S_0 is the initial state;

$\mathbb{S}_{\text{f}} \subseteq \mathbb{S}$ is the set of final states.

The elements of the automaton are defined as follows. Let

$\Sigma = \mathbb{S}_1 \times \cdots \times \mathbb{S}_m$;

$\mathbb{K} = [0 .. |obs(\mathbf{C}_1)|] \times [0 .. |obs(\mathbf{C}_2)|] \times \cdots \times [0 .. |obs(\mathbf{C}_n)|]$ denote the domain of possible index values of $OBS(\Theta)$;

$\mathbb{Q}$ denote the domain of possible queues $Queue(\Theta)$.

The *spurious active space* of $\wp(\Theta)$ is an automaton

$$\tilde{\mathcal{A}} = (\tilde{\mathbb{S}}, \mathbb{E}, \tilde{\mathbb{T}}, S_0, \mathbb{S}_{\text{f}})$$

where

$S_0 = (\sigma_0, K_0, Q_0)$, where

$\sigma_0 = (S_{01}, \ldots, S_{0p}), \sigma_0 \in \Sigma,$

$K_0 = (0 \cdots 0), K_0 \in \mathbb{K},$

$$Q_0 = (\emptyset \cdots \emptyset), Q_0 \in \mathbb{Q};$$

$$\tilde{\mathbb{S}} = \{S_0\} \cup \{S' \mid S \xrightarrow{T} S' \in \tilde{\mathbb{T}}\}, \tilde{\mathbb{S}} \subseteq \Sigma \times \mathbb{K} \times \mathbb{Q};$$

$$\mathbb{E} = \bigcup_{i=1}^{m} \mathbb{T}_i;$$

$$\mathbb{S}_f = \{(\sigma_f, K_f, Q_f) \mid \forall S_i \in \sigma_f(S_i \in \mathbb{S}_{fi}), Complete(K_f), Q_f = (\emptyset \cdots \emptyset)\};$$

$$\tilde{\mathbb{T}} : \tilde{\mathbb{S}} \times \mathbb{E} \mapsto \tilde{\mathbb{S}}.$$

Specifically, the transition function $\tilde{\mathbb{T}}$ is defined as follows:

$$N \xrightarrow{T} N' \in \tilde{\mathbb{T}},$$

where

$$N = (\sigma, K, Q), \sigma = (S_1, \ldots, S_m),$$
$$N' = (\sigma', K', Q'), \sigma' = (S_1', \ldots, S_m'),$$

if and only if the following conditions hold:

(1) *Triggerable*(T, N), where *Triggerable* is a predicate defined as follows. Let $L_\alpha \overset{\text{def}}{=} Link(\alpha)$, $T = S \xrightarrow{\alpha|\beta} S'$, $\alpha = (E_\alpha, \theta_\alpha)$. Then

$$Triggerable(T, N) \text{ iff } (Head(Q[L_\alpha]) = E_\alpha);$$

(2) *Consistent*(T, N), where *Consistent* is a predicate defined as follows. Let

$C_{\bar{i}}$ be the component relevant to T, $C_{\bar{i}} \in \mathbf{C}_{\bar{j}}$, $\mathbf{C}_{\bar{j}} \in \mathbb{P}(\Omega(\Theta))$,

$\mathbb{L}_\beta \overset{\text{def}}{=} Link(\beta) \cap \mathbb{L}(\Theta)$,

$\mathbb{L}_\beta^u \overset{\text{def}}{=} \{L_\beta \mid L_\beta \in \mathbb{L}_\beta, (\text{either } L_\beta \text{ is not saturated or } L_\beta = L_\alpha)\}$,

$\mathbb{L}_\beta^s \overset{\text{def}}{=} \mathbb{L}_\beta - \mathbb{L}_\beta^u$,

$\mathbb{L}_\beta^{so} \overset{\text{def}}{=} \{L_\beta \mid L_\beta \in \mathbb{L}_\beta^s, \text{ the saturation policy of } L_\beta \text{ is } OVERRIDE\}$,

$\mathbb{L}_\beta^{sw} \overset{\text{def}}{=} \{L_\beta \mid L_\beta \in \mathbb{L}_\beta^s, \text{ the saturation policy of } L_\beta \text{ is } WAIT\}$.

Then, *Consistent*(T, N) evaluates to true if and only if the following conditions hold:

(a) Either $\left\{ \begin{array}{l} \neg Observable(T) \text{ or} \\ C_{\bar{i}} \in Ukn(obs(\mathbf{C}_{\bar{j}})) \text{ or} \\ (E, Out) \in \beta, obs(\mathbf{C}_{\bar{j}})[K[\bar{j}] + 1] = E, \end{array} \right.$

(b) $\mathbb{L}_\beta^{sw} = \emptyset$;

(3) σ' is such that $\forall i \in [1..p]$ $\left(S_i' = \left\{ \begin{array}{ll} S' & \text{if } i = \bar{\imath} \\ S_i & \text{otherwise} \end{array} \right. \right)$;

(4) K' is such that $\forall j \in [1..n]$

$$\left(K'[j] = \left\{ \begin{array}{ll} K[j+1] & \text{if } j = \bar{\jmath}, \, Observable(T), C_{\bar{\imath}} \in Vis(obs(\mathbf{C}_{\bar{\jmath}})) \\ K[j] & \text{otherwise} \end{array} \right. \right);$$

(5) Q' is such that:

   (a) If $L_\alpha \in \mathbb{L}(\Theta)$ then $Q'[L_\alpha] = Tail(Q[L_\alpha])$,
   (b) $\forall(E, \vartheta) \in \beta, L_\beta = Link(\vartheta), L_\beta \in (\mathbb{L}_\beta^u \cup \mathbb{L}_\beta^{so})$ $(Q'[L_\beta] = Ins(Q[L_\beta], E))$,
   (c) $\forall L \in (\mathbb{L}(\Theta) - (\{L_\alpha\} \cup \mathbb{L}_\beta^u \cup \mathbb{L}_\beta^{so}))$ $(Q'[L] = Q[L])$.

A state $S \in \tilde{\mathbb{S}}$ is *convergent* in $\tilde{\mathcal{A}}$ if and only if either $S \in \mathbb{S}_f$ or there exists a transition $S \xrightarrow{T} S' \in \tilde{\mathbb{T}}$ such that S' is convergent in $\tilde{\mathcal{A}}$. A transition $S \xrightarrow{T} S' \in \tilde{\mathbb{T}}$ is *convergent* in $\tilde{\mathcal{A}}$ if and only if S' is convergent.

The active space $Act(\wp(\Theta))$ is the automaton obtained from $\tilde{\mathcal{A}}$ by selecting the states and transitions which are convergent.

Example 4.3. Consider yet another variation of Example 4.1, where we have the following diagnostic problem:

$$\wp''(\Phi) = (\Omega''(\Phi), OBS''(\Phi), \Phi_0) \tag{4.3}$$

where

$$\Omega''(\Phi) = (\mathbb{V}, \mathbb{P}''), \mathbb{P}'' = \{\{C_1\}, \{C_2\}\},$$
$$OBS''(\Phi) = (\langle b, a, b, a \rangle, \langle b, c, c \rangle).$$

What makes this example different from Example 4.1 is that the system observation is itemized, each item being inherent to a component in Φ. This causes the index K to be composed of two integers, that is

$$K = (k_1, k_2)$$

where k_1 and k_2 are relevant to the observation items $obs(\{C_1\})$ and $obs(\{C_2\})$, respectively. Consequently, a node in the active space is complete if and only if

$$K = (|obs(\{C_1\})|, |obs(\{C_2\})|) = (4, 3).$$

The relevant active space is depicted in Figure 4.4. In contrast with

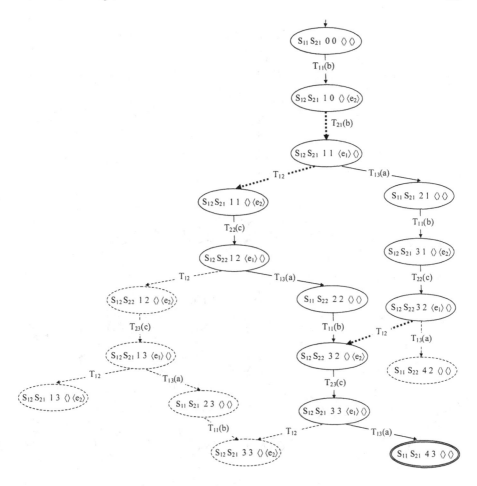

Figure 4.4. Active space relevant to the diagnostic problem (4.3).

the result displayed in Figure 4.2, two different histories are possible. This is due to the fact that, in this case, the ordering constraints among messages in $OBS''(\Phi)$ are less stringent. Intuitively, the more relaxed the ordering constraints within the observation, the larger the set of system histories consistent with the observation. ◇

Example 4.4. With reference to the active system Ψ displayed in Figure 3.3, consider the following diagnostic problem:

$$\wp(\Psi) = (\Omega(\Psi), OBS(\Psi), \Psi_0) \qquad (4.4)$$

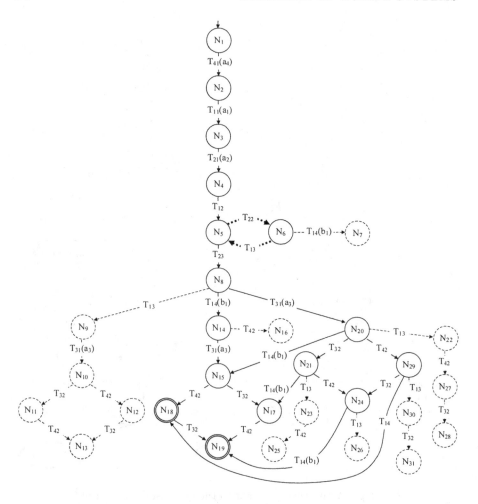

Figure 4.5.   Active space relevant to the diagnostic problem (4.4).

where

$$\Omega(\Psi) = (\mathbb{V}, \mathbb{P}), \mathbb{V} = \mathbb{C}(\Psi), \mathbb{P} = \{\{C_1, C_4\}, \{C_2, C_3\}\},$$
$$OBS(\Psi) = (\langle a_4, a_1, b_1 \rangle, \langle a_2, a_3 \rangle),$$
$$\Psi_0 = (S_{11}, S_{21}, S_{31}, S_{41}).$$

Shown in Figure 4.5 is the active space for our example. Node details are given in Table 4.1.   ◇

2.2   Universal Space

Sometimes it is convenient to generate the active space relevant to a blind diagnostic problem $\wp(\Theta) = (\Omega(\Theta), null, \Theta_0)$. In such a case, the

Table 4.1. Node details relevant to the active space displayed in Figure 4.5.

N	S_1	S_2	S_3	S_4	k_1	k_2	L_1	L_2	L_3	L_4	L_5
N_1	S_{11}	S_{21}	S_{31}	S_{41}	0	0	$\emptyset$	$\emptyset$	$\emptyset$	$\emptyset$	$\emptyset$
N_2	S_{11}	S_{21}	S_{31}	S_{42}	1	0	$\emptyset$	$\emptyset$	$\emptyset$	$\emptyset$	$\langle E_5 \rangle$
N_3	S_{12}	S_{21}	S_{31}	S_{42}	2	0	$\emptyset$	$\langle E_2 \rangle$	$\emptyset$	$\emptyset$	$\emptyset$
N_4	S_{12}	S_{22}	S_{31}	S_{42}	2	1	$\langle E_1 \rangle$	$\emptyset$	$\emptyset$	$\emptyset$	$\emptyset$
N_5	S_{13}	S_{22}	S_{31}	S_{42}	2	1	$\emptyset$	$\langle E_2 \rangle$	$\emptyset$	$\emptyset$	$\emptyset$
N_6	S_{13}	S_{22}	S_{31}	S_{42}	2	1	$\langle E_1 \rangle$	$\emptyset$	$\emptyset$	$\emptyset$	$\emptyset$
N_7	S_{11}	S_{22}	S_{31}	S_{42}	3	1	$\emptyset$	$\emptyset$	$\emptyset$	$\emptyset$	$\emptyset$
N_8	S_{13}	S_{21}	S_{31}	S_{42}	2	1	$\langle E_1 \rangle$	$\emptyset$	$\langle E_3 \rangle$	$\emptyset$	$\emptyset$
N_9	S_{13}	S_{21}	S_{31}	S_{42}	2	1	$\emptyset$	$\langle E_2 \rangle$	$\langle E_3 \rangle$	$\emptyset$	$\emptyset$
N_{10}	S_{13}	S_{21}	S_{32}	S_{42}	2	2	$\emptyset$	$\langle E_2 \rangle$	$\emptyset$	$\langle E_4 \rangle$	$\emptyset$
N_{11}	S_{13}	S_{21}	S_{31}	S_{42}	2	2	$\emptyset$	$\langle E_2 \rangle$	$\emptyset$	$\langle E_4 \rangle$	$\emptyset$
N_{12}	S_{13}	S_{21}	S_{32}	S_{41}	2	2	$\emptyset$	$\langle E_2 \rangle$	$\emptyset$	$\emptyset$	$\emptyset$
N_{13}	S_{13}	S_{21}	S_{31}	S_{41}	2	2	$\emptyset$	$\langle E_2 \rangle$	$\emptyset$	$\emptyset$	$\emptyset$
N_{14}	S_{11}	S_{21}	S_{31}	S_{42}	3	1	$\emptyset$	$\emptyset$	$\langle E_3 \rangle$	$\emptyset$	$\emptyset$
N_{15}	S_{11}	S_{21}	S_{32}	S_{42}	3	2	$\emptyset$	$\emptyset$	$\emptyset$	$\langle E_4 \rangle$	$\emptyset$
N_{16}	S_{11}	S_{21}	S_{31}	S_{41}	3	1	$\emptyset$	$\emptyset$	$\emptyset$	$\langle E_4 \rangle$	$\emptyset$
N_{17}	S_{11}	S_{21}	S_{31}	S_{42}	3	2	$\emptyset$	$\emptyset$	$\emptyset$	$\langle E_4 \rangle$	$\emptyset$
N_{18}	S_{11}	S_{21}	S_{32}	S_{41}	3	2	$\emptyset$	$\emptyset$	$\emptyset$	$\emptyset$	$\emptyset$
N_{19}	S_{11}	S_{21}	S_{31}	S_{41}	3	2	$\emptyset$	$\emptyset$	$\emptyset$	$\emptyset$	$\emptyset$
N_{20}	S_{13}	S_{21}	S_{32}	S_{42}	2	2	$\langle E_1 \rangle$	$\emptyset$	$\emptyset$	$\langle E_4 \rangle$	$\emptyset$
N_{21}	S_{13}	S_{21}	S_{31}	S_{42}	2	2	$\langle E_1 \rangle$	$\emptyset$	$\emptyset$	$\langle E_4 \rangle$	$\emptyset$
N_{22}	S_{13}	S_{21}	S_{32}	S_{42}	2	2	$\emptyset$	$\langle E_2 \rangle$	$\emptyset$	$\langle E_4 \rangle$	$\emptyset$
N_{23}	S_{13}	S_{21}	S_{31}	S_{42}	2	2	$\emptyset$	$\langle E_2 \rangle$	$\emptyset$	$\langle E_4 \rangle$	$\emptyset$
N_{24}	S_{13}	S_{21}	S_{31}	S_{41}	2	2	$\langle E_1 \rangle$	$\emptyset$	$\emptyset$	$\emptyset$	$\emptyset$
N_{25}	S_{13}	S_{21}	S_{31}	S_{41}	2	2	$\emptyset$	$\langle E_2 \rangle$	$\emptyset$	$\emptyset$	$\emptyset$
N_{26}	S_{13}	S_{21}	S_{31}	S_{42}	2	2	$\emptyset$	$\langle E_2 \rangle$	$\emptyset$	$\emptyset$	$\emptyset$
N_{27}	S_{13}	S_{21}	S_{32}	S_{41}	2	2	$\emptyset$	$\langle E_2 \rangle$	$\emptyset$	$\emptyset$	$\emptyset$
N_{28}	S_{13}	S_{21}	S_{31}	S_{41}	2	2	$\emptyset$	$\langle E_2 \rangle$	$\emptyset$	$\emptyset$	$\emptyset$
N_{29}	S_{13}	S_{21}	S_{32}	S_{41}	2	2	$\langle E_1 \rangle$	$\emptyset$	$\emptyset$	$\emptyset$	$\emptyset$
N_{30}	S_{13}	S_{21}	S_{32}	S_{41}	2	2	$\emptyset$	$\langle E_2 \rangle$	$\emptyset$	$\emptyset$	$\emptyset$
N_{31}	S_{13}	S_{21}	S_{31}	S_{41}	2	2	$\emptyset$	$\langle E_2 \rangle$	$\emptyset$	$\emptyset$	$\emptyset$

active space is called a *universal space*, and is denoted by

$$Usp(\Theta, \Theta_0).$$

Therefore, the universal space represents all the possible histories of Θ rooted in Θ_0, disregarding the generation of messages, as no constraints are given by the *null* observation. In particular, the universal space of a component C considered in isolation is the subpart of behavioral model of C reachable from the considered initial state C_0.

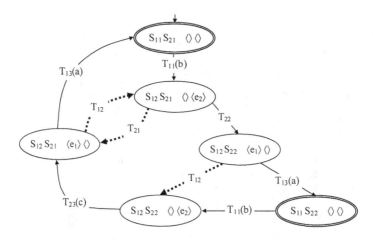

Figure 4.6.   Universal space relevant to the blind diagnostic problem of Example 4.5.

Example 4.5. Considering system Φ displayed in Figure 4.1, the universal space $Usp(\Phi, \Phi_0)$, where $\Phi_0 = (S_{11}, S_{21})$, is shown in Figure 4.6. As expected, it subsumes the active spaces depicted in Figures 4.2, 4.3, and 4.4.   ◇

3.   Diagnosis Generation

Once the reconstruction of the system reaction has been performed, the diagnostic information, which is implicitly embodied within the active space, has to be extracted so as to determine possible faulty behaviors in the reaction of the system, which is the final goal of the diagnostic process.

Formally, given an active space $Act(\wp(\Theta)) = (\mathbb{S}, \mathbb{E}, \mathbb{T}, S_0, \mathbb{S}_f)$, a path $S_0 \rightsquigarrow S_f$ in $Act(\wp(\Theta))$, where $S_f \in \mathbb{S}_f$, is a *history* of Θ. The whole (possibly unbounded) set of histories relevant to $Act(\wp(\Theta))$ is the *history set* of $\wp(\Theta)$, denoted by

$$\mathbb{H}(\wp(\Theta)).$$

A transition $S \xrightarrow{T} S' \in \mathbb{T}$ such that T is a faulty transition in a component model, is a *faulty transition* in $Act(\wp(\Theta))$ and the relevant component, a *faulty component*.

Accordingly, each history within an active space is either qualified as *faulty*, if it includes at least one faulty transition, or *nominal*, otherwise.

When considering a system history, the faulty components are those for which there exists at least a faulty transition in the history. The set of faulty components is empty if the history is nominal.

Table 4.2. Representative histories for the active space depicted in Figure 4.3.

Faulty components	History
$\emptyset$	$\langle T_{11}, T_{22}, T_{13}, T_{11}, T_{23}, T_{13} \rangle$
$\{C_1\}$	$\langle T_{11}, T_{22}, T_{12}, T_{23}, T_{13}, T_{11}, T_{22}, T_{13} \rangle$
$\{C_2\}$	$\langle T_{11}, T_{21}, T_{13}, T_{11}, T_{22}, T_{13} \rangle$
$\{C_1, C_2\}$	$\langle T_{11}, T_{21}, T_{13}, T_{11}, T_{22}, T_{12}, T_{23}, T_{13} \rangle$

Example 4.6. Considering the active space displayed in Figure 4.2, the only included history is faulty, as it embodies both faulty transitions T_{12} and T_{21}. Consequently, both C_1 and C_2 are faulty.

Instead, the history set of the active space $\wp'(\Phi)$ displayed in Figure 4.3 can be partitioned into four classes of histories, each one corresponding to a different sets of faulty components. Representative instances of these classes are provided in Table 4.2.

With reference to the active space $\wp''(\Phi)$ shown in Figure 4.4, notice that the relevant history set incorporates just two histories, namely

$$h_1(\Phi) = \langle T_{11}, T_{21}, T_{12}, T_{22}, T_{13}, T_{11}, T_{23}, T_{13} \rangle,$$
$$h_2(\Phi) = \langle T_{11}, T_{21}, T_{13}, T_{11}, T_{22}, T_{12}, T_{23}, T_{13} \rangle,$$

which involve the same set of faulty components $\{C_1, C_2\}$.

Finally, considering the active space $\wp(\Psi)$ displayed in Figure 4.5, it is easy to note that, even if the relevant history set is unbounded (due to the cycle encompassing nodes N_5 and N_6), from the viewpoint of faulty components, only two classes of histories are embodied, that correspond to the set of faulty components $\emptyset$ and $\{C_1, C_2\}$, respectively. $\diamond$

3.1 Shallow Diagnosis

A *shallow diagnosis*, or simply a *diagnosis* of a diagnostic problem $\wp(\Theta)$, denoted by $\delta(\wp(\Theta))$, is the (possibly empty) set of faulty components $C_1, \ldots, C_k$ relevant to a history in $\mathbb{H}(\wp(\Theta))$, namely

$$\delta(\wp(\Theta)) = \{C_1, \ldots, C_k\}.$$

Note that, even though the number of histories in the history set of the active space may be infinite, the number of distinct diagnoses is always finite as the number of components is finite.

If a diagnosis δ is relevant to a history h, h is said to *entail* δ, namely

$$h \models \delta.$$

Furthermore, by definition, h does not entail any other diagnosis other than δ.

The *shallow diagnostic set*, or simply the *shallow set* of $\wp(\Theta)$, denoted by $\Delta(\wp(\Theta))$, is the whole set of shallow diagnoses $\delta_1, \ldots, \delta_n$ relevant to $\wp(\Theta)$, namely

$$\Delta(\wp(\Theta)) = \{\delta_1, \ldots, \delta_n\}.$$

The notion of a shallow diagnosis is static in nature, as it merely considers the set of components that reacted abnormally during a history of the system. In particular, it neither accounts for the specific faulty events nor for the dynamics in which faulty transitions occurred.

Example 4.7. With reference to the active space displayed in Figure 4.3, the shallow set relevant to the diagnostic problem (4.2) is

$$\Delta(\wp'(\Theta)) = \{\emptyset, \{C_1\}, \{C_2\}, \{C_1, C_2\}\}.$$

Considering the active space depicted in Figure 4.4, relevant to the diagnostic problem (4.3), the corresponding shallow set is the singleton

$$\Delta(\wp''(\Theta)) = \{\{C_1, C_2\}\}.$$

Finally, the shallow set relevant to the active space shown in Figure 4.5, relevant to the diagnostic problem (4.4), is

$$\Delta(\wp(\Psi)) = \{\emptyset, \{C_1, C_2\}\}.$$

$\diamond$

3.2   Deep Diagnosis

The notion of a diagnosis introduced above is rather coarse-grained as the active space incorporates further information about faulty components. A faulty component involves a set of distinct faulty transitions in the same history and, thereby, a set of faulty events. This is why a finer-grained notion of diagnosis is introduced.

A *deep diagnosis* of a diagnostic problem $\wp(\Theta)$, denoted by $\hat{\delta}(\wp(\Theta))$, is the (possibly empty) set of pairs (C, f), where C is a component and f a faulty event generated by a transition of C within a history of $\mathbb{H}(\wp(\Theta))$, namely

$$\hat{\delta}(\wp(\Theta)) = \{(C_1, f_1), \ldots, (C_{\hat{k}}, f_{\hat{k}})\}.$$

Similarly to a shallow diagnosis, if a deep diagnosis $\hat{\delta}$ is relevant to a history h, h is said to *entail* $\hat{\delta}$, denoted by $h \models \hat{\delta}$. Furthermore, by definition, h does not entail any other deep diagnosis other than $\hat{\delta}$.

There exists a one-to-many relationship between a shallow diagnosis and a deep diagnosis, as a shallow diagnosis is obtained by projecting a deep diagnosis on components.

The *deep diagnostic set*, or simply the *deep set* of $\wp(\Theta)$, denoted by $\hat{\Delta}(\wp(\Theta))$, is the whole set of deep diagnoses $\hat{\delta}_1, \ldots, \hat{\delta}_{\hat{n}}$ relevant to $\wp(\Theta)$, namely

$$\hat{\Delta}(\wp(\Theta)) = \{\hat{\delta}_1, \ldots, \hat{\delta}_{\hat{n}}\}.$$

Accordingly, the shallow diagnostic set is obtained by restricting the deep diagnostic set on components (where duplicates are implicitly removed), that is

$$\Delta(\wp(\Theta)) = \hat{\Delta}_{\langle C(\Theta) \rangle}(\wp(\Theta)).$$

Example 4.8. With reference to the active space displayed in Figure 4.3, the deep diagnostic set relevant to the diagnostic problem (4.2) is

$$\hat{\Delta}(\wp'(\Theta)) = \{\emptyset, \{(C_1, f_1)\}, \{(C_2, f_2)\}, \{(C_1, f_1), (C_2, f_2)\}\}.$$

Considering the active space depicted in Figure 4.4, relevant to the diagnostic problem (4.3), the corresponding deep set is the singleton

$$\hat{\Delta}(\wp''(\Theta)) = \{\{(C_1, f_1), (C_2, f_2)\}\}.$$

Finally, the shallow set relevant to the active space shown in Figure 4.5, relevant to the diagnostic problem (4.4), is

$$\hat{\Delta}(\wp(\Psi)) = \{\emptyset, \{(C_1, f_1), (C_2, f_2)\}\}.$$

$\diamond$

3.3 Dynamic Diagnosis

Although more informative than shallow diagnosis, deep diagnosis lacks information both on the order in which faulty transitions have occurred and possibly repetition of them. A further notion of diagnosis is then introduced.

A *dynamic diagnosis* of a diagnostic problem $\wp(\Theta)$, denoted by $\check{\delta}(\wp(\Theta))$, is the sequence of pairs (C, f) obtained from a history in $\mathbb{H}(\wp(\Theta))$ by selecting faulty transitions, namely

$$\check{\delta}(\wp(\Theta)) = \langle (C_1, f_1), \ldots, (C_{\check{k}}, f_{\check{k}}) \rangle.$$

There exists a one-to-many relationship between a deep diagnosis and a dynamic diagnosis, as a deep diagnosis is obtained by transforming a dynamic diagnosis (which is a sequence, possibly with duplicated pairs) into a set (where duplicates are removed).

The *dynamic diagnostic set*, or simply the *dynamic set* of $\wp(\Theta)$, denoted by $\check{\Delta}(\wp(\Theta))$, is the whole set of dynamic diagnoses $\check{\delta}_1, \ldots, \check{\delta}_{\tilde{n}}$ relevant to $\wp(\Theta)$, namely

$$\check{\Delta}(\wp(\Theta)) = \{\check{\delta}_1, \ldots, \check{\delta}_{\tilde{n}}\}.$$

Accordingly, the deep diagnostic set is obtained by transforming each dynamic diagnosis in the dynamic set into a deep diagnosis (as defined above). This transformation is bound to yield two different duplicate removals:

(1) When the dynamic diagnosis is transformed into a set, duplicated pairs are removed from the resulting deep diagnosis;

(2) When dynamic diagnoses are transformed into deep diagnoses, duplicated deep diagnoses are removed from the resulting deep set.

Example 4.9. With reference to the active space displayed in Figure 4.3, the dynamic diagnostic set relevant to the diagnostic problem (4.2) will include, among others, the following dynamic diagnoses:

$$\check{\Delta}(\wp'(\Theta)) \supset \{\langle\rangle, \langle(C_2, f_2), (C_2, f_2)\rangle, \langle(C_2, f_2), (C_1, f_1), (C_1, f_1)\rangle\}.$$

Considering the active space depicted in Figure 4.4, relevant to the diagnostic problem (4.3), the corresponding dynamic set is the singleton

$$\check{\Delta}(\wp''(\Theta)) = \{\langle(C_2, f_2), (C_1, f_1)\rangle\}.$$

Finally, the dynamic set relevant to the active space shown in Figure 4.5, relevant to the diagnostic problem (4.4), includes, among others, the following dynamic diagnoses:

$$\check{\Delta}(\wp(\Psi)) \supset \{\langle\rangle, \langle(C_2, f_2), (C_1, f_1)\rangle, \langle(C_2, f_2), (C_1, f_1), (C_2, f_2), (C_1, f_1)\rangle\}.$$

$\diamond$

3.4   Diagnostic Graph

We might be interested in the following question: Is it possible to represent concisely the (possibly unbounded) set of dynamic diagnoses, just as the possibly unbounded set of histories relevant to a diagnostic problem can be represented by means of a finite graph (active space)? If so, how can we do it? To this end, we first introduce the notion of a diagnostic graph.

Let $Act(\wp(\Theta)) = (\mathbb{S}, \mathbb{E}, \mathbb{T}, S_0, \mathbb{S}_f)$ be an active space. The *diagnostic graph* of $\wp(\Theta)$, $Dgr(\wp(\Theta))$, is a finite automaton

$$Dgr(\wp(\Theta)) = (\mathbb{S}, \tilde{\mathbb{E}}, \tilde{\mathbb{T}}, S_0, \mathbb{S}_f),$$

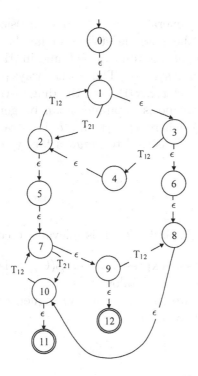

Figure 4.7. Diagnostic graph for the diagnostic problem (4.2).

where

$\tilde{\mathbb{E}} \overset{\text{def}}{=} \{T \mid T \in \mathbb{E}, \text{Faulty}(T)\} \cup \{\epsilon\}$ is the set of events;

$\tilde{\mathbb{T}} : \mathbb{S} \times \tilde{\mathbb{E}} \mapsto 2^{\mathbb{S}}$ is the transition function obtained from $\mathbb{T}$ as follows:

(1) $\forall (S \overset{T}{\to} S' \in \mathbb{T}, T \in \tilde{\mathbb{E}}) \ (S \overset{T}{\to} S' \in \tilde{\mathbb{T}})$,

(2) $\forall (S \overset{T}{\to} S' \in \mathbb{T}, T \notin \tilde{\mathbb{E}}) \ (S \overset{\epsilon}{\to} S' \in \tilde{\mathbb{T}})$.

The set of all possible paths in $Dgr(\wp(\Theta))$ is denoted by $\|Dgr(\wp(\Theta))\|$.

Example 4.10. The diagnostic graph corresponding to the active space depicted in Figure 4.3 is displayed in Figure 4.7, where nodes have been identified by integers $0 \cdots 12$. ◇

3.5 Diagnostic Expression

Given a diagnostic graph, we can distill the relevant deep diagnostic set based on the notion of a diagnostic expression.

The *diagnostic expression* of $\wp(\Theta)$, $Dex(\wp(\Theta))$, is the regular expression corresponding to the automaton $Dgr(\wp(\Theta))$. The language corresponding to $Dex(\wp(\Theta))$ is denoted by $\|Dex(\wp(\Theta))\|$.

An algorithm for generating a regular expression corresponding to the language of all the possible strings of labels associated with the transitions in a finite automaton can be found in [Hopcroft and Ullman, 1979], where ϵ is the null symbol. Thus, it is always possible to generate a diagnostic expression $Dex(\wp(\Theta))$ from the diagnostic graph $Dgr(\wp(\Theta))$.

The notion of a diagnosis entailment can be generalized as follows. Let $\mathcal{T} = \langle T_1, \ldots, T_n \rangle$, where $\forall i \in [1..n]$, T_i is a transition relevant to a component C_i. Then, $\mathcal{T}$ is said to *entail* a faulty set of components $\mathbb{C}$, denoted by

$$\mathcal{T} \models \mathbb{C},$$

if and only if

$$\mathbb{C} = \{ C \mid T \in \mathcal{T}, Faulty(T), T \text{ is relevant to component } C \}.$$

From a given diagnostic expression $Dex(\wp(\Theta))$ it is possible to extract the shallow diagnostic set as follows[5].

Let Q and Q' be two (possibly empty) sequences. Q' is said to be a *subsequence* of Q, denoted by

$$Q' \subseteq Q$$

if and only if:

(1) $\forall q \in Q'\ (q \in Q)$;

(2) The relative order among elements in Q' is the same as in Q.

The *powerset* of a sequence Q, denoted by 2^Q, is the set of subsequences of Q, namely

$$2^Q \overset{\text{def}}{=} \{ Q' \mid Q' \subseteq Q \}. \tag{4.5}$$

Let e be a regular expression and

$$\mathcal{E} = (e_1 | e_2 | \ldots | e_n)$$

the regular expression obtained from e by recursively applying the following transformations:

(1) $(e_1'|e_2'|\cdots|e_{k'}')|(e_1''|e_2''|\cdots|e_{k''}'')$ is replaced by $(e_1'|e_2'|\cdots|e_{k'}'|e_1''|\cdots|e_{k''}'')$;

(2) $e'(e_1'|e_2'|\cdots|e_k')$ is replaced by $(e'e_1'|e'e_2'|\cdots|e'e_k')$;

(3) $(e_1'|e_2'|\cdots|e_k')e'$ is replaced by $(e_1'e'|e_2'e'|\cdots|e_k'e')$;

---

[5]The generation of the deep diagnostic set follows the same pattern.

(4) $(e_1'|e_2'|\cdots|e_k')^*$ is replaced by $(e_1''|e_2''|\cdots|e_p'')$, where

$$\{e_1'', e_2'', \ldots, e_p''\} = 2^{\langle e_1', e_2', \ldots, e_k'\rangle}.$$

The *linearization* of e, denoted by $< e >$, is the set of elements in $\mathcal{E}$, namely

$$< e > \stackrel{\text{def}}{=} \{e_1, e_2, \ldots, e_n\}.$$

Theorem 4.1. *Let $Dex(\wp(\Theta))$ be a diagnostic expression of a diagnostic problem $\wp(\Theta)$. Then, the shallow set can be yielded as follows:*

$$\Delta(\wp(\Theta)) \equiv \{\delta \mid e \models \delta, e \in< Dex(\wp(\Theta)) >\}.$$

Example 4.11. The diagnostic expression relevant to the diagnostic graph corresponding to the active space displayed in Figure 4.5 is

$$Dex(\wp(\Psi)) = (T_{22}T_{13})^*,$$

and the relevant linearization

$$< Dex(\wp(\Psi)) > = \{\emptyset, T_{22}T_{13}\}$$

which, as expected based on Theorem 4.1, gives rise to the shallow set

$$\hat{\Delta}(\wp(\Psi)) = \{\emptyset, \{C_1, C_2\}\}.$$

$\diamond$

3.6   Diagnostic Hierarchy

The notions of a shallow diagnosis, a deep diagnosis, a dynamic diagnosis, as well as a history, which represent, at different abstraction levels, the diagnostic information distilled from the active space, can be accommodated within a *diagnostic hierarchy*.

The *diagnostic hierarchy* ∇ of a diagnostic problem $\wp(\Theta)$, is a 5-tuple:

$$\nabla(\wp(\Theta)) = (D, \hat{D}, \check{D}, H, \mu)$$

where

$D = \Delta(\wp(\Theta))$ is the shallow diagnostic set of $\wp(\Theta)$;

$\hat{D} = \hat{\Delta}(\wp(\Theta))$ is the deep diagnostic set of $\wp(\Theta)$;

$\check{D} = \check{\Delta}(\wp(\Theta))$ is the dynamic diagnostic set of $\wp(\Theta)$;

$H = \mathbb{H}(\wp(\Theta))$ is the history set of $\wp(\Theta)$;

μ is the polymorphic mapping function defined as follows:

(1) $h \mapsto \check{\delta} \in \mu$, $h \in H$, $\check{\delta} \in \check{D}$, iff the restriction of h on the faulty events and relevant components equals $\check{\delta}$,

(2) $\check{\delta} \mapsto \hat{\delta} \in \mu$, $\check{\delta} \in \check{D}$, $\hat{\delta} \in \hat{D}$, iff $Set(\check{\delta}) = \hat{\delta}$,

(3) $\hat{\delta} \mapsto \delta \in \mu$, $\hat{\delta} \in \hat{D}$, $\delta \in D$, iff the restriction of $\hat{\delta}$ on components equals δ.

4.   Summary

- *Reconstruction*: the process of generating the sound and complete set of system histories which are consistent with a given diagnostic problem. Such (possibly unbounded) set of histories is intensionally (concisely) represented by a graph called active space.

- *Active Space*: a finite automaton whose language represents the whole set of histories that are consistent with a given diagnostic problem. Each history is a path from the initial state of the system to a final (quiescent) system state.

- *Shallow Diagnosis*: the set of faulty components relevant to a given system history. The whole set of shallow diagnoses relevant to a diagnostic problem is the shallow (diagnostic) set of the diagnostic problem.

- *Deep Diagnosis*: the set of faults relevant to a given system history, each fault being an association between a component and a faulty event. The whole set of deep diagnoses relevant to a diagnostic problem is the deep (diagnostic) set of the diagnostic problem.

- *Dynamic Diagnosis*: the sequence of faults relevant to a given system history, each fault being an association between a component and a faulty event. The whole (possibly unbounded) set of dynamic diagnoses relevant to a diagnostic problem is the dynamic (diagnostic) set of the diagnostic problem.

- *Diagnostic Graph*: the nondeterministic automaton obtained from an active space by substituting nominal transitions labels with ϵ (null) labels. The resulting language implicitly represents the dynamic diagnostic set of the relevant diagnostic problem.

- *Diagnostic Expression*: the regular expression, on the alphabet of the domain of faulty transitions relevant to an active space, expressing the language of the diagnostic graph corresponding to the given active space.

- *Diagnostic Hierarchy*: a mapping between the different diagnosis abstractions relevant to a given diagnostic problem. Specifically, histories are mapped onto dynamic diagnoses, dynamic diagnoses onto deep diagnoses, and deep diagnoses onto shallow diagnoses.

Appendix A
Algorithms

This appendix provides two algorithms, expressed in pseudo-code, relevant to simulation-based diagnosis. Algorithm A.2 implements the reconstruction of the system behavior based on a given diagnostic problem. Algorithm A.3 computes the deep diagnostic set of a given diagnostic problem.

Algorithm A.2. (*Reconstruct*) The *Reconstruct* function generates a graph-based representation of the active space $Act(\wp(\Theta))$ relevant to the input diagnostic problem (the notation of the algorithm is based on the formal specification of an active space provided in Section 2.1). To this end, it makes use of the recursive function *RStep*, whose definition is embedded within *Reconstruct*, between the parameter specification and the body of *Reconstruct*, the latter starting at Line 55. At Line 56, the parameters of the first call of *RStep* are instantiated, namely σ_0, K, Q, and $\aleph$. *RStep* is first called at Line 57.

Considering the body of *RStep* (Lines 1–54), possible transitions among those leaving from the current node N are searched for within the nested loop at Line 3. Lines 4–10 set the variables introduced in the definition of active space, Point 2. If the transition entails an output event on a saturated link with saturation mode *WAIT*, it is discarded (Line 12). Lines 14–16 checks the consistency of the possible message associated with the candidate transition with the index of the observation: if inconsistent, the transition is discarded. Line 17 makes a copy of the fields of the current node into the new variables σ', K', and Q'. The new state reached by the candidate transition is updated at Line 18. If the transition is observable and the relevant component is visible by the observer, the observation index is incremented at Line 20. At Line 23, the triggering event α is dequeued from the relevant link (only if the latter is an internal event of Θ). The queues relevant to the output events of the candidate transition are updated at Lines 25–27. At Line 28, the new node N' is created. If N' was not visited already (Line 35), *RStep* is recursively called at Line 36, and an edge from N to N' is possibly created (Line 42).

The loop terminates at Line 44. At this point, three cases are possible:

(1) N is final (Line 45): N is marked as consistent and returned (Lines 46–47);

(2) N is not final, but an edge from it to a new node was created (Line 48): N is simply returned (Line 49);

(3) N is not final and no edge was created from it (Line 50): N is marked as inconsistent and a null value is returned (Line 52).

Lines 59–62 in *Reconstruct* remove from the graph $\tilde{\Re}$ the possible dangling nodes and edges which are not part of any path from the root of $\tilde{\Re}$ to a final node. These anomalous sub-graphs of $\tilde{\Re}$ correspond to cycles in which all the nodes are marked as unknown. However, consistent silent cycles possibly remain still in $\tilde{\Re}$ after the pruning.

Lines 63–68 establish the isomorphism between the graph representation of the reconstruction $\Re$ and its elements $\mathbb{S}$, $\mathbb{E}$, $\mathbb{T}$, S_0, and S_f. The resulting reconstruction is eventually returned at Line 69.

function *Reconstruct*$(\wp(\Theta))$:

                The active space $Act(\wp(\Theta)) = (\mathbb{S}, \mathbb{E}, \mathbb{T}, S_0, S_f), \mathbb{S} \subseteq \Sigma \times \mathbb{K} \times \mathbb{Q}$

   input

      $\wp(\Theta) = (\Omega(\Theta), OBS(\Theta), \Theta_0)$: A diagnostic problem for system Θ;

    function *RecStep*$(\sigma, K, Q, \textbf{out}\ \aleph)$: The root of a sub-graph of $Act(\wp(\Theta))$

      input

         $\sigma = (S_1, \ldots, S_m) \in \Sigma$,

         $K \in \mathbb{K}$,

         $Q = Queue(\Theta)$: a queue of Θ,

         $\aleph \subseteq \Sigma \times \mathbb{K} \times \mathbb{Q}$: the set of nodes visited up to the current call of *RecStep*

      output

         $\aleph$: The updated set of visited nodes;

      begin {RecStep}

1.   Insert $N = (\sigma, K, Q)$ into $\aleph$ and mark it as *unknown*;

2. $\mathbb{T}_{cand} := \{T \mid T = S \xrightarrow{\alpha|\beta} S', T$ relevant to a component in $\Theta, S \in \sigma,$
                $\alpha = (E_\alpha, \theta_\alpha), (\alpha \in Ready(Q)$ or $Link(\alpha) \notin \mathbb{L}(\Theta))\}$;

3.   **for each** $T = S \xrightarrow{\alpha|\beta} S' \in \mathbb{T}_{cand}$ **do**

4. $C_{\bar{i}} :=$ the component relevant to T, $C_{\bar{i}} \in \mathbf{C}_{\bar{j}}$, $\mathbf{C}_{\bar{j}} \in \mathbb{P}(\Omega(\Theta))$;

5. $L_\alpha := Link(\alpha)$;

6. $\mathbb{L}_\beta := Link(\beta) \cap \mathbb{L}(\Theta)$;

7. $\mathbb{L}_\beta^u := \{L_\beta \mid L_\beta \in \mathbb{L}_\beta,$ (either L_β is not saturated or $L_\beta = L_\alpha)\}$;

8. $\mathbb{L}_\beta^s := \mathbb{L}_\beta - \mathbb{L}_\beta^u$;

9. $\mathbb{L}_\beta^{so} := \{L_\beta \mid L_\beta \in \mathbb{L}_\beta^s,$ the saturation policy of L_β is *OVERRIDE*$\}$;

10. $\mathbb{L}_\beta^{sw} := \{L_\beta \mid L_\beta \in \mathbb{L}_\beta^s,$ the saturation policy of L_β is *WAIT*$\}$;

11.   **if** $\mathbb{L}_\beta^{sw} \neq \emptyset$ **then**

12.   **goto** 3

13.   **end-if**;

14.   **if** $Observable(T), (E, Out) \in \beta$ **and** $C_{\bar{i}} \in \mathbb{V}(\Omega(\Theta))$ **and**
          $obs(\mathbf{C}_{\bar{j}})[K[\bar{j}] + 1] \neq E$ **then**

15.   **goto** 3

16.   **end-if**;

17. $\sigma' := \sigma$; $K' := K$; $Q' := Q$;

18. $\sigma'[\bar{i}] = S'$;

19.   **if** $Observable(T)$ **and** $C_{\bar{i}} \in \mathbb{V}(\Omega(\Theta))$ **then**

20. $K'[\bar{j}] := K'[\bar{j}] + 1$

21.   **end-if**;

22.   **if** $L_\alpha \in \mathbb{L}(\Theta)$ **then**

```
23.          Q'[L_α] := Tail(Q[L_α])
24.        end-if;
25.        for each event (E, ϑ) ∈ β, L_β = Link(ϑ), L_β ∈ (𝕃_β^u ∪ 𝕃_β^so) do
26.          Q'[L_β] := Ins(Q[L_β], E)
27.        end-for;
28.        N' := (σ', K', Q');
29.        if N' ∈ ℵ then
30.          if N' is marked as inconsistent then
31.              N'_ℜ := nil
32.          else
33.              N'_ℜ := N'
34.          end-if
35.        else
36.          N'_ℜ := RecStep(σ', K', D', ℵ)
37.        end-if;
38.        if N'_ℜ ≠ nil then
39.          if N is marked as unknown and N' as consistent then
40.              Mark N as consistent
41.          end-if;
42.          Create an edge N ⟶^T N'
43.        end-if
44.      end-for;
45.      if N = (σ, K, (∅ ··· ∅)) and Complete(K) then
46.        Mark N as consistent;
47.        return N
48.      elsif At least one edge was created at Line 42 then
49.          return N
50.      else
51.        Mark N as inconsistent;
52.          return nil
53.      end-if
54.    end {RecStep};

55.  begin {Reconstruct}
56.    σ_0 := (S_{01}, ..., S_{0p}); K := (0 ··· 0); Q := (∅ ··· ∅); ℵ := ∅;
57.    N_0 := RecStep(σ_0, K, D, ℵ);
58.    ℜ̃ := the graph rooted in N_0;
59.    for each N ∈ ℵ such that N is marked as unknown do
60.      if ∄ a path N ↝ (σ_f, K_f, (∅ ··· ∅)) in ℜ̃ such that Complete(K_f) then
61.        Remove N from ℜ̃;
62.    Remove from ℜ̃ all the dangling edges;
63.    𝕊 := the set of nodes in ℜ̃;
64.    𝔼 := the set of labels of edges in ℜ̃;
65.    𝕋 := the set of edges in ℜ̃;
66.    S_0 := the root of ℜ̃;
67.    𝕊_f ⊆ 𝕊 := the set of nodes N_f = (σ_f, K_f, (∅ ··· ∅)) such that Complete(K_f);
68.    ℜ := (𝕊, 𝔼, 𝕋, S_0, 𝕊_f);
69.    return ℜ
70.  end {Reconstruct}.
```

Algorithm A.3. (*Diagnose*) The *Diagnose* function takes as input a diagnostic problem and generates as output the relevant deep diagnostic set. To do so, at Line 2, it first makes up the active space of the diagnostic problem by calling the *Reconstruct* function. Then, it produces the relevant diagnostic expression (Line 3) and the corresponding linearization (Line 4). Based on the latter, the deep set is generated at Line 5 and returned at Line 6.

> function $Diagnose(\wp(\Theta))$: The deep diagnostic set $\hat{\Delta}(\wp(\Theta))$
> **input**
> $\wp(\Theta)$: A diagnostic problem;
> 1. **begin** {Diagnose}
> 2. $Act(\wp(\Theta)) := Reconstruct(\wp(\Theta))$;
> 3. $Dex(\wp(\Theta)) :=$ the diagnostic expression of $\wp(\Theta)$;
> 4. $\{e_1, \ldots, e_k\} :=$ the linearization $< Dex(\wp(\Theta)) >$;
> 5. $\hat{\Delta}(\wp(\Theta)) := \{\hat{\delta}_1, \ldots, \hat{\delta}_k\}$ where $\forall i \in [1 .. k]$
> $(\hat{\delta}_i = \{(C, f) \mid T \in e_i, T$ relevant to C, f is the fault event of $T\})$;
> 6.   **return** $\hat{\Delta}(\wp(\Theta))$
> 7. **end** {Diagnose}.

Appendix B
Proofs of Theorems

Theorem 4.1 *Let $Dex(\wp(\Theta))$ be a diagnostic expression of a diagnostic problem $\wp(\Theta)$. Then, the shallow set can be yielded as follows:*

$$\Delta(\wp(\Theta)) \equiv \{\delta \mid e \models \delta, e \in< Dex(\wp(\Theta)) >\}.$$

Proof. Based on the definitions of diagnosis and diagnostic expression, we may assert that:

(1) $\forall h \in \mathbb{H}(\wp(\Theta)), h \models \delta$ $(h' \in \|Dgr(\wp(\Theta))\|, h' \models \delta)$;

(2) $\forall h \in \|Dgr(\wp(\Theta))\|, h \models \delta$ $(h' \in \mathbb{H}(\wp(\Theta)), h' \models \delta)$.

Intuitively, such an isomorphism shows that the diagnostic information in $Dgr(\wp(\Theta))$ is equivalent to that in $Act(\wp(\Theta))$, that is, the set of diagnoses entailed by the whole set of paths in the two graphs is the same. Since $Dex(\wp(\Theta))$ represents the whole set of strings of the language relevant to $Dgr(\wp(\Theta))$, namely $\|Dex(\wp(\Theta))\| \equiv \|Dgr(\wp(\Theta))\|$, we may claim that $Dex(\wp(\Theta))$ carries the same diagnostic information as $Act(\wp(\Theta))$. Thus, we have to show that the linearization of $Dex(\wp(\Theta))$ carries the same diagnostic information as $Dex(\wp(\Theta))$. To this end, denoting with $Set(Q)$ the transformation of the sequence Q into a set (with removal of duplicates), it suffices proving that

$$\forall e \in \|Dex(\wp(\Theta))\|, e \models \delta \ (e' \in< Dex(\wp(\Theta)) >, e' \models \delta, Set(e') = Set(e))$$

and vice versa. In other terms, we have to show that:

(1) If $e \in \|Dex(\wp(\Theta))\|$ then $e' \in< Dex(\wp(\Theta)) >, Set(e') = Set(e)$;

(2) If $e \in< Dex(\wp(\Theta)) >$ then $e' \in \|Dex(\wp(\Theta))\|, Set(e') = Set(e)$.

To prove the above entailments, note that, according to the definition of linearization, among the four transformations, only the last one actually changes the language of the regular expression, that is:

$$(e'_1|e'_2|\cdots|e'_k)^* \text{ is replaced by } (e''_1|e''_2|\cdots|e''_p),$$

where

$$\{e''_1, e''_2, \ldots, e''_p\} = 2^{\langle e'_1, e'_2, \cdots, e'_k \rangle}.$$

However, what matters is that in such a transformation the diagnostic information of the two expressions is the same. The proof of the latter statement amounts to showing that:

(1) If $e \in \|(e_1'|e_2'|\cdots|e_k')^*\|$ then $e' \in \{e_1'', e_2'', \ldots, e_p''\}, Set(e) = Set(e')$;

(2) If $e \in \{e_1'', e_2'', \ldots, e_p''\}$ then $e' \in \|(e_1'|e_2'|\cdots|e_k')^*\|, Set(e) = Set(e')$.

In fact, on the one hand, if $e \in \|(e_1'|e_2'|\cdots|e_k')^*\|$ then e is composed of one or several of zero or more elements in $\{e_1', \cdots, e_k'\}$. This implies that, owing to duplicate removal, $Set(e)$ incorporates zero or more elements in $\{e_1', \cdots, e_k'\}$, that is, $Set(e) \subseteq \{e_1', \cdots, e_k'\}$. Based on the definition of sequence powerset given in Formula (4.5), there exists $e_i'' \in \{e_1'', \cdots, e_p''\}$ such that $Set(e_i'') = Set(e)$.

On the other hand, if $e \in \{e_1'', e_2'', \ldots, e_p''\}$ then the set of elements in e is a subset of $\{e_1', \cdots, e_k'\}$, which implies that $e \in \|(e_1'|e_2'|\cdots|e_k')^*$, where $Set(e) = Set(e')$.

The proof of Theorem 4.1 stems from the fact that, whenever applying each of the four transformations defined in the linearization, the language relevant to the new regular expression, even if different, keeps maintaining the same diagnostic information, as just shown. □

Chapter 5

MODULAR DIAGNOSIS

Abstract

In monolithic diagnosis, the reconstruction of the system behavior is carried out in a single step (monolithically) based on the simulation of the system reaction that is consistent with the given diagnostic problem. However, this is not in general the best way to perform reconstruction, both because the search space, besides the system observation, is constrained by the component models only and because no support for parallel computation is provided. A better way to proceed is to recursively break down the diagnostic problem into several smaller subproblems, each one relevant to a subpart of the system, namely a *cluster*. The subproblem associated with a cluster is obtained by projecting the observer, the observation, and the initial state on the cluster, which in general results in an open subsystem. Such a decomposition is specified by a *reconstruction graph* that, in its simplest form, is a tree that mirrors the decomposition of the original diagnostic problem. Starting from the leaves of the graph, reconstruction is carried out for lower-level clusters, while the resulting active spaces are joint to obtain the active space relevant to upper-level clusters. This is iterated until the root of the reconstruction graph is reached: the resulting active space is equivalent to the active space we would obtain by means of a monolithic reconstruction, as detailed in Chapter 4. In its more general form, the reconstruction graph is a DAG, where each node may represent several clusters that share the same diagnostic subproblem. This way, the method allows the parallel reconstructions of clusters, whilst supporting reuse of intermediate active spaces.

1.   Introduction

In Chapter 4 we have introduced a reconstruction technique that generates the active space relevant to a diagnostic problem in a monolithic way. When the diagnostic problem involves a large system, the reconstruction is bound to become extremely complex in time and space. Such a complexity stems from the monolithic nature of the reconstruction, where the diagnostic problem is faced as a whole and the active space is generated in one shot.

This situation is common in several computing areas. For example, in database systems [Elmasri and Navathe, 1994, Ullman and Widom, 1997, Atzeni et al., 1999], a relational query typically involves the join of several large relations, namely

$$\bowtie_p (R_1, R_2, \ldots, R_n),$$

where R_i are relations and $\bowtie$ is the join relational operator with predicate p. The semantics of the query is as follows. Consider a tuple t belonging to the Cartesian product of the n relations,

$$t \in (R_1 \times R_2 \times \cdots \times R_n).$$

Tuple t belongs to the result if and only if predicate p is satisfied for t, namely $p(t)$ evaluates to true. If we answer the query this way, we need to consider a number N of tuples corresponding to the (possibly huge) cardinality of the above Cartesian product, namely

$$N = |R_1 \times R_2 \times \cdots \times R_n| = |R_1| * |R_2| * \cdots * |R_n|.$$

Although it is virtually possible to answer the query this way, relational database systems try to find a strategy that allows the correct answer in a more efficient way. The basic idea is to apply local selections on single relations in order to reduce the number of tuples considered by the join, namely

$$\bowtie_{p'} (\sigma_{p_1}(R_1), \sigma_{p_2}(R_2), \ldots, \sigma_{p_n}(R_n)),$$

where σ is the selection relational operator, p_i a restriction of predicate p on relation R_i, and p' the new join predicate involving conditions spanning over different relations. Since the cardinality of the intermediate relations generated by selections may be shrunk considerably, the number of tuples considered by the successive join is bound to become much smaller.

This strategy can be generalized in such a way that the original query is substituted by a more articulated (yet equivalent) query represented by a tree called *query plan*, where nodes are relational operators and leaves are relations[1].

---

[1] Typically, the morphology of the query plan is dictated not only by heuristic rules (such as moving downward the local selections) but also by statistical knowledge relevant to the size of the relations within the database.

Query optimization in relational databases is conceptually similar to the problem of behavior reconstruction in active systems. Histories correspond to tuples, component models to relations, diagnostic problems to queries, and active spaces to answers[2]. Specifically, a diagnostic problem can be seen as the 'join' of the component models where the role of 'predicate' is played by the observation.

Modular diagnosis is the task of generating the active space relevant to a diagnostic problem by means of several intermediate steps, based on a structure that parallels the query plan of relational databases, called the reconstruction graph. Similarly to a query plan, where nodes are associated with intermediate relations, nodes of the reconstruction graph are associated with intermediate active spaces relevant to a projection of the diagnostic problem on a subpart of the system, namely a cluster. Whatever the reconstruction graph, the active space associated with the root equals the active space obtained monolithically.

In the remainder of the chapter, Section 2 introduces the notion of a cluster. Section 3 defines the concept of a decomposition of a system into a set of clusters. Formal properties of subsumption and monotonicity are introduced. The technique of modular reconstruction is given in Section 4, while the compositional (extended) definition of an active space is provided in Section 5. The notions of a diagnostic problem decomposition and a reconstruction graph are outlined in Section 6. The introduced concepts are summarized in Section 7. Relevant pseudo-coded algorithms and proofs of theorems are reported in Appendix A and B, respectively.

2. Cluster

In monolithic diagnosis the reconstruction of the system behavior is targeted to the entire system, that is considered as a whole during simulation. By contrast, modular diagnosis requires reconstruction to be applicable to subparts of the systems. Thus, we need to refine the notion of a subsystem introduced in Chapter 4.

A subsystem ξ of an active system Θ is called a *cluster* of Θ. Generally speaking, a cluster is open, as it may include dangling on-terminals (see Section 4 of Chapter 3) resulting from the isolation of a component from the neighboring components that are not within the cluster. The mode in which the cluster behaves is implicitly defined by the models of the components and the links within the cluster. We may define a cluster

---

[2] A difference between the realms of databases and diagnosis consists of the fact that the number of histories within active spaces may be unbounded, while relations always incorporate a finite number of tuples.

model as the network of component and link models, thereby abstracting from actual components and links, as follows.

Let λ be a domain of labels, $\mathbb{M}$ a domain of component models, and $\mathbb{M}^\mathbb{L}$ a domain of link models. A *cluster model* is a directed graph

$$M_\xi = (\mathbb{C}_\xi, \mathbb{L}_\xi, \mathbb{D}_\xi)$$

where

$\mathbb{C}_\xi \subseteq \lambda \times \mathbb{M}$ is the set of nodes;

$\mathbb{L}_\xi \subseteq \mathbb{O}_L \times \mathbb{I}_L \times \mathbb{M}^\mathbb{L}$ is the set of edges, where

$$\mathbb{O}_L = \{C.O \mid (C, M) \in \mathbb{C}_\xi, O \in \mathbb{O}(M)\},$$
$$\mathbb{I}_L = \{C.I \mid (C, M) \in \mathbb{C}_\xi, I \in \mathbb{I}(M)\};$$

$\mathbb{D}_\xi$ is the set of dangling terminals (see Section 3.4).

Furthermore, the following conditions hold:

(1) M_ξ is connected;

(2) Terminals of component models are not overloaded, that is, if

$$(C_i.O_i, C_j.I_j, M_k^L) \in \mathbb{L}_\xi$$

then $\forall C_m \neq C_i, \forall C_l \neq C_j, \forall M^L \in \mathbb{M}^\mathbb{L}$ the following conditions hold:

$$(C_i.O_i, C_l.I_l, M^L) \notin \mathbb{L}_\xi, (C_m.O_m, C_j.I_j, M^L) \notin \mathbb{L}_\xi.$$

If $\mathbb{C}_\xi$ is a singleton, then M_ξ is an *atomic cluster model*. A *cluster*

$$\xi = (\mathbb{C}, \mathbb{L}, \mathbb{D})$$

is an instantiation of a cluster model M_ξ, where $\mathbb{C}$ is the set of components instantiating the pairs (C, M) in M_ξ, $\mathbb{L}$ the set of links instantiating the relationships between terminals of component models in M_ξ, and $\mathbb{D}$ the resulting dangling terminals.

3.   Decomposition

Modular diagnosis requires the decomposition of the diagnostic problem into subproblems applied to subparts of the system. The problem decomposition is reflected into the system decomposition. More generally, we need to define the notion of a cluster decomposition, as follows.

Let Θ be the system incorporating ξ. The *frontier* of ξ is the set of links defined as follows:

$$Front(\xi) = \{L \mid L \in (\mathbb{L}(\Theta) - \mathbb{L}(\xi)), L = (C.O \Rightarrow C'.I'), C' \in \mathbb{C}(\xi)\}.$$

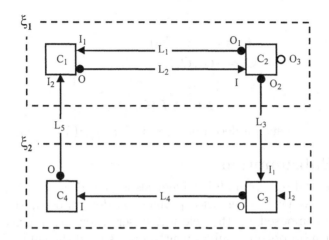

Figure 5.1. Clusters ξ_1 and ξ_2 of system Ψ (see Figure 3.4).

A *decomposition*

$$\Xi(\xi) = \{\xi_1, \dots, \xi_n\}$$

of a cluster ξ is a set of disjoint clusters ξ_i such that $\{\mathbb{C}(\xi_1), \dots, \mathbb{C}(\xi_n)\}$ is a partition of $\mathbb{C}(\xi)$.

3.1 Interface

The decomposition of a cluster loses the connections among subclusters. On the other hand, in order for modular diagnosis to be effective, we need to account for these connections sooner or later. Thus, we need to define the notion of an *interface* of a decomposition as follows.

The interface of a decomposition $\Xi(\xi)$, is the set of links connecting components belonging to different clusters in $\Xi(\xi)$, namely

$$Interf(\Xi(\xi)) = \{L \mid L \in \mathbb{L}(\xi), L = (C_1.O \Rightarrow C_2.I), C_1 \in \mathbb{C}(\xi_i),$$
$$C_2 \in \mathbb{C}(\xi_j), i \neq j, \xi_i \in \Xi(\xi), \xi_j \in \Xi(\xi)\}.$$

Example 5.1. Shown in Figure 5.1 is a decomposition of system Ψ (depicted in Figure 3.4) into two clusters, namely

$$\Xi(\Psi) = \{\xi_1, \xi_2\}$$

where

$$\xi_1 = (\mathbb{C}_1, \mathbb{L}_1, \mathbb{D}_1),$$
$$\mathbb{C}_1 = \{C_1, C_2\},$$
$$\mathbb{L}_1 = \{L_1, L_2\},$$
$$\mathbb{D}_1 = \{I_2(C_1), O_2(C_2), O_3(C_2)\},$$

and

$$\xi_2 = (\mathbb{C}_2, \mathbb{L}_2, \mathbb{D}_2),$$
$$\mathbb{C}_2 = \{C_3, C_4\},$$
$$\mathbb{L}_2 = \{L_4\},$$
$$\mathbb{D}_2 = \{I_1(C_3), I_2(C_3), O(C_4)\}.$$

The interface of such a decomposition is $Interf(\Xi(\Psi)) = \{L_3, L_5\}$.   ◇

3.2   Subsumption

An essential role in modular diagnosis is played by the notion of subsumption. When a diagnostic problem is decomposed, the temporal constraints imposed by the relevant observation are relaxed so as to cope with subproblems. Subsequently, once solved the subproblems, the resulting active spaces are merged and a new (larger) active space is made up. However, in order to preserve the soundness of the diagnostic method, relaxation of observations are supposed to fulfill specific subsumption patterns.

Let

$$OBS(\xi) = (obs(\mathbf{C}_1), \ldots, obs(\mathbf{C}_n)),$$
$$OBS'(\xi) = (obs'(\mathbf{C}_1), \ldots, obs'(\mathbf{C}_n)),$$

be two observations of cluster ξ relevant to observers $\Omega(\xi) = (\mathbb{V}, \mathbb{P})$ and $\Omega'(\xi) = (\mathbb{V}', \mathbb{P}')$, respectively. $OBS(\xi)$ *subsumes* $OBS'(\xi)$, denoted by

$$OBS(\xi) \succeq OBS'(\xi),$$

if and only if

$$\mathbb{P} = \mathbb{P}' = \{\mathbf{C}_1, \ldots, \mathbf{C}_n\},$$
$$\mathbb{V} \supseteq \mathbb{V}',$$
$$\forall i \in [1 .. n] \ \left(obs'(\mathbf{C}_i) = obs_{\langle \mathbb{V}' \cap \mathbf{C}_i \rangle}(\mathbf{C}_i)\right).$$

Example 5.2. Consider system Ψ of Figure 3.4. Since a system is a cluster, we may apply the notion of subsumption to observations of Ψ. Let $\Omega(\Psi) = (\mathbb{V}, \mathbb{P})$ and $\Omega'(\Psi) = (\mathbb{V}', \mathbb{P}')$ be two observers relevant to observations $OBS(\Psi)$ and $OBS'(\Psi)$, respectively, where

$$\mathbb{V} = \{C_1, C_2, C_3, C_4\},$$
$$\mathbb{V}' = \{C_1, C_3\},$$
$$\mathbb{P} = \mathbb{P}' = \{\{C_1, C_4\}, \{C_2, C_3\}\},$$
$$OBS(\Psi) = (\langle a_4, a_1, b_1 \rangle, \langle a_2, a_3 \rangle),$$
$$OBS'(\Psi) = (\langle a_1, b_1 \rangle, \langle a_3 \rangle).$$

Clearly,

$$OBS(\Psi) \succeq OBS'(\Psi),$$

insofar as

$$\mathbb{P} = \mathbb{P}',$$
$$(\mathbb{V} = \{C_1, C_2, C_3, C_4\}) \supseteq (\mathbb{V}' = \{C_1, C_3\}),$$
$$obs'(\mathbf{C}_1) = \langle a_1, b_1 \rangle = obs_{\langle \mathbb{V}' \cap \mathbf{C}_1 \rangle}(\mathbf{C}_1),$$
$$obs'(\mathbf{C}_2) = \langle a_3 \rangle = obs_{\langle \mathbb{V}' \cap \mathbf{C}_2 \rangle}(\mathbf{C}_2).$$

$\diamond$

Proposition 5.1. *The subsumption relationship is transitive, that is,*

$$OBS(\xi) \succeq OBS'(\xi), OBS'(\xi) \succeq OBS''(\xi) \models OBS(\xi) \succeq OBS''(\xi).$$

The notion of subsumption can be naturally extended to a diagnostic problem as follows. Let

$$\wp(\xi) = (\Omega(\xi), OBS(\xi), \xi_0),$$
$$\wp'(\xi) = (\Omega'(\xi), OBS'(\xi), \xi_0)$$

be two diagnostic problems relevant to cluster ξ. $\wp(\xi)$ is said to subsume $\wp'(\xi)$, denoted

$$\wp(\xi) \succeq \wp'(\xi),$$

if and only if

$$OBS(\xi) \succeq OBS'(\xi).$$

3.3   Monotonicity

In order to provide formal support to the soundness of the modular approach, we need to relate the notion of subsumption between the observation relevant to a diagnostic problem and the set of observations obtained by decomposing the latter into a set of subproblems. Specifically, the temporal constraints imposed by the original observation are required to subsume the union of the constraints imposed by its fragmented pieces.

Let $\Xi(\xi) = \{\xi_1, \dots, \xi_m\}$ be a decomposition of a cluster ξ. An observation $OBS(\xi)$ is *monotonic* with respect to a set of observations $\mathbf{OBS}(\Xi(\xi)) = \{OBS(\xi_1), \dots, OBS(\xi_m)\}$, denoted by

$$OBS(\xi) \unrhd \mathbf{OBS}(\Xi(\xi)),$$

if and only if

$$\forall j \in [1 .. m] \left(OBS_{\langle \xi_j \rangle}(\xi) \succeq OBS(\xi_j) \right). \tag{5.1}$$

Example 5.3. With reference to the decomposition of system Ψ shown in Figure 5.1, consider the following observations:

$$OBS(\Psi) = (\langle a_4, a_1, b_1 \rangle, \langle a_2, a_3 \rangle),$$
$$OBS(\xi_1) = (\langle a_1, b_1 \rangle, \langle \rangle),$$
$$OBS(\xi_2) = (\langle \rangle, \langle a_3 \rangle),$$

with relevant observers

$$\Omega(\Psi) = (\{C_1, C_2, C_3, C_4\}, \{\{C_1, C_4\}, \{C_2, C_3\}\},$$
$$\Omega(\xi_1) = (\{C_1\}, \{\{C_1\}, \{C_2\}\},$$
$$\Omega(\xi_2) = (\{C_3\}, \{\{C_3\}, \{C_4\}\}.$$

Clearly,

$$OBS(\Psi) \trianglerighteq \{OBS(\xi_1), OBS(\xi_2)\},$$

insofar as

$$(OBS_{\langle \xi_1 \rangle}(\xi) = (\langle a_1, b_1 \rangle, \langle a_2 \rangle)) \succeq OBS(\xi_1),$$
$$(OBS_{\langle \xi_2 \rangle}(\xi) = (\langle a_4 \rangle, \langle a_3 \rangle)) \succeq OBS(\xi_2).$$

$\diamondsuit$

As with subsumption, the notion of monotonicity can be extended to a diagnostic problem as follows. Let

$$\wp(\xi) = (\Omega(\xi), OBS(\xi), \xi_0)$$

be a diagnostic problem, and

$$\mathbf{P}(\Xi(\xi)) = \{\wp(\xi_1), \dots, \wp(\xi_m)\}$$

a set of diagnostic problems relevant to decomposition $\Xi(\xi)$. $\wp(\xi)$ is said to be monotonic with respect to $\mathbf{P}(\Xi(\xi))$, denoted

$$\wp(\xi) \trianglerighteq \mathbf{P}(\Xi(\xi)),$$

if and only if

$$\forall j \in [1 .. m] \ \left(\wp_{\langle \xi_j \rangle}(\xi) \succeq \wp(\xi_j) \right). \tag{5.2}$$

Theorem 5.1. *Let $OBS(\xi)$ be an observation of a cluster $\xi = (\mathbb{C}, \mathbb{L}, \mathbb{D})$,*

$$\Xi(\xi) = \{\xi_1, \dots, \xi_m\}$$

a decomposition of ξ, and

$$\mathbf{OBS}(\Xi(\xi)) = \{OBS(\xi_1), \dots, OBS(\xi_m)\}$$

a set of observations of clusters in $\Xi(\xi)$ such that

$$OBS(\xi) \trianglerighteq \mathbf{OBS}(\Xi(\xi)).$$

Then,

$$Ukn(OBS(\xi)) \subseteq \bigcup_{j=1}^{m} Ukn(OBS(\xi_j)).$$

Example 5.4. With reference to Example 5.3 where

$$OBS(\Psi) \trianglerighteq \{OBS(\xi_1), OBS(\xi_2)\},$$

we have, as expected by Theorem 5.1,

$$(Ukn(OBS(\Psi)) = \emptyset) \subseteq (Ukn(OBS(\xi_1)) \cup Ukn(OBS(\xi_2)) = \{C_2, C_4\}).$$

$\diamond$

4. Modular Reconstruction

Chapter 4 has presented a technique for reconstructing the reaction of a system Θ based on a given diagnostic problem

$$\wp(\Theta) = (\Omega(\Theta), OBS(\Theta), \Theta_0),$$

whose goal is to generate the active space $Act(\wp(\Theta))$, that is, the automaton representing all possible histories of Θ that explain $\wp(\Theta)$. In doing so, the technique performs a simulation of the reactive behavior of Θ, starting at Θ_0, by applying the triggerable transitions at the current state of Θ based on the behavioral models of the involved components. Such a technique is called monolithic since simulation is applied to the system as a whole.

However, monolithic reconstruction may result inappropriate in real, large-scale active systems, where the search space is bound to become excessively wide. Furthermore, monolithic reconstruction does not support parallel computation: the active space is made up by a single process, namely the reconstruction algorithm.

Parallel computation can be achieved by breaking down the diagnostic problem relevant to system Θ into a set of subproblems, each of which is pertinent to a cluster of Θ. Basically, as detailed in Section 3 of Chapter 4, the system is decomposed into a set of disjoint clusters, each cluster ξ being associated with the diagnostic problem resulting from the restriction of $\wp(\Theta)$ on ξ. That is, given a decomposition

$$\Xi(\Theta) = \{\xi_1, \dots, \xi_m\},$$

the original diagnostic problem is transformed into a set of diagnostic problems

$$\mathbf{P}(\Xi(\Theta)) = \{\wp(\xi_1), \ldots, \wp(\xi_m)\}$$

where

$$\forall i \in [1 .. m] \ (\wp(\xi_i) = \wp_{\langle \xi_i \rangle}(\Theta)).$$

Such a decomposition can be carried out recursively: a diagnostic problem relevant to a cluster can be broken down into other subproblems corresponding to a decomposition of the cluster (a generalization of this approach is presented in Section 6.1).

Therefore, the active system is considered no longer as a whole, but rather as the composition of several clusters and relevant interfaces. Basically, the modular reconstruction follows two steps:

(1) The active space relevant to the diagnostic subproblem $\wp(\xi_i)$ is generated by its own (possibly in parallel);

(2) The resulting active spaces are appropriately joined to obtain the active space relevant to the original diagnostic problem $\wp(\Theta)$.

The final graph obtained by merging the active spaces $Act(\wp(\xi_i))$ is equivalent to the active space yielded by a monolithic reconstruction of $\wp(\Theta)$ (see Proposition 5.2).

The second step (join) is required because the decomposition is bound to lose constraints both in the observation (temporal ordering among messages) and in the topology (links within the interface are not considered). Such constraints are forced by merging the active spaces based on the observation of Θ and the simulation of the behavior of the links within the interface.

Since we already know how to monolithically reconstruct the system behavior relevant to a cluster, we only need to clarify how the join mechanism works. Thus, assume to have built (monolithically) the set of active spaces

$$\mathbf{A} = \{Act(\wp(\xi_1)), \ldots, Act(\wp(\xi_m))\}.$$

We may view $Act(\wp(\xi_i))$ as the (constrained) behavioral model of cluster ξ_i, exactly as the active space $Act(\Theta)$ can be thought of as the (constrained) behavioral model of system Θ. Thus, the join step can be seen as the (monolithic) reconstruction of the behavior of Θ based on the (constrained) model of the clusters in $\Xi(\Theta)$.

With this in mind, the join algorithm mirrors monolithic reconstruction. Specifically, it performs a simulation of the system behavior by applying the triggerable transitions from the current state of Θ based on the active spaces in **A**. In order for a transition to be triggerable, the following conditions must be true:

(1) The transition is among those leaving one of the cluster states in the current node of the active space;

(2) If the input event of the transition comes from a link L of the interface of $\Xi(\Theta)$, the relevant input must be the first event (ready) in the queue of events within L;

(3) If the transition generates a message, the latter must be consistent with the ordering relationship among messages in $OBS(\Theta)$;

(4) If the transition generates an event on a link L in $Interf(\Xi(\Theta))$, it must be consistent with the management policy of L.

As in monolithic reconstruction, if no transition is triggerable, backtracking is carried out. In order to maintain information on the current state of the reconstruction process, each node N of the active space is required to store three different pieces of information:

$$N = (\sigma, K, Q),$$

where

σ is the record of states of the active spaces relevant to the clusters;

K is the index of the observation, that is, a record of integers denoting the messages already generated up to the current node;

Q is a queue of $Interf(\Theta)$, that is, the set of queues of messages within the links in the interface of Θ.

A node of the active space is final when:

(1) The index K is *complete*, that is, all the messages in $OBS(\Theta)$ have been generated;

(2) Each state in σ is final in the corresponding (local) active space;

(3) Q is empty.

The initial node

$$N_0 = (\sigma_0, K_0, Q_0)$$

is such that

(1) The elements in σ_0 are the initial states of the local active spaces;

(2) All elements in K_0 are zero;

(3) All links in Q are empty.

When the join is based on a linear observation (K is simply an integer), it can be described by Algorithm 5.1.

Algorithm 5.1. (*Join*)

1. $\aleph = \{N_0\}$ (N_0 is unmarked); $\mathcal{E} := \emptyset$;

2. Repeat Steps 3 through 5 until all nodes in $\aleph$ are marked;

3. Get an unmarked node $N = (\sigma, K, Q)$ in $\aleph$;

4. For each $i \in [1 .. m]$, for each edge leaving a state in σ (within a local active space) marked by a transition T, if T is triggerable, that is, when its triggering event is relevant to a link L in $Interf(\Xi(\Theta))$, if such an event is available in L and T is consistent with both $OBS(\Theta)$ and the link policy (when T generates output events on links in $Interf(\Xi(\Theta))$), do the following steps:

   (a) Create a copy $N' = (\sigma', K', Q')$ of N;
   (b) Set $\sigma'[i]$ to the state reached by the edge marked by T;
   (c) If T is observable (a message is generated), then increment K' by one;
   (d) If the triggering event E of T is relevant to a link L_j in $Interf(\Xi(\Theta))$, then remove E from $Q'[j]$;
   (e) Insert the internal output events of T into the relevant queues in Q';
   (f) If $N' \notin \aleph$ then insert N' into $\aleph$; (N' is unmarked)
   (g) Insert edge $N \xrightarrow{T} N'$ into $\mathcal{E}$;

5. Mark N;

6. Remove from $\aleph$ all the nodes and from $\mathcal{E}$ all the edges that are not on a path from the initial state N_0 to a final state in $\aleph$.   ◇

Example 5.5. Consider Example 4.4, where a monolithic reconstruction relevant to the diagnostic problem

$$\wp(\Psi) = (\Omega(\Psi), OBS(\Psi), \Psi_0)$$

was carried out based on the following assumptions:

$$\Omega(\Psi) = (\mathbb{V}, \mathbb{P}), \text{ where } \mathbb{V} = \mathbb{C}(\Psi), \mathbb{P} = \{\{C_1, C_4\}, \{C_2, C_3\}\},$$
$$OBS(\Psi) = (\langle a_4, a_1, b_1 \rangle, \langle a_2, a_3 \rangle),$$
$$\Psi_0 = (S_{11}, S_{21}, S_{31}, S_{41}).$$

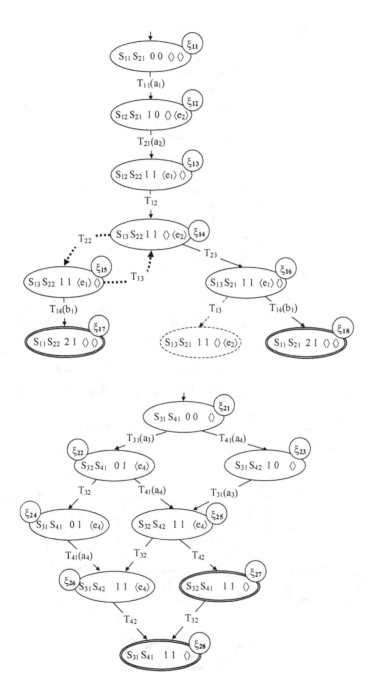

Figure 5.2. Active spaces $Act(\wp(\xi_1))$ (top) and $Act(\wp(\xi_2))$ (bottom) (see Example 5.5).

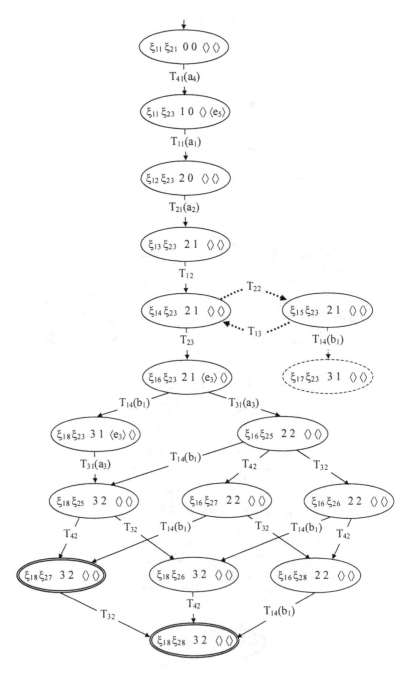

Figure 5.3. Active space for system Ψ resulting from the join of the active spaces displayed in Figure 5.2 (see Example 5.5).

We now create the same active space in a modular way. To this end, with reference to clusters ξ_1 and ξ_2 shown in Figure 5.1, we first generate the active spaces relevant to the diagnostic subproblems obtained by restricting $\wp(\Psi)$ on such clusters (see Section 8.1 of Chapter 4), respectively, that is,

$$\wp(\xi_1) = \wp_{\langle\xi_1\rangle}(\Psi) = (\Omega(\xi_1), OBS(\xi_1), \xi_{1_0}),$$

where

$$\Omega(\xi_1) = (\{C_1, C_2\}, \{\{C_1\}, \{C_2\}\})$$
$$OBS(\xi_1) = (\langle a_1, b_1 \rangle, \langle a_2 \rangle),$$
$$\xi_{1_0} = (S_{11}, S_{21}),$$

and

$$\wp(\xi_2) = \wp_{\langle\xi_2\rangle}(\Psi) = (\Omega(\xi_2), OBS(\xi_2), \xi_{2_0}),$$

where

$$\Omega(\xi_2) = (\{C_3, C_4\}, \{\{C_4\}, \{C_3\}\})$$
$$OBS(\xi_2) = (\langle a_4 \rangle, \langle a_3 \rangle),$$
$$\xi_{2_0} = (S_{31}, S_{41}).$$

The monolithic reconstructions relevant to the diagnostic subproblems $\wp(\xi_1)$ and $\wp(\xi_2)$ are displayed respectively on the top and on the bottom of Figure 5.2. Nodes of the corresponding active spaces have been marked by $\xi_{11} \cdots \xi_{18}$ and $\xi_{21} \cdots \xi_{28}$, respectively.

Based on such local active spaces, the generation of the system behavior relevant to the diagnostic problem $\wp(\Psi)$ is carried out in accordance with Algorithm 5.1. The result is shown in Figure 5.3. Such an active space is equivalent to the active space displayed in Figure 4.5, obtained in Example 4.4 by the monolithic approach. In particular, it incorporates the same history set $\mathbb{H}(\wp(\Theta))$. ◇

5. Compositional Definition of Active Space

The notion of active space based on modular reconstruction can be formalized as follows. Let $\wp(\xi) = (\Omega(\xi), OBS(\xi), \xi_0)$ be a diagnostic problem for cluster ξ. An active space of $\wp(\xi)$ is an automaton

$$Act(\wp(\xi)) = (\mathbb{S}, \mathbb{E}, \mathbb{T}, S_0, \mathbb{S}_f)$$

where

$\mathbb{S}$ is the set of states;

$\mathbb{E}$ is the set of events;

$\mathbb{T}$ is the transition function, $\mathbb{T} : \mathbb{S} \times \mathbb{E} \mapsto \mathbb{S}$;

S_0 is the initial state;

$\mathbb{S}_f \subseteq \mathbb{S}$ is the set of final states.

The elements of the automaton are defined as follows.

(1) If ξ incorporates a single component C with model

$$M_C = (\mathbb{S}_C, \mathbb{E}_{\text{in}C}, \mathbb{I}_C, \mathbb{E}_{\text{out}C}, \mathbb{O}_C, \mathbb{T}_C)$$

and the observer of $OBS(\xi)$ is blind, then

$\mathbb{S} = \mathbb{S}_C$;

$\mathbb{E} = \mathbb{T}_C$;

$\mathbb{T} : \mathbb{S} \times \mathbb{E} \mapsto \mathbb{S}$;

$S_0 = \xi_0$;

$\mathbb{S}_f = \{S_f \in \mathbb{S}_C$ such that there exists a path $S_0 \rightsquigarrow S_f$ in $M_C\}$.

In this case the active space is called *atomic*.

(2) If $\Xi(\xi) = \{\xi_1, \ldots, \xi_m\}$ is a decomposition of ξ and

$$\mathbf{A} = \{Act(\wp(\xi_1)), \ldots, Act(\wp(\xi_m))\}$$

is a set of active spaces relevant to $\Xi(\xi)$, then

$$Act(\wp(\xi)) = \mathcal{J}_{\wp(\xi)}(\mathbf{A})$$

where $\mathcal{J}$ is the *join operator* defined in Section 5.1.

5.1   Join Operator

Let

$OBS(\xi) = (obs(\mathbf{C}_1), \ldots, obs(\mathbf{C}_n))$ be an observation of a cluster ξ,

$\Xi(\xi) = \{\xi_1, \ldots, \xi_m\}$ a decomposition of ξ,

$\mathbf{A} = \{Act(\wp(\xi_1)), \ldots, Act(\wp(\xi_m))\}$ a set of relevant active spaces where

$$\forall i \in [1 .. m] \ (Act(\wp(\xi_i)) = (\mathbb{S}_i, \mathbb{E}_i, \mathbb{T}_i, S_{0_i}, \mathbb{S}_{f_i})),$$
$$OBS(\xi) \trianglerighteq \{OBS(\xi_1), \ldots, OBS(\xi_m)\},$$

$\Sigma = \mathbb{S}_1 \times \cdots \times \mathbb{S}_m,$

$\mathbb{K}$ the domain of possible index values of $OBS(\xi)$;

$\mathbb{Q}$ the domain of possible queues in $Interf(\Xi(\xi))$.

The spurious active space obtained by applying the spurious join $\tilde{\mathcal{J}}$ of $\mathbf{A}$ based on the diagnostic problem $\wp(\xi)$, denoted by $\tilde{\mathcal{J}}_{\wp(\xi)}(\mathbf{A})$, is an automaton

$$\tilde{\mathcal{J}}_{\wp(\xi)}(\mathbf{A}) = (\tilde{\mathbb{S}}, \mathbb{E}, \tilde{\mathbb{T}}, S_0, \mathbb{S}_f)$$

where

$S_0 = (\sigma_0, K_0, Q_0)$, where

$\quad \sigma_0 = (S_{0_1}, \ldots, S_{0_m}), \sigma_0 \in \Sigma,$
$\quad K_0 = (0 \cdots 0), K_0 \in \mathbb{K},$
$\quad Q_0 = (\emptyset \cdots \emptyset), Q_0 \in \mathbb{Q};$

$\tilde{\mathbb{S}} = \{S_0\} \cup \{S' \mid S \xrightarrow{T} S' \in \tilde{\mathbb{T}}\}, \tilde{\mathbb{S}} \subseteq \Sigma \times \mathbb{K} \times \mathbb{Q};$

$\mathbb{E} = \bigcup_{i=1}^{m} \mathbb{T}_i;$

$\mathbb{S}_f = \{(\sigma_f, K_f, Q_f) \mid \forall S_i \in \sigma_f \, (S_i \in \mathbb{S}_{f_i}), Complete(K_f), Q_f = (\emptyset \cdots \emptyset)\};$

$\tilde{\mathbb{T}} : \tilde{\mathbb{S}} \times \mathbb{E} \mapsto \tilde{\mathbb{S}}.$

Specifically, the transition function $\tilde{\mathbb{T}}$ is defined as follows:

$$N \xrightarrow{T} N' \in \tilde{\mathbb{T}},$$

where

$N = (\sigma, K, Q), \sigma = (S_1, \ldots, S_m),$

$N' = (\sigma', K', Q'), \sigma' = (S'_1, \ldots, S'_m),$

if and only if the following conditions hold:

(1) *Triggerable*(T, N), where *Triggerable* is a predicate defined as follows.
Let $L_\alpha = Link(\alpha), T = S \xrightarrow{\alpha|\beta} S', \alpha = (E_\alpha, \theta_\alpha)$. Then

$$Triggerable(T, N) \text{ iff } (L_\alpha \notin Interf(\Xi(\xi)) \text{ or } Head(Q[L_\alpha]) = E_\alpha);$$

(2) *Consistent*(T, N), where *Consistent* is a predicate defined as follows.
Let

$$T \in \mathbb{E}_{\bar{\imath}},$$

$C_{\bar{\imath}}$ be the component relevant to T, $C_{\bar{\imath}} \in \mathbf{C}_{\bar{\jmath}}$, $\mathbf{C}_{\bar{\jmath}} \in \mathbb{P}(\Omega(\Theta))$,

$\mathbb{L}_\beta = Link(\beta) \cap Interf(\Xi(\xi))$,

$\mathbb{L}_\beta^{\mathrm{u}} = \{L_\beta \mid L_\beta \in \mathbb{L}_\beta, (\text{either } L_\beta \text{ is not saturated or } L_\beta = L_\alpha)\}$,

$\mathbb{L}_\beta^{\mathrm{s}} = \mathbb{L}_\beta - \mathbb{L}_\beta^{\mathrm{u}}$,

$\mathbb{L}_\beta^{\mathrm{so}} = \{L_\beta \mid L_\beta \in \mathbb{L}_\beta^{\mathrm{s}}, \text{ the saturation policy of } L_\beta \text{ is } OVERRIDE\}$,

$\mathbb{L}_\beta^{\mathrm{sw}} = \{L_\beta \mid L_\beta \in \mathbb{L}_\beta^{\mathrm{s}}, \text{ the saturation policy of } L_\beta \text{ is } WAIT\}$.

Then, $Consistent(T, N)$ evaluates to true if and only if the following conditions hold:

(a) Either $\begin{cases} \neg Observable(T) \text{ or} \\ C_{\bar{\imath}} \in Ukn(obs(\mathbf{C}_{\bar{\jmath}})) \text{ or} \\ (E, Out) \in \beta, obs(\mathbf{C}_{\bar{\jmath}})[K[\bar{\jmath}] + 1] = E, \end{cases}$

(b) $\mathbb{L}_\beta^{\mathrm{sw}} = \emptyset$;

(3) σ' is such that $\forall i \in [1 .. m] \left(S_i' = \begin{cases} S' & \text{if } i = \bar{\imath} \\ S_i & \text{otherwise} \end{cases} \right)$;

(4) K' is such that $\forall j \in [1 .. n]$

$$\left(K'[j] = \begin{cases} K[j + 1] & \text{if } j = \bar{\jmath}, Observable(T), C_{\bar{\imath}} \in Vis(obs(\mathbf{C}_{\bar{\jmath}})) \\ K[j] & \text{otherwise} \end{cases} \right);$$

(5) Q' is such that:

(a) If $L_\alpha \in Interf(\Xi(\xi))$ then $Q'[L_\alpha] = Tail(Q[L_\alpha])$,

(b) $\forall (E, \vartheta) \in \beta, L_\beta = Link(\vartheta), L_\beta \in (\mathbb{L}_\beta^{\mathrm{u}} \cup \mathbb{L}_\beta^{\mathrm{so}}) (Q'[L_\beta] = Ins(Q[L_\beta], E))$.

(c) $\forall L \in (Interf(\Xi(\xi)) - (\{L_\alpha\} \cup \mathbb{L}_\beta^{\mathrm{u}} \cup \mathbb{L}_\beta^{\mathrm{so}})) (Q'[L] = Q[L])$.

The join $\mathcal{J}$ of $\mathbf{A}$ based on the diagnostic problem $\wp(\xi)$ is the automaton obtained from $\tilde{\mathcal{J}}_{\wp(\xi)}(\mathbf{A})$ by selecting the states and transitions which are convergent, that is, those connected with a final state.

5.2   Canonical Reconstruction

The monolithic reconstruction presented in Chapter 4 can be seen as a degenerate case of modular reconstruction, where only atomic active spaces are involved.

Let $\wp(\xi)$ be a diagnostic problem, where $\mathbb{C}(\xi) = \{C_1, \ldots, C_n\}$ and $\xi_0 = (C_{10}, \ldots, C_{n0})$. Let

$$\mathbf{A}(\xi) = \{Act(\wp(C_1)), \ldots, Act(\wp(C_n))\}$$

be the set of atomic active spaces relevant to components in ξ. The *canonical reconstruction* of $\wp(\xi)$, denoted by $\Re(\wp(\xi))$, is defined as follows:

$$\Re(\wp(\xi)) = \mathcal{J}_{\wp(\xi)}(\mathbf{A}(\xi)).$$

Proposition 5.2. *Let* $\wp(\xi) = (\Omega(\xi), OBS(\xi), \xi_0)$ *be a diagnostic problem for cluster* ξ, $\Xi(\xi) = \{\xi_1, \ldots, \xi_m\}$ *a decomposition of* ξ,

$$\mathbf{A}_\xi = \{Act(\wp(\xi_1)), \ldots, Act(\wp(\xi_m))\}$$

a set of relevant active spaces where

$$\forall i \in [1 .. m] \ (\wp(\xi_i) = (\Omega(\xi_i), OBS(\xi_i), \xi_{i_0})),$$

such that ξ_0 *is the composition of* $\xi_{1_0} \cdots \xi_{m_0}$ *and*

$$OBS(\xi) \trianglerighteq \{OBS(\xi_1), \ldots, OBS(\xi_m)\}.$$

Let $\mathbf{A}_C$ *the set of atomic active spaces of components in* ξ. *Then,*

$$\mathcal{J}_{\wp(\xi)}(\mathbf{A}_\xi) \equiv \Re(\wp(\xi)). \tag{5.3}$$

Proposition 5.2 provides a formal basis for the modular reconstruction of the system reaction. As a matter of fact, Equation (5.3) allows the system behavior reconstruction to be made up in several steps, each of which yields a new active space based on a relevant diagnostic subproblem. The recursive application of Equation (5.3) leads to the concept of a reconstruction graph given below.

6. Problem Decomposition

Section 4 has introduced the basic notion of a modular reconstruction in its simplest form: the diagnostic problem $\wp(\Theta)$ is decomposed into a set of diagnostic subproblems $\wp(\xi_i)$ relevant to a decomposition of Θ into clusters ξ_i. Each $\wp(\xi_i)$ is obtained as a projection of $\wp(\Theta)$ on ξ_i.

This modular paradigm can be generalized in four directions:

(1) The diagnostic problem can be recursively split into subproblems, thereby being represented as a tree, where the root corresponds to the original problem and nodes to subproblems; in particular, leaves represent the lowest-level subproblems that are coped with monolithically.

(2) Let $\wp(\xi)$ be a diagnostic (sub)problem and

$$\mathbf{P}(\Xi(\xi)) = \{\wp(\xi_1), \ldots, \wp(\xi_m)\}$$

the set of diagnostic subproblems relevant to a decomposition of $\Xi(\xi)$. Each $\wp(\xi_i)$ is not required to be the projection of $\wp(\xi)$ on ξ_i but, rather, to be subsumed by $\wp(\xi)$. In other words, the projection (restriction) relationship is relaxed into subsumption. More concisely, $\wp(\xi)$ is only supposed to be monotonic with respect to $\mathbf{P}(\Xi(\xi))$, namely

$$\wp(\xi) \unrhd \mathbf{P}(\Xi(\xi)).$$

(3) The third generalization requires the definition of a diagnostic problem matching. Let

$$\wp(\xi) = (\Omega(\xi), OBS(\xi), \xi_0),$$
$$\wp(\xi') = (\Omega(\xi'), OBS(\xi'), \xi_0'),$$

be two diagnostic problems relevant to different clusters ξ and ξ', respectively; $\wp(\xi)$ *matches* $\wp(\xi')$, denoted

$$\wp(\xi) \, \lozenge \, \wp(\xi'),$$

if and only if

(a) The model of ξ equals the model of ξ',

(b) $\xi_0 = \xi_0'$,

(c) $\Omega(\xi) = \Omega(\xi')$.

Then, two matching diagnostic subproblems can be merged into a single diagnostic subproblem which is an abstraction of them. This way, the decomposition tree is bound to become a DAG, where a node may be abstract, thereby referring to several clusters. Note that the matching relationship can be applied to abstract nodes as well.

(4) Instead of being associated with component models (atomic active spaces), the leaves of the graph correspond to universal spaces which are assumed to have been somehow previously computed (an atomic active space is in fact a universal space for the relevant component).

The above generalizations are substantiated by the notion of a reconstruction graph.

6.1   Reconstruction Graph

A *reconstruction graph* is a DAG where each node corresponds to a diagnostic (sub)problem associated with a set of clusters, called the *extension*. If the latter is a singleton, the diagnostic problem is concrete,

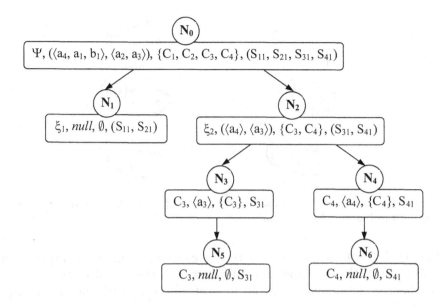

Figure 5.4. Reconstruction graph for diagnostic problem $\wp(\Psi)$ (see Example 5.6).

as it refers to the corresponding cluster. Otherwise, the diagnostic problem is abstract, insofar as it is relevant to each of the clusters within the extension.

In particular, the root of the graph is associated with the diagnostic problem of a system Θ, while leaves correspond to given universal spaces. From the reconstruction viewpoint, the solution of a diagnostic problem is the relevant active space. Thus, solutions of diagnostic problems associated with leaf nodes are supposed available, while those for non-leaf nodes are to be computed.

The reconstruction process follows the reconstruction plan starting from the leaves. Then, while moving upward in the graph, the reconstruction process enforces observation constraints and interface constraints on progressively growing clusters, obtained by aggregating smaller clusters up to the root, which corresponds to the whole system along with its observation. As such, the active space associated with the root of the reconstruction plan is the solution of the original diagnostic problem.

Example 5.6. Example 5.5 has shown a modular reconstruction for the diagnostic problem coped with monolithically in Example 4.4. Now we solve the same diagnostic problem by means of the reconstruction graph shown in Figure 5.4. Each node of the graph is labeled by essential information, including the cluster at hand (since the graph is in fact a

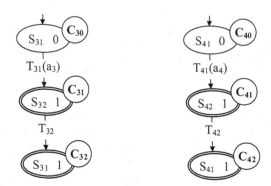

Figure 5.5. Active spaces relevant to nodes N_3 and N_4 of the reconstruction graph displayed in Figure 5.4 (see Example 5.6).

tree, each diagnostic problem is concrete), the relevant observation , the visible set of the corresponding observer, and the initial state.

The original diagnostic problem $\wp(\Psi) = (\Omega(\Psi), OBS(\Psi), \Psi_0)$ associated with N_0, where

$$\Omega(\Psi) = (\mathbb{V}, \mathbb{P}), \text{ where } \mathbb{V} = \mathbb{C}(\Psi), \mathbb{P} = \{\{C_1, C_4\}, \{C_2, C_3\}\},$$
$$OBS(\Psi) = (\langle a_4, a_1, b_1 \rangle, \langle a_2, a_3 \rangle),$$
$$\Psi_0 = (S_{11}, S_{21}, S_{31}, S_{41}),$$

has been decomposed into two diagnostic problems, one relevant to ξ_1 and the other to ξ_2, where the former is represented by the universal space $Usp(\xi_1, (S_{11}, S_{21}))$ depicted on the top of Figure 5.6, while the latter is further decomposed into nodes N_3 and N_4, corresponding to components C_3 and C_4, respectively. However, even if relevant to components, N_3 and N_4 are non-leaf nodes, as they correspond to the diagnostic problems

$$\wp(C_3) = (\Omega(C_3), \langle a_3 \rangle, S_{31}),$$
$$\wp(C_4) = (\Omega(C_4), \langle a_4 \rangle, S_{41}),$$

which are solved by applying the join operator on the single universal spaces associated with nodes N_5 and N_6, as shown in Figure 5.5 (such a join is in fact a selection, that is, a specialization of the relevant universal space, where the history set is subjected to a restriction constrained by the component observation).

Then, the active space relevant to N_2 is generated as the join of the active spaces associated with N_3 and N_4, as shown on the bottom of Figure 5.6. Finally, the active space relevant to the original diagnostic problem for Ψ is yielded as the join of the universal space of N_1 (top of

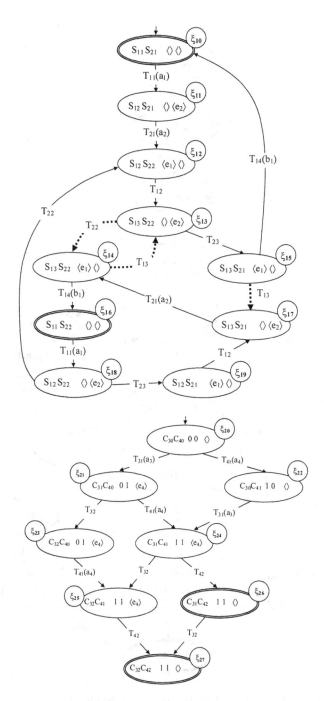

Figure 5.6. Active spaces for nodes N_1 (top) and N_2 (bottom) of the reconstruction graph displayed in Figure 5.4 (see Example 5.6).

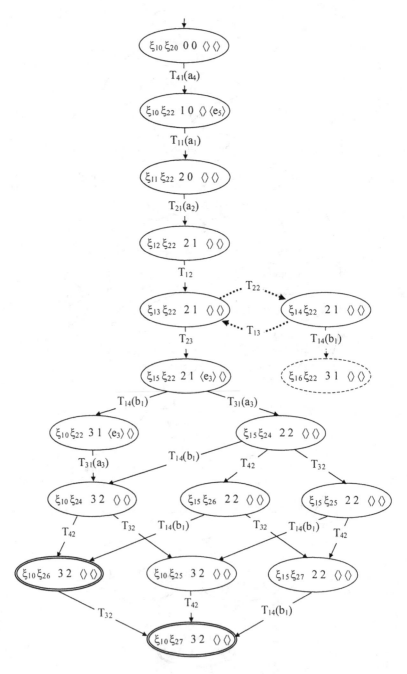

Figure 5.7. Active space for the root N_0 of the reconstruction graph displayed in Figure 5.4, obtained by joining the active spaces in Figures 5.5 (see Example 5.6).

Figure 5.6) and the active space of N_2 (bottom of Figure 5.6), as shown in Figure 5.7. As expected, the result equals the active space shown in Figure 5.3, which was determined by means of a different modular reconstruction. ◇

6.2 Formal Definition of a Reconstruction Graph

We can now formalize the notion of a reconstruction graph as follows. Let

$\wp(\Theta) = (\Omega(\Theta), OBS(\Theta), \Theta_0)$ be a diagnostic problem for a system Θ,

$\Xi(\Theta)$ a decomposition of Θ,

M_Θ the model of Θ,

Σ the domain of clusters ξ defined as follows:

$$\Sigma = \{\xi \mid \Xi(\xi) \in (2^{\Xi(\Theta)} - \emptyset)\},$$

$\mathbb{M}$ the domain of the models M of the clusters in Σ,

$\mathbf{\Omega}(\mathbb{M})$ the domain of observers for models in $\mathbb{M}$,

OBS$(\mathbb{M})$ the domain of observations $OBS(M)$,

$\mathbb{M}_0$ the domain of initial states of cluster models in $\mathbb{M}$,

$\mathcal{P}(\mathbb{M})$ the domain of diagnostic problems $\wp(M) = (\Omega(M), OBS(M), M_0)$, $\Omega(M) \in \mathbf{\Omega}(\mathbb{M})$.

A *reconstruction graph* based on $\wp(\Theta)$ and $\Xi(\Theta)$, is a directed acyclic graph,

$$Rgraph(\wp(\Theta), \Xi(\Theta)) = (\mathbb{N}, \mathbb{E}, N_0)$$

where

$\mathbb{N} \subseteq \mathcal{P}(\mathbb{M}) \times (2^\Sigma - \emptyset)$ is the set of nodes $N = (\wp(M), \mathbb{Z})$, where $\mathbb{Z}$ is the *extension* of N,

$\mathbb{E} : \mathbb{N} \mapsto 2^{[\mathbb{N}]}$ is the set of edges, where $2^{[\mathbb{N}]}$ is the domain of multisets over $\mathbb{N}$, and

N_0 is the root,

such that, denoting with $Succ(N)$ the multiset of successive nodes of N, namely

$$Succ(N) = [N' \mid N' \in \mathbb{N}, N \to N' \in \mathbb{E}],$$

with $Prec(N)$ the multiset of preceding nodes of N, namely

$$Prec(N) = [N' \mid N' \in \mathbb{N}, N' \to N \in \mathbb{E}],$$

and with $Leaf(N)$ the boolean function defined as follows:

$$Leaf(N) \stackrel{\text{def}}{=} \begin{cases} true & \text{if } Succ(N) = \emptyset \\ false & \text{otherwise} \end{cases},$$

the following conditions hold:

(1) $N_0 = (\wp(\Theta), \{\Theta\}), \wp(\Theta) = (\Omega(\Theta), OBS(\Theta), \Theta_0)$;

(2) $\forall(\wp(M), Z) \in \mathbb{N}, \forall \xi \in Z$, M is the cluster model of ξ;

(3) $\forall N \in \mathbb{N}, Leaf(N), N = (\wp(M), Z), \wp(M) = (\Omega(M), OBS(M), M_0)$, we have

$$OBS(M) = null, \forall \xi \in Z \, (\xi \in \Xi(\Theta));$$

(4) $\forall N \in \mathbb{N}, N = (\wp(M), Z), \wp(M) = (\Omega(M), OBS(M), M_0)$, such that

$\neg Leaf(N), Succ(N) = [N_1, \ldots, N_m]$,
$\forall i \in [1..m] \, (N_i = (\wp(M_i), Z_i), \wp(M_i) = (\Omega(M_i), OBS(M_i), M_{i_0}))$,

the following conditions hold:

(a) $M_0 = M_{0_1} \oplus M_{0_2} \oplus \cdots \oplus M_{0_m}$,
(b) $\forall \xi \in Z \, (\Xi(\xi) = \{\xi_1, \ldots, \xi_m\}, \forall i \in [1..m](\xi_i \in Z_i))$,
(c) $\forall \xi \in Z, \forall \xi' \in Z, \xi \neq \xi', (\Xi(\xi) \cap \Xi(\xi') = \emptyset)$,
(d) $\wp(M) \trianglerighteq \{\wp(M_1), \ldots, \wp(M_m)\}$;

(5) $\forall N \in (\mathbb{N} - \{N_0\})$ the following conditions hold:

(a) $\forall \xi \in Z(N) \, (N' \in Prec(N), \xi' \in Z(N'), \xi' \supseteq \xi)$,
(b) $\forall N_1 \in Prec(N), \forall N_2 \in Prec(N), \forall \xi_1 \in Z(N_1), \forall \xi_2 \in Z(N_2), \xi_1 \neq \xi_2 \, (\mathbb{C}(\xi_1 \cap \xi_2) = \emptyset)$.

It is possible to associate with each node $N \in \mathbb{N}$ a natural number representing the *level* of N, defined as follows:

$$Level(N) = \begin{cases} 0 & \text{if } Leaf(N) \\ Level(N') + 1 & \text{if the following condition is true:} \end{cases}$$

$$N' \in Succ(N), \nexists N'' \, (N'' \in Succ(N), Level(N'') > Level(N')).$$

Clearly, the following property holds:

$$\forall N, N \in \mathbb{N}, N \neq N_0 \, (Level(N) < Level(N_0)).$$

It is possible as well to associate with each node $N = (\wp(M), \mathbb{Z}) \in \mathbb{N}$, $\wp(M) = (\Omega(M), OBS(M), M_0)$, an active space by means of the *Active* recursive function defined as follows:

$$Active(N) = \begin{cases} Usp(M, M_0) & \text{if } Leaf(N) \\ \mathcal{J}_{\wp(M)}(Active(N_1), \ldots, Active(N_m)) & \text{otherwise,} \end{cases}$$

(5.4)

where $Succ(N) = [N_1, \ldots, N_m]$.

Theorem 5.2. *Let* $Rgraph(\wp(\Theta), \Xi(\Theta)) = (\mathbb{N}, \mathbb{E}, N_0)$ *be a reconstruction graph. Then, the following properties hold:*

(1) $\forall N \in \mathbb{N}, N = (\wp(M), \mathbb{Z}), \wp(M) = (\Omega(M), OBS(M), M_0), \forall \xi \in \mathbb{Z}$

$$(OBS_{\langle \xi \rangle}(\Theta) \succeq OBS(M));$$

(2) $\forall N \in (\mathbb{N} - \{N_0\}) \ (|\mathbb{Z}(N)| = \sum_{N' \in Prec(N)} |\mathbb{Z}(N')|);$

(3) $Active(N_0) \equiv \Re(\wp(\Theta)).$

Corollary 5.1. *Let* $(\mathbb{N}, \mathbb{E}, N_0)$ *and* $(\mathbb{N}', \mathbb{E}', N_0')$ *be two different reconstruction graphs relevant to the same diagnostic problem* $\wp(\Theta)$. *Then*

$$Active(N_0) \equiv Active(N_0').$$

The corollary represents the formal basis for the preservation of the soundness and completeness of the diagnostic method. In other words, whatever the recursive decomposition of a diagnostic problem, the resulting active space is always the same. Thus, the choice of a particular reconstruction graph is bound to influence the complexity (efficiency) of the reconstruction process, rather than the result.

7. Summary

- *Cluster*: a connected subpart of a system incorporating a set of components and all the links among them. The behavioral model of a cluster is implicitly defined by the models of its components and links. A system is a special case of cluster.

- *Decomposition*: the mode in which a cluster is fragmented into subparts whose components represent a partition of the components of the cluster. It is the topological notion that supports modular diagnosis.

- *Interface*: the set of links connecting different subclusters relevant to a decomposition.

- *Subsumption*: a relation between two observations, which provides support to the soundness and completeness of modular reconstruction. Intuitively, an observation subsumes another observation when the former enforces all the temporal constraints of the latter.

- *Monotonicity*: the extension of the subsumption relationship between an observation of a diagnostic problem and the set of observations relevant to a decomposition of the diagnostic problem: the temporal constraints of the observation are required to include the union of the temporal constraints of its fragments.

- *Modular reconstruction*: the process of making up the active space relevant to a diagnostic problem by joining the active spaces relevant to a decomposition of the problem into subproblems.

- *Canonical reconstruction*: a degenerate case of modular reconstruction, in which the decomposition of the system is atomic, that is, where each subsystem is a component (it is in fact a monolithic reconstruction).

- *Reconstruction graph*: a DAG representing the mode in which a diagnostic problem is recursively decomposed into subproblems. It supports modular reconstruction. Starting from leaf nodes and traversing the graph upward, increasingly larger active spaces are generated by joining smaller active spaces. Whatever the reconstruction graph, the resulting active space (associated with the root) is always the same.

Appendix A
Algorithms

This appendix provides two algorithms, expressed in pseudo-code, relevant to modular diagnosis. Based on a given diagnostic (sub)problem, Algorithm A.2 implements the join of a set of active spaces relevant to a cluster decomposition. It is a variation of function *Reconstruct* (see Chapter 4), in accordance with the compositional definition of active space provided in Section 5.5.

Algorithm A.3 makes up the active space of a given diagnostic problem, starting from a set of active spaces relevant to a decomposition of the system.

Algorithm A.2. (*Join*) The *Join* function generates a graph-based representation of the active space $Act(\wp(\Theta))$ relevant to the input diagnostic problem (the notation of the algorithm is based on the compositional specification of an active space provided in Section 5). It makes use of the recursive function *JStep*, whose definition is embedded within *Join*, between the parameter specification and the body of *Join*, the latter starting at Line 55. At Line 56, the parameters of the first call of *JStep* are instantiated, namely σ_0, K, Q, and $\aleph$. *JStep* is first called at Line 57.

Considering the body of *JStep* (Lines 1–54), possible transitions among those leaving from the current node N are searched for within the nested loop at Line 3. A candidate transition is the mark T of an edge leaving a node in σ, the current record of states of the input active spaces. If the input event of T is relevant to a link within the interface of the set of clusters, then such an event is required to be ready in Q.

Lines 4–10 set the variables introduced in the compositional definition of active space, Point 2. If the transition entails an output event on a saturated link with saturation mode *WAIT*, it is discarded (Line 12).

Lines 14–16 checks the consistency of the possible message associated with the candidate transition with the index of the observation: if inconsistent, the transition is discarded.

Line 17 makes a copy of the fields of the current node into the new variables σ', K', and Q'. The new state reached by the candidate transition is updated at Line 18. If the transition is observable and the relevant component is visible by the observer, the observation index is incremented at Line 20.

At Line 23, the triggering event α is dequeued from the relevant link (only if the latter is within the interface). The queues relevant to the output events of the candidate transition are updated at Lines 25–27.

At Line 28, the new node N' is created. If N' was not visited already (Line 35), *JStep* is recursively called at Line 36, and an edge from N to N' is possibly created (Line 42).

The loop terminates at Line 44. At this point, three cases are possible:

(1) N is final (Line 45): N is marked as consistent and returned (Lines 46–47);

(2) N is not final, but an edge from it to a new node was created (Line 48): N is simply returned (Line 49);

(3) N is not final and no edge was created from it (Line 50): N is marked as inconsistent and a null value is returned (Line 52).

Lines 59–62 in *Join* remove from the graph $\tilde{\mathcal{J}}$ the possible dangling nodes and edges which are not part of any path from the root of $\tilde{\mathcal{J}}$ to a final node. These anomalous subgraphs of $\tilde{\mathcal{J}}$ correspond to cycles in which all the nodes are marked as unknown. However, consistent silent cycles possibly remain still in $\tilde{\mathcal{J}}$ after the pruning.

Lines 63–68 establish the isomorphism between the graph representation of the reconstruction $\mathcal{J}$ and its elements $\mathbb{S}$, $\mathbb{E}$, $\mathbb{T}$, S_0, and $\mathbb{S}_f$. The resulting reconstruction is eventually returned at Line 69.

function *Join*($\mathbf{A}$, $\wp(\xi)$):
$\qquad$ The join $\mathcal{J}_{\wp(\xi)}(\mathbf{A}) = (\mathbb{S}, \mathbb{E}, \mathbb{T}, S_0, \mathbb{S}_f)$, $\mathbb{S} \subseteq \Sigma \times \mathbb{K} \times \mathbb{Q}$

$\quad$ **input**
$\qquad$ $\mathbf{A} = \{Act(\wp(\xi_i)) \mid Act(\wp(\xi_i)) = (\mathbb{S}_i, \mathbb{E}_i, \mathbb{T}_i, S_{0_i}, \mathbb{S}_{f_i}), i \in [1 .. m]\}$:
$\qquad\qquad$ The active spaces of a decomposition $\Xi(\xi) = \{\xi_1, \dots, \xi_m\}$,
$\qquad\qquad$ $\wp(\xi) = (\Omega(\xi), OBS(\xi), \xi_0)$: A diagnostic problem for cluster ξ;

function *JStep*($\sigma, K, Q,$ out $\aleph$): The root of a subgraph of $Act(\wp(\Theta))$
$\quad$ **input**
$\qquad$ $\sigma = (S_1, \dots, S_m) \in \Sigma$,
$\qquad$ $K \in \mathbb{K}$,
$\qquad$ $Q = Queue(\xi)$: a queue of ξ,
$\qquad$ $\aleph \subseteq \Sigma \times \mathbb{K} \times \mathbb{Q}$: the set of nodes visited up to the current call of *JStep*
$\quad$ **output**
$\qquad$ $\aleph$: The updated set of visited nodes;
$\quad$ **begin** {JStep}
1. $\qquad$ Insert $N = (\sigma, K, Q)$ into $\aleph$ and mark it as *unknown*;
2. $\qquad$ $\mathbb{T}_{cand} := \{S_i \xrightarrow{T} S_i' \in \mathbb{T}_i \mid i \in [1 .. m], T = S \xrightarrow{\alpha|\beta} S', \alpha = (E_\alpha, \theta_\alpha),$
$\qquad\qquad\qquad\qquad\qquad$ $(\alpha \in Ready(Q)$ **or** $Link(\alpha) \notin Interf(\Xi(\xi)))\}$;
3. $\qquad$ **for each** $S_{\bar{i}} \xrightarrow{T} S_{\bar{i}}' \in \mathbb{T}_{cand}, T = S \xrightarrow{\alpha|\beta} S'$ **do**
4. $\qquad\qquad$ $C_{\bar{i}} :=$ the component relevant to T, $C_{\bar{i}} \in \mathbf{C}_{\bar{j}}, \mathbf{C}_{\bar{j}} \in \mathbb{P}(\Omega(\xi))$;
5. $\qquad\qquad$ $L_\alpha := Link(\alpha)$;
6. $\qquad\qquad$ $\mathbb{L}_\beta := Link(\beta) \cap Interf(\Xi(\xi))$;
7. $\qquad\qquad$ $\mathbb{L}_\beta^u := \{L_\beta \mid L_\beta \in \mathbb{L}_\beta,$ (either L_β is not saturated or $L_\beta = L_\alpha)\}$;
8. $\qquad\qquad$ $\mathbb{L}_\beta^s := \mathbb{L}_\beta - \mathbb{L}_\beta^u$;
9. $\qquad\qquad$ $\mathbb{L}_\beta^{so} := \{L_\beta \mid L_\beta \in \mathbb{L}_\beta^s,$ the saturation policy of L_β is $OVERRIDE\}$;
10. $\qquad\quad$ $\mathbb{L}_\beta^{sw} := \{L_\beta \mid L_\beta \in \mathbb{L}_\beta^s,$ the saturation policy of L_β is $WAIT\}$;
11. $\qquad\quad$ **if** $\mathbb{L}_\beta^{sw} \neq \emptyset$ **then**
12. $\qquad\qquad$ **goto** 3
13. $\qquad\quad$ **end-if**;
14. $\qquad\quad$ **if** $Observable(T), (E, Out) \in \beta$ **and** $C_{\bar{i}} \in Vis(OBS(\xi))$ **and**

```
                    obs(Cⱼ̃)[K[j̃] + 1] ≠ E then
15.             goto 3
16.         end-if;
17.         σ' := σ; K' := K; Q' := Q;
18.         σ'[ī] = S'ᵢ̃;
19.         if Observable(T) and Cī̃ ∈ Vis(OBS(ξ)) then
20.             K'[j̃] := K'[j̃] + 1
21.         end-if;
22.         if Lα ∈ Interf(Ξ(ξ)) then
23.             Q'[Lα] := Tail(Q[Lα])
24.         end-if;
25.         for each event (E, ϑ) ∈ β, Lβ = Link(ϑ), Lβ ∈ (𝕃ᵤβ ∪ 𝕃ˢᵒβ) do
26.             Q'[Lβ] := Ins(Q[Lβ], E)
27.         end-for;
28.         N' := (σ', K', Q');
29.         if N' ∈ ℵ then
30.             if N' is marked as inconsistent then
31.                 N'ⱼ̃ := nil
32.             else
33.                 N'ⱼ̃ := N'
34.             end-if
35.         else
36.             N'ⱼ̃ := JStep(σ', K', Q', ℵ)
37.         end-if;
38.         if N'ⱼ̃ ≠ nil then
39.             if N is marked as unknown and N' as consistent then
40.                 Mark N as consistent
41.             end-if;
42.             Create an edge N —T→ N'
43.         end-if
44.     end-for;
45.     if N = (σ, K, (∅···∅)) and ∀Sᵢ ∈ σ (Sᵢ ∈ 𝕊fᵢ) and Complete(K) then
46.         Mark N as consistent;
47.         return N
48.     elsif At least one edge was created at Line 42 then
49.             return N
50.     else
51.         Mark N as inconsistent;
52.         return nil
53.     end-if
54. end {JStep};

55. begin {Join}
56.     σ₀ := (S₀₁, ..., S₀ₘ); K := (0···0); Q := (∅···∅); ℵ := ∅;
57.     N₀ := JStep(σ₀, K, Q, ℵ);
58.     𝒥̃ := the graph rooted in N₀;
59.     for each N ∈ ℵ such that N is marked as unknown do
60.         if ∄ a path N ⤳ (σf, Kf, (∅···∅)) in 𝒥̃ such that
                        ∀Sᵢ ∈ σf (Sᵢ ∈ 𝕊fᵢ) and Complete(Kf) then
```

61.   Remove N from $\tilde{\mathcal{J}}$;
62.   Remove from $\tilde{\mathcal{J}}$ all the dangling edges;
63. $\mathbb{S} :=$ the set of nodes in $\tilde{\mathcal{J}}$;
64. $\mathbb{E} :=$ the set of labels of edges in $\tilde{\mathcal{J}}$;
65. $\mathbb{T} :=$ the set of edges in $\tilde{\mathcal{J}}$;
66. $S_0 :=$ the root of $\tilde{\mathcal{J}}$;
67. $\mathbb{S}_f \subseteq \mathbb{S} :=$ the set of nodes $N_f = (\sigma_f, K_f, (\emptyset \cdots \emptyset))$ such that
$$\forall S_i \in \sigma_f \ (S_i \in \mathbb{S}_{f_i}) \text{ and } Complete(K_f);$$
68. $\mathcal{J} := (\mathbb{S}, \mathbb{E}, \mathbb{T}, S_0, \mathbb{S}_f);$
69.   **return** $\mathcal{J}$
70. **end** {Join}.

Algorithm A.3. (*ModRec*) The *ModRec* function generates the active space relevant to a diagnostic problem. The diagnostic problem comes with a set of active spaces relevant to a decomposition of the system. When no prior computation has been carried out, such active spaces are atomic, that is, the models of the components within the system. Otherwise, they are the result of previous computation: intuitively, they can be viewed as the (possibly constrained) models of the clusters relevant to the decomposition.

First, a reconstruction graph must be generated (Line 2). However, the algorithm provides no support for such a generation. Then, the available active spaces associated with the lowest-level nodes of the graph ($Level(N) = 0$) are inserted into the accumulator Acc (Line 3).

The active spaces relevant to the upper nodes are yielded within the loop embraced by Lines 5–12. At each iteration, the level is incremented (Line 6) and, for each node N of that level, the set $\mathbf{A}'$ of active spaces associated with the successive nodes of N is built (Line 8).

At this point, the active space relevant to N can be computed by the join of the active spaces in $\mathbf{A}'$ based on the diagnostic problem $\wp(M(N))$ (Line 9). Then, the newly created active space A_N is added to Acc (Line 10).

The loop terminates when the active space relevant to the root N_0 of the reconstruction graph is reached (Line 12). Such an active space is the result returned at Line 13.

function $\boldsymbol{ModRec}(\wp(\Theta), \mathbf{A}(\Xi(\Theta)))$: The active space $Act(\wp(\Theta))$
input
$\wp(\Theta) = (\Omega(\Theta), OBS(\Theta), \Theta_0)$: A diagnostic problem for system Θ,
$\mathbf{A}(\Xi(\Theta)) = \{Act(\wp(\xi_1)), \ldots, Act(\wp(\xi_m))\}$:
the active spaces of a decomposition of Θ, $\Xi(\Theta) = \{\xi_1, \ldots, \xi_m\}$;
1. **begin** {ModRec}
2.   Create a reconstruction graph $Rgraph(\wp(\Theta), \Xi(\Theta)) = (\mathbb{N}, \mathbb{E}, N_0)$;
3. $Acc := \{Act(\wp(M(N)) \mid N \in \mathbb{N}, Level(N) = 0\}$;
4. $level := 0$;
5.   **repeat**
6. $level := level + 1$;
7.   **for each** $N \in \mathbb{N}, \ Level(N) = level$ **do**
8. $\mathbf{A}' := \{A' \mid A' \in Acc, A' \text{ relevant to } N', N' \in Succ(N)\}$;
9. $A_N := Join(\mathbf{A}', \wp(M(N)))$;
10. $Acc := Acc \cup \{A_N\}$
11.   **end-for**

12.   **until** $level = Level(N_0)$;
13.   **return** the active space relevant to N_0
14. **end** {ModRec}.

Appendix B
Proofs of Theorems

Theorem 5.1 *Let $OBS(\xi)$ be an observation of a cluster $\xi = (\mathbb{C}, \mathbb{L}, \mathbb{D})$,*

$$\Xi(\xi) = \{\xi_1, \ldots, \xi_m\}$$

a decomposition of ξ, and

$$\mathbf{OBS}(\Xi(\xi)) = \{OBS(\xi_1), \ldots, OBS(\xi_m)\}$$

a set of observations of clusters in $\Xi(\xi)$ such that

$$OBS(\xi) \unrhd \mathbf{OBS}(\Xi(\xi)). \tag{B.1}$$

Then,

$$Ukn(OBS(\xi)) \subseteq \bigcup_{j=1}^{m} Ukn(OBS(\xi_j)).$$

The proof of this theorem is based on the following lemma:

Lemma 5.1.1 *Let*

$$OBS(\xi) = (obs(\mathbf{C}_1), \ldots, obs(\mathbf{C}_n)),$$
$$OBS'(\xi) = (obs'(\mathbf{C}_1), \ldots, obs'(\mathbf{C}_n))$$

be two observations of cluster ξ. Then,

$$OBS(\xi) \succeq OBS'(\xi) \models Ukn(OBS(\xi)) \subseteq Ukn(OBS'(\xi)).$$

Proof. Let $\mathbb{V}$ and $\mathbb{V}'$ be the visible sets of the observers of the two observations, respectively. From the definition of subsumption, we have

$$\mathbb{V} \supseteq \mathbb{V}',$$

therefore,

$$\mathbb{C} - \mathbb{V} \subseteq \mathbb{C} - \mathbb{V}',$$

which, according to the definition of unknown set (Formula (3.3)), implies

$$Ukn(OBS(\xi)) \subseteq Ukn(OBS'(\xi)),$$

which concludes the proof of Lemma 5.1.1.

Proof of Theorem 5.1.   Since, according to relation (B.1),

$$OBS(\xi) \trianglerighteq \mathbf{OBS}(\Xi(\xi)) = \{OBS(\xi_1), \ldots, OBS(\xi_m)\},$$

from the definition of subsumption, Condition (5.1), we have

$$\forall j \in [1..m]\ (OBS_{\langle \xi_j \rangle}(\xi) \succeq OBS(\xi_j))$$

which, by virtue of Lemma 5.1.1, can be written as

$$\forall j \in [1..m](Ukn(OBS_{\langle \xi_j \rangle}(\xi)) \subseteq Ukn(OBS(\xi_j))).$$

Since

$$\bigcup_{j=1}^{m} Ukn(OBS_{\langle \xi_j \rangle}(\xi)) = Ukn(OBS(\xi)),$$

from the above containment relationships it follows that

$$Ukn(OBS(\xi)) \subseteq \bigcup_{j=1}^{m} Ukn(OBS(\xi_j)),$$

which concludes the proof of the theorem. $\square$

Theorem 5.2 *Let* $Rgraph(\wp(\Theta), \Xi(\Theta)) = (\mathbb{N}, \mathbb{E}, N_0)$ *be a reconstruction graph. Then, the following properties hold:*

(1) $\forall N \in \mathbb{N}, N = (\wp(M), \mathbb{Z}), \wp(M) = (\Omega(M), OBS(M), M_0), \forall \xi \in \mathbb{Z}$

$$(OBS_{\langle \xi \rangle}(\Theta) \succeq OBS(M));$$

(2) $\forall N \in (\mathbb{N} - \{N_0\})\ (|\mathbb{Z}(N)| = \sum_{N' \in Prec(N)} |\mathbb{Z}(N')|);$

(3) $Active(N_0) \equiv \Re(\wp(\Theta)).$

Proof.   We prove each of the three properties separately.

Proof of Property (1).   Note that this property trivially holds for every leaf node and for the root N_0. Thus, we have to prove it for a generic node

$$N \in (\mathbb{N} - (\{N_0\} \cup \{N_f \mid Leaf(N_f)\})).$$

From Condition (4d) of the definition of reconstruction graph, we have

$$\wp(M) \trianglerighteq \{\wp(M_1), \ldots, \wp(M)\},$$

where $Succ(N) = [N_1, \ldots, N_s], \forall j \in [1..s]\ (N_i = (\wp(M_j), \mathbb{Z}_j))$. That is, by definition of subsumption,

$$\forall j \in [1..s](\wp_{\langle C(M_j) \rangle}(M) \succeq OBS(M_j)),$$

in other words, by definition of monotonicity,

$$OBS(M) \trianglerighteq \{OBS(M_1), \ldots, OBS(M_s)\},$$

in other terms,

$$\forall j \in [1..s](OBS_{\langle C(M_j) \rangle}(M) \succeq OBS(M_j)).$$

The subsumption relationship means that the restriction of the observation relevant to a node in $Prec(N)$ subsumes the observation relevant to N. Since the subsumption relationship is transitive (see Proposition 5.1), Property (1) is proved.

Proof of Property (2). Let $Prec(N) = [N_1, \ldots, N_p]$. Considering the definition of reconstruction graph, from Conditions (4b) and (5b) it follows that for each node N_i in $Prec(N)$ and for each cluster ξ' in $\mathbb{Z}(N_i)$ there exists a different cluster ξ in $\mathbb{Z}(N)$. Thus, on the one hand,

$$\mathbb{Z}(N) \supseteq \sum_{i=1}^{n} \mathbb{Z}(N_i).$$

On the other, from Conditions (5a) and (5b) we may state that for each cluster ξ in $\mathbb{Z}(N)$ there exists one and only one cluster ξ' in the extension of a preceding node $N' \in Prec(N)$ which incorporate ξ. This implies that

$$\mathbb{Z}(N) \subseteq \sum_{i=1}^{n} \mathbb{Z}(N_i).$$

From the above containment relationships it follows that

$$|\mathbb{Z}(N)| = \sum_{i=1}^{n} \mathbb{Z}(N_i) = \sum_{N' \in Prec(N)} |\mathbb{Z}(N')|.$$

which concludes the proof of Property (2).

Proof of Property (3). The proof is based on the following lemma:

Lemma 5.2.1 *For each* $N \in \mathbb{N}, \neg Leaf(N)$, $N = (\wp(M), \mathbb{Z})$, *where*

$$\wp(M) = (\Omega(M), OBS(M), M_0),$$

the following property holds:

$$Active(N) \equiv \Re(\wp(M)).$$

Proof. The proof of Lemma 5.21 is by induction on the nodes.

Basis. Based on the definition of *Active* and universal space *Usp* given in Equation (5.4) and Section 2.2, respectively, if $Leaf(N)$, then

$$Active(N) = Usp(M, M_0) = Act(null, M_0) = Act(\wp(M))$$

Induction. We assume that, for each $N' \in Succ(N)$, the following property holds:

$$Active(N') = Act(\wp(M')),$$

where $N' = (\wp(M'), \mathbb{Z}')$, $\wp(M') = (\Omega(M'), OBS(M'), M'_0)$. Based on the definition of *Active*, if $\neg Leaf(N)$, $Succ(N) = [N_1, \ldots, N_s]$, then

$$Active(N) = \mathcal{J}_{\wp(M)}(Active(N_1), \ldots, Active(N_s))$$

where

$$M_0 = M_{0_1} \oplus M_{0_2} \oplus \cdots \oplus M_{0_s}, \qquad (B.2)$$

$$OBS(M) \trianglerighteq \{OBS(M_1), \ldots, OBS(M_s)\}. \qquad (B.3)$$

Based on definition of compound active space, we may rewrite the above formula as

$$Active(N) = Act(\wp(M)).$$

The conclusion of the induction is based on Theorem 5.3, which allows us to replace $Act(\wp(M))$ with $\Re(\wp(M))$, thereby proving Lemma 5.2.1.

Based on Lemma 5.2.1, if $N = N_0$ then

$$Active(N_0) = \Re(\wp(M_\Theta)) = \Re(\wp(\Theta)),$$

which concludes the proof of Property (3), as well as the proof of Theorem 5.2.   □

Corollary 5.1 *Let $(\mathbb{N}, \mathbb{E}, N_0)$ and $(\mathbb{N}', \mathbb{E}', N_0')$ be two different reconstruction graphs relevant to the same diagnostic problem $\wp(\Theta)$. Then*

$$Active(N_0) \equiv Active(N_0').$$

Proof.   Since, based on Theorem 5.2,

$$Active(N_0) \equiv \Re(\wp(\Theta)),$$

$$Active(N_0') \equiv \Re(\wp(\Theta)),$$

we conclude that $Active(N_0) \equiv Active(N_0')$.   □

III

POLYMORPHIC SYSTEMS

Chapter 6

SIMULATION-BASED DIAGNOSIS

Abstract

The diagnostic techniques presented in Chapters 4 and 5, either monolithic or modular, are applicable to a class of DESs with asynchronous behavior, where the system reaction is a sequence of component transitions. Thus, it is not possible for two or more component transitions to be fired in parallel. By contrast, in a synchronous DES all the triggerable transitions are fired in parallel. The reaction is a sequence of system transitions, each being a group of component transitions. A polymorphic DES integrates both synchronous and asynchronous behavior. The type of behavior is established by the type of the links wherein events are ready. A link can be either synchronous or asynchronous. At each point of the system reaction, transitions triggered by events that are ready on synchronous links have higher priority than transitions triggered by events on asynchronous links. Thus, if there is a synchronous link with a ready event, the system transition will be synchronous (including all the triggerable component transitions), otherwise it will be asynchronous (including just a component transition). Both monolithic and modular diagnostic approaches are extended for polymorphic systems. In particular, since a system transition may involve several observable component transitions, the signature of the system reaction is a sequence of sets of messages. However, the observation relevant to a diagnostic problem is still structured as a collection of items, each being a sequence of messages. The distillation of the diagnostic information from the active space follows the same pattern adopted for active systems.

1.   Introduction

Chapters 4 and 5 have introduced two variants of a diagnostic technique for active systems, namely monolithic and modular diagnosis.

Although the modular technique can be viewed as a generalization of the monolithic one, the nature of the system to be diagnosed is the same. In particular, the system is asynchronous and, consequently, a system transition corresponds to a component transition and at most one message is generated (when the transition is observable). Each component transition is either triggered by an event ready on one of its input terminals or coming from the external world.

By contrast, in synchronous DESs (see Chapter 2), a system transition consists of several component transitions, specifically, the whole set of transitions that are triggerable in the relevant state. Therefore, ready events are consumed in parallel and several messages can be generated in one shot.

One may argue that the diagnostic techniques for asynchronous systems can be straightforwardly extended to cope with synchronous systems as well.

Instead of this, we prefer generalizing the class of considered DESs, so that a system integrates both synchronous and asynchronous behavior. Then, we extend the technique to cope with this new class of DESs.

A system that integrates both synchronous and asynchronous behavior is called polymorphic. The reaction of a polymorphic system is a sequence of system transitions, each of which is either synchronous or asynchronous. When asynchronous, the system transition involves a single component transition as assumed in previous chapters. When synchronous, the system transition encompasses (in general) several component transitions that are activated in parallel, like in synchronous DESs.

What distinguishes the two different behaviors is the kind of links on which ready events are stored. In a polymorphic system, links can be either synchronous or asynchronous. The ready events on synchronous links are considered as a whole: all relevant triggerable transitions are fired in parallel.

A formal result of this chapter is the equivalence of the monolithic and modular reconstruction in the generalized realm of polymorphic systems.

The chapter is organized as follows. Section 2 introduces the notion of a polymorphic system. Monolithic and modular reconstruction techniques for polymorphic systems are presented in Sections 3 and 4, respectively. The formal definition of a polymorphic system is provided in Section 5. Section 6 contains a summary of the introduced concepts. A pseudo-coded algorithm relevant to the generalized reconstruction technique is outlined in Appendix A. Proofs of relevant theorems, including the reconstruction equivalence theorem, are given in Appendix B.

2. Polymorphic System

From the modeling point of view, a polymorphic system is very close to an active system as far as component models and system topology are concerned. The peculiarity of a polymorphic system lies in the modeling of links. In Chapter 3, a link model was defined as a 4-tuple, with the implicit assumption that all links were asynchronous. In order to support the integration of synchronous and asynchronous behavior, in a polymorphic system a link can be either classified as synchronous or asynchronous. Consequently, a link model M_L is characterized by a further attribute that makes such a distinction.

Specifically, a link model in a polymorphic system is a 5-tuple,

$$M_L = (Y, I, O, \chi, \pi),$$

where:

$Y \in \{SYNC, ASYNC\}$ is the *type*;

I is the *input terminal*;

O is the *output terminal*;

χ is the *capacity*;

π is the *saturation policy*.

When $Y = ASYNC$, no constraints are provided on the remaining attributes. Otherwise, when $Y = SYNC$, the following constraints hold:

(1) $\chi = 1$;

(2) $\pi = WAIT$.

The vice versa does not hold in general, that is, it may be the case that $\chi = 1$, $\pi = WAIT$, and $Y = ASYNC$. In other words, when an event E is ready on a synchronous link L, no other events can be queued in L until E is consumed.

At most one synchronous link is directed toward a component C, that is, at most one input terminal of C can be connected with a synchronous link.

Like active systems, a polymorphic system Π is a triple $(\mathbb{C}, \mathbb{L}, \mathbb{D})$ of components, links, and dangling terminals. However, the on-terminals $\mathbb{D}_{on} \subseteq \mathbb{D}$ must be qualified either as synchronous or asynchronous, that is,

$$\mathbb{D}_{on} = \mathbb{D}_{on}^s \cup \mathbb{D}_{on}^a,$$

where

$\mathbb{D}_{on}^{s}$ is the set of *synchronous on-terminals*;

$\mathbb{D}_{on}^{a}$ is the set of *asynchronous on-terminals*.

Once connected to links, component terminals inherit the qualification of the relevant link, that is, a component terminal connected with a link L is either synchronous or asynchronous if and only if L is synchronous or asynchronous, respectively.

Example 6.1. Shown on the top of Figure 6.1 is the topology of a polymorphic system Π, obtained as a variation of system Ψ depicted in Figure 3.3, where links L_1 and L_3 (represented by dotted arrows) are synchronous. Thus, since the component models are still the same, the only difference of Π with respect to Ψ lies in the synchronism of such links. Formally, $\Pi = (\mathbb{C}, \mathbb{L}, \mathbb{D})$ where

$\mathbb{C} = \{C_1, C_2, C_3, C_4\}$;

$\mathbb{L} = \{L_1, L_2, L_3, L_4, L_5\}$;

$\mathbb{D} = \{O_3(C_2), I_2(C_3)\}$;

$\mathbb{D}_{on} = \mathbb{D}_{on}^{a} = \{I_2\}$;

$\mathbb{D}_{on}^{s} = \emptyset$;

$\mathbb{D}_{off} = \{O_3\}$.

Since input terminal I_2 belongs to the set $\mathbb{D}_{on}^{a}$ of asynchronous on-terminals, I_2 is assumed to be connected with an asynchronous link external to Π.   ◇

3.   Monolithic Reconstruction

The reconstruction of the system behavior based on a diagnostic problem involving a polymorphic system Π is still represented by an automaton (active space) where each path from the root to a final state is a possible system history. However, due to synchronous transitions, a transition in the active space is in general labeled by a set of component transitions rather than a single (asynchronous) transition.

To generate the active space, the reconstruction technique performs a simulation of the reactive behavior of Π, starting from the initial state Π_0, by applying the triggerable system transitions from the current state of Π based on the behavioral models of the involved components.

However, the rules governing the selection of the system transition differ from those applied to asynchronous systems as follows.

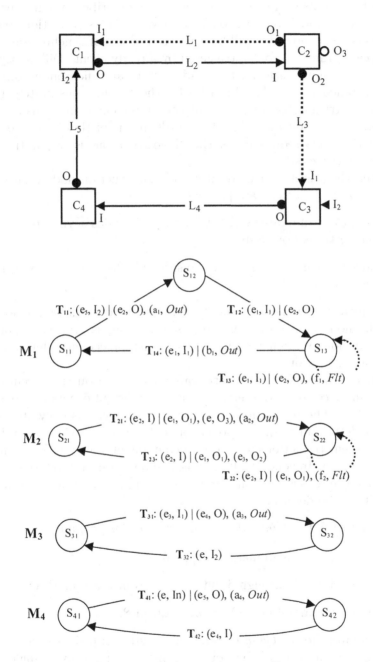

Figure 6.1. Polymorphic system Π.

If no event is ready at a synchronous terminal, only an asynchronous component transition can be considered, as described in Chapter 4. In such a case, candidates will consist of asynchronous transitions only.

Otherwise, candidates will be composed of sets of synchronous transitions, each transition being triggered by a ready event. Differently from the asynchronous case, all the ready events at synchronous terminals are consumed in parallel. In order for the synchronous system transition to be triggerable, all the involved component transitions are to be triggerable on their own. In other words, if one of these does not fulfill the triggerability requirements, the whole synchronous system transition cannot be triggered.

When the system is closed, at each reconstruction node (state of the active space), two cases are possible:

(1) The set of candidate system transitions includes asynchronous (component) transitions only;

(2) The set of candidate system transitions includes sets of synchronous transitions only.

Consequently, it is not possible that the set of candidates include both asynchronous transitions and sets of synchronous transitions because a synchronous system transition has higher precedence than an asynchronous transition.

It may sound odd that several candidate synchronous system transitions can be considered, seeing that all the relevant events are consumed in parallel. The reason for such an apparent inconsistency stems from the nondeterminism of component models: when an event may trigger two different transitions at a certain component state, both alternatives must be given. In general, all possible combinations are to be considered.

When the reconstruction is monolithic and based on a linear observation, it can be described by Algorithm 6.1.

Algorithm 6.1. (*Monolithic reconstruction for polymorphic systems*)

1. $\aleph := \{N_0\}$ (N_0 is unmarked); $\mathcal{E} := \emptyset$;

2. Repeat Steps 3 through 6 until all nodes in $\aleph$ are marked;

3. Get an unmarked node $N = (\sigma, K, Q)$ in $\aleph$;

4. Compute the set *Cand* of candidate system transitions (a candidate is in general a set of component transitions, possibly a singleton);

5. For each $\tau = \{T_1, \ldots, T_k\}$ in *Cand*, where $\forall i \in [1\,..\,k]$, T_i is relevant to component C_i, do the following steps:

(a) Create a copy $N' = (\sigma', K', Q')$ of N;

(b) For each component transition $T_i \in \tau$ do the following steps:

   (i) Set $\sigma'[i]$ to the state reached by T_i;

   (ii) If T_i is observable (a message is generated), then increment K' by one;

   (iii) If the triggering event E of T_i is relevant to an internal link L_j, then remove E from $Q'[j]$;

   (iv) Insert the internal output events of T_i into the relevant queues in Q';

(c) If $N' \notin \aleph$ then insert N' into $\aleph$; (N' is unmarked)

(d) Insert edge $N \xrightarrow{\tau} N'$ into $\mathcal{E}$;

6. Mark N;

7. Remove from $\aleph$ all the nodes and from $\mathcal{E}$ all the edges that are not on a path from the initial state N_0 to a final state in $\aleph$. ◇

Example 6.2. With reference to the active system Π displayed in Figure 6.1, consider the following diagnostic problem:

$$\wp(\Pi) = (\Omega(\Pi), OBS(\Pi), \Pi_0) \tag{6.1}$$

where

$$\Omega(\Pi) = (\mathbb{V}, \mathbb{P}), \mathbb{V} = \mathbb{C}(\Pi), \mathbb{P} = \{\{C_1, C_4\}, \{C_2, C_3\}\},$$
$$OBS(\Pi) = (\langle a_4, a_1, b_1 \rangle, \langle a_2, a_3 \rangle),$$
$$\Pi_0 = (S_{11}, S_{21}, S_{31}, S_{41}).$$

Note that $\wp(\Pi)$ mirrors the diagnostic problem $\wp(\Psi)$ of Example 4.4. The resulting active space is shown in Figure 6.2. Each node is identified by the set of component states (first four fields), the observation index (fifth and sixth fields), and the link states (last five fields). Each edge is marked by one or several component transitions. Specifically, node

$$N = (S_{13}, S_{21}, S_{31}, S_{42}, 2, 1, \langle e_1 \rangle, \langle \rangle, \langle e_3 \rangle, \langle \rangle, \langle \rangle)$$

is left by two edges marked by system transitions τ_1 and τ_2, where

$$\tau_1 = \{T_{14}, T_{31}\},$$
$$\tau_2 = \{T_{13}, T_{31}\}.$$

This is due to, on the one hand, the simultaneous presence of ready events e_1 and e_3 on synchronous links L_1 and L_3, respectively, on the

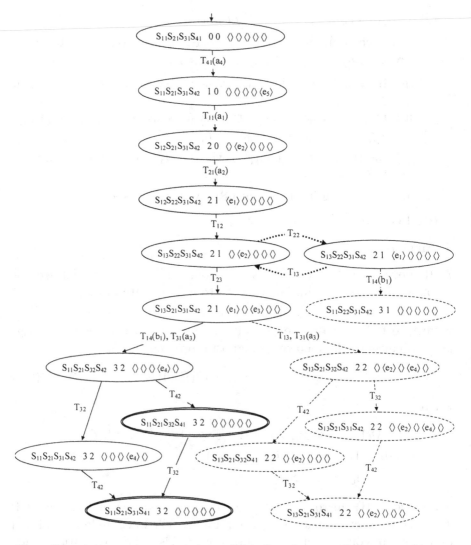

Figure 6.2. Active space, generated monolithically, relevant to diagnostic problem 6.1 (see Example 6.2).

other, the nondeterminism in state S_{13}, where event e_1 is bound to trigger either T_{13} or T_{14}. A possible system history is

$$h(\wp(\Pi)) = \langle \{T_{41}\}, \{T_{11}\}, \{T_{21}\}, \{T_{12}\}, \{T_{23}\}, \{T_{14}, T_{31}\}, \{T_{42}\} \rangle,$$

which includes all asynchronous transitions but one, $\tau = \{T_{14}, T_{31}\}$. $\diamond$

4.   Modular Reconstruction

Chapter 5 has presented a modular approach to the reconstruction problem. Basically, the original diagnostic problem for a system Θ is divided into a set of diagnostic subproblems relevant to a decomposition of Θ. Once reconstructed the system behavior corresponding to each subproblem, the final active space is generated by joining the local active spaces relevant to the clusters in the decomposition.

The modular technique can be naturally adapted to cope with the domain of polymorphic systems, although some peculiarities determined by the synchronous behavior must be considered, both in reconstruction and join steps.

4.1   Partial Active Space Generation

When a diagnostic problem $\wp(\Pi)$ is broken down into a set $\mathbf{P}$ of diagnostic subproblems relevant to a decomposition $\Xi(\Pi) = \{\xi_1, \ldots, \xi_m\}$,

$$\mathbf{P}(\Xi(\Pi)) = \{\wp(\xi_1), \ldots, \wp(\xi_m)\},$$

each cluster ξ_i is in general open, thereby including a set of dangling on-terminals. When no synchronous input on-terminals are included, the reconstruction process as described by Algorithm 6.1 can determine the set *Cand* of candidate transitions without any assumption on dangling input on-terminals[1]. By contrast, when the cluster ξ_i embodies a non-empty set $\mathbf{I}$ of synchronous input on-terminals, things are different. Specifically, in order to fulfill completeness, Step 4 of Algorithm 6.1 is expected to account for all the assumptions on the states of the links $\mathbf{L}$ connecting terminals in $\mathbf{I}$. In fact, since a candidate synchronous transition is composed of all the component transitions relevant to synchronous input terminals for which an event is ready, the lack of knowledge about links in $\mathbf{L}$ requires the algorithm to generate a candidate system transition for each different assumption.

Considering a cluster $\xi \in \Xi(\Pi)$, the computation of *Cand* in Step 4 of Algorithm 6.1 is described by Algorithm 6.2.

Specifically, at Line 1, the set $\mathbf{T}^{\text{a}}_{\text{cand}}$ of candidate asynchronous system transitions is generated, each system transition being a singleton.

Lines 2-4 yield the set $\tilde{\mathbf{T}}_{\text{s}}$ of internal (partial) candidate synchronous system transitions, considering the ready events within links internal to ξ. Lines 5-7 determine the set of external candidate system transitions, considering the events ready at dangling on-terminals. However, due

---

[1] Step 4 of the algorithm is generic in nature, inasmuch as no support is provided to the computation of *Cand*.

to the lack of knowledge about such external events, the actual set of candidate synchronous system transitions $\mathbf{T}^s_{cand}$ is computed by merging an internal candidate with a (possibly empty) subset of an external candidate, so as to cover all possibilities (Line 8).

Finally, at Line 9, non-triggerable system transitions are removed from $\mathbf{T}^s_{cand}$. The actual set *Cand* of candidate system transitions is determined at Line 10. In particular, if $\mathbf{T}^s_{cand}$ incorporates a system transition involving a component transition triggered by an event within ξ, then $\mathbf{T}^s_{cand}$ is the set of candidates. Otherwise, since no event is ready within a link of ξ and it is not possible to establish the existence of ready events at dangling synchronous on-terminals, the set of candidates is the union of the triggerable synchronous and asynchronous system transitions.

Algorithm 6.2. (*Computation of the candidate system transitions*)

1. $\mathbf{T}^a_{cand} := \{\{T\} \mid T$ is a triggerable asynchronous transition for a component in $\xi\}$;

2. $\mathbf{T}_s :=$ the set of synchronous transitions triggered by an event at a non-dangling terminal of ξ;

3. $\hat{\mathbf{T}}_s := \{\mathbb{T}_1, \ldots, \mathbb{T}_p\}$ where $\forall i \in [1 \mathinner{..} p](\mathbb{T}_i = \{T \mid T \in \mathbf{T}_s, T$ relevant to a component C_i of $\xi\})$;

4. $\tilde{\mathbf{T}}_s := \mathbb{T}_1 \times \cdots \times \mathbb{T}_p$;

5. $\mathbf{T}'_s :=$ the set of synchronous transitions triggered by an event at a dangling on-terminal of ξ;

6. $\hat{\mathbf{T}}'_s := \{\mathbb{T}'_1, \ldots, \mathbb{T}'_{p'}\}$ where $\forall i \in [1 \mathinner{..} p'](\mathbb{T}'_i = \{T' \mid T' \in \mathbf{T}'_s, T'$ relevant to a component C_i of $\xi\})$;

7. $\tilde{\mathbf{T}}'_s := \mathbb{T}'_1 \times \cdots \times \mathbb{T}'_{p'}$;

8. $\mathbf{T}^s_{cand} := \{\tau \mid \tau = \tilde{\tau} \cup \tilde{\tau}'', \tilde{\tau} \in \tilde{\mathbf{T}}_s, \tilde{\tau}'' \subseteq \tilde{\tau}', \tilde{\tau}' \in \tilde{\mathbf{T}}'_s\}$;

9. Remove from $\mathbf{T}^s_{cand}$ the system transitions that are not triggerable;

10. If $\mathbf{T}^s_{cand}$ involves a component transition triggered by an event ready within a link of ξ then *Cand* $:= \mathbf{T}^s_{cand}$ else *Cand* $:= \mathbf{T}^a_{cand} \cup \mathbf{T}^s_{cand}$. $\diamond$

4.2   Join of Partial Active Spaces

As for asynchronous systems, once generated the active spaces relevant to the diagnostic subproblems, a join operation is required to make up the active space relevant to the original diagnostic problem. In this step, both the temporal constraints on the system observation and the constraints on the interface are to be enforced. What is peculiar to the join of active spaces for a polymorphic system is the enforcement of the synchronism, where applicable. As highlighted above, the completeness of the reconstruction for a cluster $\xi \in \Xi(\Pi)$ requires the algorithm to account for all the possible assumptions at the frontier of ξ, particularly at input synchronous dangling on-terminals. In the join step it is possible to solve such an uncertainty as much as it is for asynchronous systems.

When the join is based on a linear observation, it can be described by Algorithm 6.3.

Algorithm 6.3. (*Join of active spaces for polymorphic systems*)

1. $\aleph := \{N_0\}$ (N_0 is unmarked); $\mathcal{E} := \emptyset$;

2. Repeat Steps 3 through 6 until all nodes in $\aleph$ are marked;

3. Get an unmarked node $N = (\sigma, K, Q)$ in $\aleph$;

4. Compute the set *Cand* of candidate system transitions;

5. For each $\tau \in Cand$ do the following steps:

   (a) Create a copy $N' = (\sigma', K', Q')$ of N;

   (b) Let $\mathbf{T} = \{\tau_1, \ldots, \tau_m\}$ be the set of $\tau_i \subseteq \tau$ relevant to clusters $\xi_i \in \Xi(\Pi)$, respectively;

   (c) For each $i \in [1 .. m]$, such that $\tau_i \neq \emptyset$, set $\sigma'[i]$ to the state reached by τ_i in the partial active space relevant to ξ_i;

   (d) Increment K by the number of observable component transitions in τ;

   (e) For each component transition $T \in \tau$ do the following steps:

       (i) If the triggering event E of T is relevant to a link L_j in *Interf*$(\Xi(\Pi))$, then remove E from $Q'[j]$;

       (ii) Insert the internal output events of T into the relevant queues in Q';

   (f) If $N' \notin \aleph$ then insert N' into $\aleph$; (N' is unmarked)

   (g) Insert edge $N \xrightarrow{T} N'$ into $\mathcal{E}$;

6. Mark N;

7. Remove from $\aleph$ all the nodes and from $\mathcal{E}$ all the edges that are not on a path from the initial state N_0 to a final state in $\aleph$. $\diamond$

Two remarks are worthwhile. On the one hand, Step 4 of the above algorithm provides no explanation on how *Cand* is actually computed.

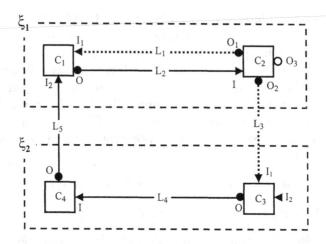

Figure 6.3. Clusters ξ_1 and ξ_2 of system Π (see Figure 6.1).

This can be envisaged by looking at Algorithm 6.2 and conceptually replacing component models with partial active spaces, which can be thought of as the constrained behavioral models of the relevant clusters. In particular, synchronous cluster transitions are merged appropriately so as to guarantee that all triggerable transitions are fired in parallel.

On the other, the computation of the field σ' (Step 5.c) is carried out after grouping component transitions in τ by subsets relevant to clusters of the decomposition. In fact, τ is in general formed by the composition of several cluster transitions (specifically, when synchronism is applicable), each of which moves the reconstruction state to a new node that is the new value of the corresponding element in σ'.

Example 6.3. Consider Example 6.2, where a monolithic reconstruction relevant to the diagnostic problem

$$\wp(\Pi) = (\Omega(\Pi), OBS(\Pi), \Pi_0) \tag{6.2}$$

was carried out based on the following assumptions:

$$\Omega(\Pi) = (\mathbb{V}, \mathbb{P}), \mathbb{V} = \mathbb{C}(\Pi), \mathbb{P} = \{\{C_1, C_4\}, \{C_2, C_3\}\},$$
$$OBS(\Pi) = (\langle a_4, a_1, b_1 \rangle, \langle a_2, a_3 \rangle),$$
$$\Pi_0 = (S_{11}, S_{21}, S_{31}, S_{41}).$$

We now make up the reconstruction of the system reaction by means of the modular technique. With reference to clusters ξ_1 and ξ_2 shown in Figure 6.3, we need to generate the active spaces relevant to the diagnostic subproblems obtained by restricting $\wp(\Pi)$ on such clusters,

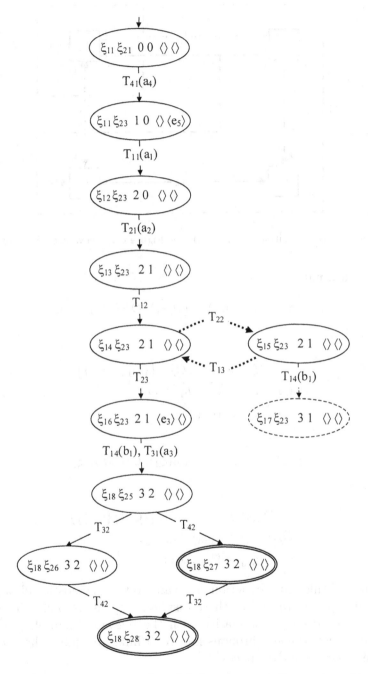

Figure 6.4. Active space, generated modularly, relevant to diagnostic problem (6.2) (see Example 6.3).

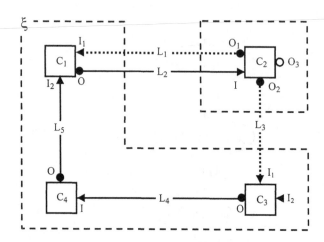

Figure 6.5. Decomposition of system Π (see Figure 6.1) relevant to Example 6.4.

respectively, namely,

$$\wp(\xi_1) = \wp_{\langle \xi_1 \rangle}(\Pi) = (\Omega(\xi_1), OBS(\xi_1), \xi_{1_0}),$$

where

$$\Omega(\xi_1) = (\{C_1, C_2\}, \{\{C_1\}, \{C_2\}\})$$
$$OBS(\xi_1) = (\langle a_1, b_1 \rangle, \langle a_2 \rangle),$$
$$\xi_{1_0} = (S_{11}, S_{21}),$$

and

$$\wp(\xi_2) = \wp_{\langle \xi_2 \rangle}(\Pi) = (\Omega(\xi_2), OBS(\xi_2), \xi_{2_0}),$$

where

$$\Omega(\xi_2) = (\{C_3, C_4\}, \{\{C_4\}, \{C_3\}\})$$
$$OBS(\xi_2) = (\langle a_4 \rangle, \langle a_3 \rangle),$$
$$\xi_{2_0} = (S_{31}, S_{41}).$$

The monolithic reconstructions relevant to the diagnostic subproblems $\wp(\xi_1)$ and $\wp(\xi_2)$ are in fact those displayed in Figure 5.2. This coincidence stems from the similarity with the relevant Example 5.5. As a matter of fact, the synchronous nature of links L_1 and L_3 has no effect on these partial active spaces[2].

---

[2]This is only a coincidence. Generally speaking, synchronous links introduce different reconstructions even for subproblems.

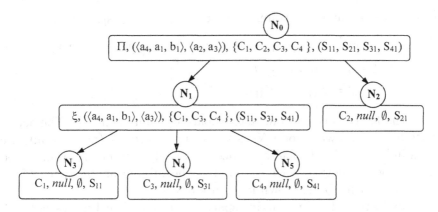

Figure 6.6. Reconstruction graph for diagnostic problem of Example 6.4.

Based on such partial active spaces, the generation of the active space relevant to the original diagnostic problem $\wp(\Pi)$ is carried out as described by Algorithm 6.3. The result is depicted in Figure 6.4, which, as expected, is equivalent to the active space generated monolithically and displayed in Figure 6.2 (spurious nodes are irrelevant to the equivalence). Also note that, although the partial active spaces are identical to those determined in Example 5.5 (Figure 5.2), the final active space differs from the corresponding one depicted in Figure 5.3. In particular, the number of states has shrunk (from 16 in Figure 5.3 to 11 in Figure 6.4). Such a difference comes with no surprise, as Π differs from Ψ in the topology, specifically, the nature of links L_1 and L_3. ◇

Example 6.4. Example 6.3 has shown a modular reconstruction for the diagnostic problem coped with monolithically in Example 6.2. Now, with reference to the decomposition of Π depicted in Figure 6.5, we solve the same diagnostic problem by means of the reconstruction graph shown in Figure 6.6. Each node of the graph is labeled by the identifier of the cluster at hand, the relevant observation, the visible set of the corresponding observer, and the initial state. Accordingly, cluster ξ embodies components C_1, C_3, and C_4, and links L_4 and L_5.

The diagnostic problem $\wp(\Pi) = (\Omega(\Pi), OBS(\Pi), \Pi_0)$ associated with the root N_0, where

$$\Omega(\Pi) = (\mathbb{V}, \mathbb{P}), \text{ where } \mathbb{V} = \mathbb{C}(\Pi), \mathbb{P} = \{\{C_1, C_4\}, \{C_2, C_3\}\},$$
$$OBS(\Pi) = (\langle a_4, a_1, b_1 \rangle, \langle a_2, a_3 \rangle),$$
$$\Pi_0 = (S_{11}, S_{21}, S_{31}, S_{41}),$$

has been decomposed into two diagnostic problems, one relevant to ξ and the other to C_2 (see Figure 6.5), where the former is a monolithic reconstruction involving components C_1, C_3 and C_4, specifically,

$$\Omega(\xi) = (\mathbb{V}, \mathbb{P}), \text{ where } \mathbb{V} = \mathbb{C}(\xi), \mathbb{P} = \{\{C_1, C_4\}, \{C_3\}\},$$
$$OBS(\xi) = (\langle a_4, a_1, b_1 \rangle, \langle a_3 \rangle),$$
$$\xi_0 = (S_{11}, S_{31}, S_{41}),$$

while the latter is the atomic active space of component C_2. The active space associated with node N_1 is shown in Figure 6.7 (node details are given in Table 6.1). The active space relevant to the root N_0 is obtained by joining the active space of N_1 and the atomic active space of C_2 associated with N_2, as shown in Figure 6.8.

As expected, the three active spaces represented in Figures 6.2, 6.4, and 6.8, which have been determined monolithically, modularly, and based on a reconstruction graph, respectively, are equivalent. ◇

5. Formalization of a Polymorphic Active Space

The compositional definition of active space given in Section 5 of Chapter 5 can be generalized for polymorphic systems as follows.

Let $\wp(\xi) = (\Omega(\xi), OBS(\xi), \xi_0)$ be a diagnostic problem for a polymorphic cluster ξ. An *active space* of $\wp(\xi)$ is an automaton

$$Act(\wp(\xi)) = (\mathbb{S}, \mathbb{E}, \mathbb{T}, S_0, \mathbb{S}_f)$$

where

$\mathbb{S}$ is the set of states;

$\mathbb{E}$ is the set of events;

$\mathbb{T}$ is the transition function, $\mathbb{T} = \mathbb{T}^a \cup \mathbb{T}^s$ such that $\mathbb{T}^a \cap \mathbb{T}^s = \emptyset$, where

$\mathbb{T}^a$ is the asynchronous transition function,

$\mathbb{T}^s$ is the synchronous transition function;

S_0 is the initial state;

$\mathbb{S}_f \subseteq \mathbb{S}$ is the set of final states.

The elements of the automaton are defined as follows.

(1) If ξ incorporates a single component C with model

$$M_C = (\mathbb{S}_C, \mathbb{E}_{in_C}, \mathbb{I}_C, \mathbb{E}_{out_C}, \mathbb{O}_C, \mathbb{T}_C)$$

and the observer of $OBS(\xi)$ is blind, then

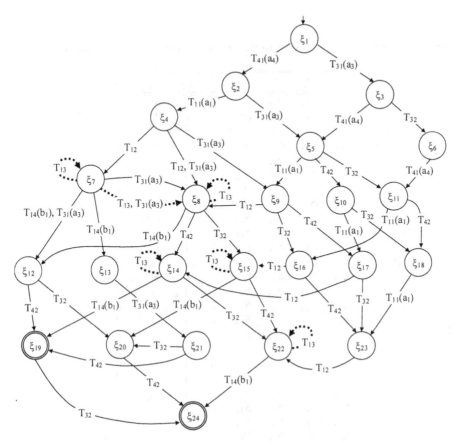

Figure 6.7. Active space relevant to node N_1 of the reconstruction graph displayed in Figure 6.6 (node details are shown in Table 6.1).

$\mathbb{S} = \mathbb{S}_C;$

$\mathbb{E} = \mathbb{T}_C;$

$\mathbb{T}^a : \mathbb{S} \times \mathbb{E} \mapsto \mathbb{S};$

$\mathbb{T}^s : \mathbb{S} \times \mathbb{E} \mapsto \mathbb{S};$

$S_0 = \xi_0;$

$\mathbb{S}_f = \{ S_f \in \mathbb{S}_C \mid \text{there exists a path } S_0 \rightsquigarrow S_f \text{ in } M_C \}.$

Furthermore, the following condition holds. Let

$$T^* = S \xrightarrow{T} S' \in \mathbb{T}, T = S \xrightarrow{\alpha \mid \beta} S' \in \mathbb{T}_C, \alpha = (E, \vartheta).$$

Then,

$$T^* \in \left\{ \begin{array}{ll} \mathbb{T}^a & \text{if either } Link(\vartheta) \text{ is asynchronous or } \vartheta = In \\ \mathbb{T}^s & \text{otherwise.} \end{array} \right.$$

Table 6.1. Node details relevant to the active space displayed in Figure 6.7.

ξ	S_1	S_3	S_4	k_1	k_2	L_4	L_5
ξ_1	S_{11}	S_{21}	S_{31}	0	0	$\emptyset$	$\emptyset$
ξ_2	S_{11}	S_{21}	S_{32}	1	0	$\emptyset$	$\langle e_5 \rangle$
ξ_3	S_{11}	S_{22}	S_{31}	0	1	$\langle e_4 \rangle$	$\emptyset$
ξ_4	S_{12}	S_{21}	S_{32}	2	0	$\emptyset$	$\emptyset$
ξ_5	S_{11}	S_{22}	S_{32}	1	1	$\langle e_4 \rangle$	$\langle e_5 \rangle$
ξ_6	S_{11}	S_{21}	S_{31}	0	1	$\langle e_4 \rangle$	$\emptyset$
ξ_7	S_{13}	S_{21}	S_{32}	2	0	$\emptyset$	$\emptyset$
ξ_8	S_{13}	S_{22}	S_{32}	2	1	$\langle e_4 \rangle$	$\emptyset$
ξ_9	S_{12}	S_{22}	S_{32}	2	1	$\langle e_4 \rangle$	$\emptyset$
ξ_{10}	S_{11}	S_{22}	S_{31}	1	1	$\emptyset$	$\langle e_5 \rangle$
ξ_{11}	S_{11}	S_{21}	S_{32}	1	1	$\langle e_4 \rangle$	$\langle e_5 \rangle$
ξ_{12}	S_{11}	S_{22}	S_{32}	3	1	$\langle e_4 \rangle$	$\emptyset$
ξ_{13}	S_{11}	S_{21}	S_{32}	3	0	$\emptyset$	$\emptyset$
ξ_{14}	S_{13}	S_{22}	S_{31}	2	1	$\emptyset$	$\emptyset$
ξ_{15}	S_{13}	S_{21}	S_{32}	2	1	$\langle e_4 \rangle$	$\emptyset$
ξ_{16}	S_{12}	S_{21}	S_{32}	2	1	$\langle e_4 \rangle$	$\emptyset$
ξ_{17}	S_{12}	S_{22}	S_{31}	2	1	$\emptyset$	$\emptyset$
ξ_{18}	S_{11}	S_{21}	S_{31}	1	1	$\emptyset$	$\langle e_5 \rangle$
ξ_{19}	S_{11}	S_{22}	S_{31}	3	1	$\emptyset$	$\emptyset$
ξ_{20}	S_{11}	S_{21}	S_{32}	3	1	$\langle e_4 \rangle$	$\emptyset$
ξ_{21}	S_{11}	S_{22}	S_{32}	3	1	$\langle e_4 \rangle$	$\emptyset$
ξ_{22}	S_{13}	S_{21}	S_{31}	2	1	$\emptyset$	$\emptyset$
ξ_{23}	S_{12}	S_{21}	S_{31}	2	1	$\emptyset$	$\emptyset$
ξ_{24}	S_{11}	S_{21}	S_{31}	3	1	$\emptyset$	$\emptyset$

In this case the active space is called *atomic*.

(2) If $\Xi(\xi) = \{\xi_1, \ldots, \xi_m\}$ is a decomposition of ξ and

$$\mathbf{A} = \{Act(\wp(\xi_1)), \ldots, Act(\wp(\xi_m))\}$$

is a set of active spaces relevant to $\Xi(\xi)$, then

$$Act(\wp(\xi)) = \mathcal{J}_{\wp(\xi)}(\mathbf{A})$$

where $\mathcal{J}$ is the (polymorphic) *join operator* defined in the next section. In this case the active space is called *compound*.

5.1 Polymorphic Join Operator

The polymorphic join operator, that extends the join operator defined in Chapter 5, is formalized as follows.

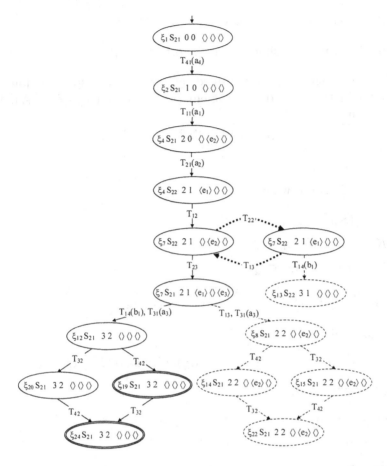

Figure 6.8. Active space for the root N_0 of the reconstruction graph displayed in Figure 6.6, obtained by joining the active space of Figure 6.7 and the atomic active space of C_2 (see Example 6.4).

Let

$$OBS(\xi) = (obs(\mathbf{C}_1), \ldots, obs(\mathbf{C}_n)) \text{ be an observation of } \xi,$$

$\Xi(\xi)$ a decomposition of ξ,

$\mathbf{A} = \{Act(\wp(\xi_1)), \ldots, Act(\wp(\xi_m))\}$ a set of relevant active spaces such that

$$\forall i \in [1 .. m] \; (Act(\wp(\xi_i)) = (\mathbb{S}_i, \mathbb{E}_i, \mathbb{T}_i, S_{0_i}, \mathbb{S}_{f_i})),$$
$$OBS(\xi) \trianglerighteq \{OBS(\xi_1), \ldots, OBS(\xi_m)\},$$

$$\Sigma = \mathbb{S}_1 \times \cdots \times \mathbb{S}_m,$$

$\mathbb{K}$ the domain of possible index values of $OBS(\xi)$;

$\mathbb{Q}$ the domain of possible queues in $Interf(\Xi(\xi))$.

The spurious active space obtained by applying the spurious join $\tilde{\mathcal{J}}$ of $\mathbf{A}$ based on the diagnostic problem $\wp(\xi)$, denoted by $\tilde{\mathcal{J}}_{\wp(\xi)}(\mathbf{A})$, is an automaton

$$\tilde{\mathcal{J}}_{\wp(\xi)}(\mathbf{A}) = (\tilde{\mathbb{S}}, \mathbb{E}, \tilde{\mathbb{T}}, S_0, \mathbb{S}_f)$$

where

$S_0 = (\sigma_0, K_0, Q_0)$, where

$\quad \sigma_0 = (S_{0_1}, \ldots, S_{0_m}), \sigma_0 \in \Sigma,$

$\quad K_0 = (0 \cdots 0), K_0 \in \mathbb{K},$

$\quad Q_0 = (\emptyset \cdots \emptyset), Q_0 \in \mathbb{Q};$

$\tilde{\mathbb{S}} = \{S_0\} \cup \{S' \mid S \xrightarrow{\tau} S' \in \tilde{\mathbb{T}}\}, \tilde{\mathbb{S}} \subseteq \Sigma \times \mathbb{K} \times \mathbb{Q};$

$\mathbb{E} = \bigcup_{i=1}^m \mathbb{T}_i;$

$\mathbb{S}_f = \{(\sigma_f, K_f, Q_f) \mid \forall S_i \in \sigma_f \, (S_i \in \mathbb{S}_{f_i}), Complete(K_f), Q_f = (\emptyset \cdots \emptyset)\};$

$\tilde{\mathbb{T}} = \tilde{\mathbb{T}}^a \cup \tilde{\mathbb{T}}^s$, where

$\quad \tilde{\mathbb{T}}^a \cap \tilde{\mathbb{T}}^s = \emptyset,$

$\quad \tilde{\mathbb{T}}^a \supseteq \mathbb{T}^a,$

$\quad \tilde{\mathbb{T}}^s \supseteq \mathbb{T}^s,$

$\quad \tilde{\mathbb{T}}^a : \tilde{\mathbb{S}} \times \mathbb{E} \mapsto \tilde{\mathbb{S}},$

$\quad \tilde{\mathbb{T}}^s : \tilde{\mathbb{S}} \times 2^{\mathbb{E}} \mapsto \tilde{\mathbb{S}}.$

Specifically, the transition function $\tilde{\mathbb{T}}$ is defined as follows:

$$N \xrightarrow{\tau} N' \in \tilde{\mathbb{T}},$$

where

$N = (\sigma, K, Q), \sigma = (S_1, \ldots, S_m),$

$N' = (\sigma', K', Q'), \sigma' = (S'_1, \ldots, S'_m),$

$\tau = \{T_1, \ldots, T_p\},$

if and only if the following conditions hold:

(1) $\tau \in Dom(N)$, where $Dom(N)$ is defined as follows. Let *Async*, *Dummy*, and *Sync* be functions of a state S belonging to an active space $A = Act(\wp(\eta))$, defined as follows:

$$Async(S) = \{\{T\} \mid S \xrightarrow{T} S' \in \mathbb{T}^a(A)\},$$

$$Dummy(S) = \begin{cases} \{\emptyset\} & \text{if } \{\tau' \mid S \xrightarrow{\tau'} S' \in \mathbb{T}^s(A), \\ & \quad (T = \tau' \text{ or } T \in \tau'), Link(T) \in \mathbb{L}(\eta)\} = \emptyset \\ \emptyset & \text{otherwise,} \end{cases}$$

$$Sync(S) = \begin{cases} Dummy(S) \cup \{\tau' \mid S \xrightarrow{\tau'} S' \in \mathbb{T}^s(A)\} & \text{if } A \text{ is compound} \\ \{\emptyset\} \cup \{\{T\} \mid S \xrightarrow{T} S' \in \mathbb{T}^s(A)\} & \text{otherwise.} \end{cases}$$

Let $\mathbf{S}_1 \boxtimes \mathbf{S}_2$ denote the *aggregative composition* of two sets of sets, defined as follows:

$$\mathbf{S}_1 \boxtimes \mathbf{S}_2 = \{\mathbf{S} \mid \mathbf{S} = \mathbb{S}_1 \cup \mathbb{S}_2, (\mathbb{S}_1, \mathbb{S}_2) \in \mathbf{S}_1 \times \mathbf{S}_2\}.$$

Let *ADom*, *Kernel*, and *SDom* be functions of N, defined as follows:

$$ADom(N) = \bigcup_{i=1}^{m} Async(S_i),$$

$$Kernel(N) = \{L \mid L \in Interf(\Xi(\xi)), L \text{ is synchronous}, Q[L] \neq \emptyset\},$$

$$SDom(N) = \{\tau' \mid \tau' \in (Sync(S_1) \boxtimes \cdots \boxtimes Sync(S_m)), \\ \forall L \in Kernel(N)(\exists T \in \tau'(Link(T) = L))\}.$$

Then,

$$Dom(N) = \begin{cases} SDom(N) & \text{if } \emptyset \notin SDom(N) \\ ADom(N) \cup (SDom(N) - \{\emptyset\}) & \text{otherwise} \end{cases};$$

(2) *Triggerable*(τ, N), where *Triggerable* is a predicate defined as follows. Let

$$\mathbb{L}_\alpha = \{L_\alpha \mid L_\alpha = Link(\alpha), T \in \tau, T = S \xrightarrow{\alpha|\beta} S', \\ \alpha = (E_\alpha, \theta_\alpha), L_\alpha \in Interf(\Xi(\xi))\}.$$

Then,

$$Triggerable(\tau, N) \Leftrightarrow \forall L_\alpha \in \mathbb{L}_\alpha, L_\alpha = Link(\alpha), \alpha = (E_\alpha, \theta_\alpha) \\ (Head(Q[L_\alpha]) = E_\alpha);$$

(3) *Consistent*(τ, N), where *Consistent* is a predicate defined as follows. Let

$$\mathbb{L}_\beta = \{L_\beta \mid L_\beta = Link(B), B \in \beta, T = S \xrightarrow{\alpha|\beta} S' \in \tau, L_\beta \in Interf(\Xi(\xi))$$

$$\mathbb{L}_\beta^u = \{L_\beta \mid L_\beta \in \mathbb{L}_\beta, (\text{either } L_\beta \text{ is not saturated or } L_\beta \in \mathbb{L}_\alpha)\},$$

$$\mathbb{L}_\beta^s = \mathbb{L}_\beta - \mathbb{L}_\beta^u,$$

$$\mathbb{L}_\beta^{so} = \{L_\beta \mid L_\beta \in \mathbb{L}_\beta^s, \text{ the saturation policy of } L_\beta \text{ is } OVERRIDE\},$$

$$\mathbb{L}_\beta^{sw} = \{L_\beta \mid L_\beta \in \mathbb{L}_\beta^s, \text{ the saturation policy of } L_\beta \text{ is } WAIT\},$$

$Obs(\tau) = \{m_1, \ldots, m_r\}$ denote the set of messages relevant to the observable transitions in τ,

$Comp(Obs(\tau))$ denote the set of components relevant to $Obs(\tau)$,

$\mathbf{C}_j^\tau = \mathbf{C}_j \cap Comp(Obs(\tau)) \cap Vis(OBS(\xi)), j \in [1..n]$,

$obs(\mathbf{C})[x..y]$ denote the set of messages between indexes x and y,

$\mu_j = obs(\mathbf{C}_j)[(K[j]+1)..(K[j]+|\mathbf{C}_j^\tau|)]$.

Then,

$$Consistent(\tau, N) \Leftrightarrow \left\{ \begin{array}{l} \forall j \in [1..n], \mathbf{C}_j^\tau \neq \emptyset \ (\mu_j = Obs_{\langle \mathbf{C}_j^\tau \rangle}(\tau)) \\ \mathbb{L}_\beta^{sw} = \emptyset \end{array} \right. ;$$

(4) σ' is such that, denoting with

$$Group(\tau, \Xi(\xi)) = \{\tau_i \mid i \in [1..m], \tau_i \subseteq \tau \text{ relevant to components in } \xi_i\},$$

the following condition holds:

$$\forall i \in [1..m] \left(S_i' = \left\{ \begin{array}{ll} S_{i'} & \text{if } \tau_i \in Group(\tau, \Xi(\xi)), \tau_i \neq \emptyset, S_i \xrightarrow{\tau_i} S_{i'} \in \mathbb{T}_i \\ S_i & \text{otherwise} \end{array} \right. \right.$$

(5) K' is such that $\forall j \in [1..n] \ (K'[j] = K[j] + |\mathbf{C}_j^\tau|)$;

(6) Q' is such that:

   (a) $\forall L_\alpha \in \mathbb{L}_\alpha \ (Q'[L_\alpha] = Tail(Q[L_\alpha]))$,

   (b) $\forall (E, \vartheta) \in \beta, L_\beta = Link(\vartheta), L_\beta \in (\mathbb{L}_\beta^u \cup \mathbb{L}_\beta^{so}) \ (Q'[L_\beta] = Ins(Q[L_\beta], E))$,

   (c) $\forall L \in (Interf(\Xi(\xi)) - (\mathbb{L}_\alpha \cup \mathbb{L}_\beta^u \cup \mathbb{L}_\beta^{so})) \ (Q'[L] = Q[L])$.

The join $\mathcal{J}$ of $\mathbf{A}$ based on the diagnostic problem $\wp(\xi)$ is the automaton obtained from $\tilde{\mathcal{J}}_{\wp(\xi)}(\mathbf{A})$ by selecting the states and transitions which are convergent, that is, those connected to a final state.

Proposition 6.1. *Let N be a node of a compound active space. Then, the following entailments hold:*

$$\tau \in Sync(N) \Rightarrow \tau \in SDom(N);$$
$$\tau \in Async(N) \Rightarrow \tau \in ADom(N).$$

The notion of a canonical reconstruction $\Re(\wp(\xi))$ defined in Section 5.2 of Chapter 5 can be inherited as is within the framework of polymorphic systems. Due to formal reasons, it is worthwhile introducing the definition of a restriction of a polymorphic active space as follows.

Let $A = \Re(\wp(\xi)) = (\mathbb{S}, \mathbb{E}, \mathbb{T}, S_0, \mathbb{S}_f)$ be a polymorphic active space yielded by the canonical reconstruction of the diagnostic problem

$$\wp(\xi) = (\Omega(\xi), OBS(\xi), \xi_0).$$

The *restriction* of A on a cluster $\check{\xi} \subseteq \xi$, denoted by $A_{\langle \check{\xi} \rangle}$, is the automaton $\check{A} = (\check{\mathbb{S}}, \check{\mathbb{E}}, \check{\mathbb{T}}, \check{S}_0, \check{\mathbb{S}}_f)$ where:

$\check{\mathbb{S}} = \mathbb{S}_{\langle \check{\xi} \rangle}$ is the restriction of $\mathbb{S}$ on $\check{\xi}$, defined as follows. For each $S = (\sigma, K, D)$ in $\mathbb{S}$, $\check{S} = S_{\langle \check{\xi} \rangle} = (\check{\sigma}, \check{K}, \check{D})$ belongs to $\check{\mathbb{S}}$, where

> $\check{\sigma} \subseteq \sigma$ is the projection on the set of states relevant to the components in $\mathbb{C}(\check{\xi})$,
>
> $\check{K}$ is the index of the observation $OBS_{\langle \check{\xi} \rangle}(\xi)$, whose value follows Condition (5) of the definition of polymorphic active space, specifically, if $\check{S} \xrightarrow{\check{\tau}} \check{S}' \in \check{\mathbb{T}}$, then for each j in $[1..\check{n}]$, where $\check{n}$ is the number of observation items of $OBS(\check{\xi})$, we have:
>
> $$\check{K}'[j] = \check{K}[j] + |\mathbf{C}_j^{\check{\tau}}|,$$
>
> $\check{Q} \subseteq Q$ is the set of dangling sets relevant to the links in $\mathbb{L}(\check{\xi})$;

$\check{\mathbb{E}} = \mathbb{E}_{\langle \check{\xi} \rangle}$ is the restriction of $\mathbb{E}$ on $\check{\xi}$, that is, the set of events relevant to the models of components in $\mathbb{C}(\check{\xi})$;

$\check{\mathbb{T}} = \mathbb{T}_{\langle \check{\xi} \rangle}$ is the restriction of $\mathbb{T}$ on $\check{\xi}$, defined as follows: for each $S \xrightarrow{\tau} S'$ in $\mathbb{T}$, $\check{S} \xrightarrow{\check{\tau}} \check{S}'$ belongs to $\mathbb{T}'$, where

> $\check{S} = S_{\langle \check{\xi} \rangle}$, $\check{S}' = S'_{\langle \check{\xi} \rangle}$,
>
> $\check{\tau} = \tau_{\langle \check{\xi} \rangle}$, $\check{\tau} \subseteq \tau$, $\check{\tau} \neq \emptyset$, is the set of transitions in τ which are relevant to components in $\mathbb{C}(\check{\xi})$;

$\check{S}_0 = S_{0\langle \check{\xi} \rangle}$ is the restriction of S_0 on components in $\mathbb{C}(\check{\xi})$;

$\check{\mathbb{S}}_{\mathrm{f}} = \{\check{S}_f \mid \check{S}_{\mathrm{f}} = S_{\langle\check{\xi}\rangle}, S \in \mathbb{S}, \mathit{Complete}(K(\check{S}_{\mathrm{f}})), Q(\check{S}_{\mathrm{f}}) = (\emptyset \cdots \emptyset)\}.$

The restriction of A on a diagnostic problem $\wp(\check{\xi}) = (\Omega(\check{\xi}), \mathit{OBS}(\check{\xi}), \check{\xi}_0)$, where $\wp(\check{\xi}) \preceq \wp_{\langle\check{\xi}\rangle}(\xi)$, equals the restriction $A_{\langle\check{\xi}\rangle}$ but K, which is the index of $\mathit{OBS}(\check{\xi})$, rather than the index of $\mathit{OBS}_{\langle\check{\xi}\rangle}(\xi)$.

Proposition 6.2. *Let A' be the restriction of an active space $\Re(\wp(\xi))$ on either a cluster ξ' or a diagnostic problem $\wp(\xi')$. Then, each state and each transition in A' is convergent.*

Finally, Proposition 5.2 can be rephrased in the context of polymorphic systems by the next theorem.

Theorem 6.1. *Let $\wp(\xi) = (\Omega(\xi), \mathit{OBS}(\xi), \xi_0)$ be a diagnostic problem for a polymorphic cluster ξ, $\Xi(\xi) = \{\xi_1, \ldots, \xi_m\}$ a decomposition of ξ,*

$$\mathbf{A}_\xi = \{\mathit{Act}(\wp(\xi_1)), \ldots, \mathit{Act}(\wp(\xi_m))\}$$

a set of relevant active spaces where

$$\forall i \in [1 \mathbin{..} m] \; (\wp(\xi_i) = (\Omega(\xi_i), \mathit{OBS}(\xi_i), \xi_{i_0})),$$

such that ξ_0 is the composition of $\xi_{1_0} \cdots \xi_{m_0}$ and

$$\mathit{OBS}(\xi) \unrhd \{\mathit{OBS}(\xi_1), \ldots, \mathit{OBS}(\xi_m)\}.$$

Let $\mathbf{A}_C$ the set of atomic active spaces of components in ξ. Then,

$$\mathcal{J}_{\wp(\xi)}(\mathbf{A}_\xi) \equiv \Re(\wp(\xi)).$$

Theorem 6.1 provides a formal basis for the modular reconstruction of the reaction of a polymorphic system, possibly based on a reconstruction plan, as shown in the previous examples.

6.   Summary

- *Polymorphic System*: a discrete-event system that integrates synchronous and asynchronous behavior. Polymorphic systems generalizes the class of active systems introduced in Chapter 3.

- *Synchronous Link*: a link whose capacity and saturation policy are constrained to be 1 and *WAIT*, respectively, such that an event ready on it is consumed before any other event ready on asynchronous links.

- *Synchronous Component Transition*: a component transition triggered by an event ready on a synchronous link.

- *Synchronous System Transition*: the set of component transitions that are triggered by the events ready on the synchronous links of the polymorphic system. All such transitions are fired in parallel.

- *Polymorphic Active Space*: a finite automaton whose language represents the whole set of histories that are consistent with a given diagnostic problem for a polymorphic system. Each history is a path from the initial state of the system to a final (quiescent) system state. In general, each edge is marked by a set of component transitions that makes up a system transition. A polymorphic active space can be yielded either monolithically or based on a reconstruction plan.

Appendix A
Algorithms

This appendix generalizes the *Join* algorithm presented in Chapter 5 for modular diagnosis. Based on a given diagnostic problem, Algorithm A.4 implements the polymorphic join of a set of active spaces relevant to a cluster decomposition. The generation of the active space based on a given reconstruction plan, as specified by algorithm *ModRec* in Chapter 5, remains unchanged for polymorphic systems provided that the call to *Join* at Line 9 be replaced by a call to *Pjoin*.

Algorithm A.4. (*Pjoin*) The *Pjoin* function generates a graph-based representation of the active space $Act(\wp(\Theta))$ relevant to the input diagnostic problem (the notation of the algorithm is based on the formal definition of polymorphic active space provided in Section 5). It makes use of the recursive function *PStep*, whose definition is embedded within *Pjoin*, between the parameter specification and the body of *Pjoin*, the latter starting at Line 54. At Line 55, the parameters of the first call to *PStep* are instantiated, namely σ_0, K, Q, and $\aleph$. *PStep* is first called at Line 56.

Considering the body of *JStep* (Lines 1–53), at Line 2 the domain *Dom* of system transitions from the current node N is determined. Then, each of them is considered in the loop starting at Line 3. A candidate system transition is the mark τ of an edge leaving a node in σ, the current record of states of the input active spaces.

Lines 4–9 set the variables introduced in the formal definition of polymorphic active space, Point 3. All the input events of component transitions in τ relevant to a link within the interface of the set of clusters are required to be ready in Q. Similarly, all the output events of the transitions in τ are required not to be directed towards a saturated link with saturation mode *WAIT* (Lines 10–11).

Lines 13–17 check the consistency of the possible messages associated with the transitions in τ with the index of the observation: if inconsistent, the transition is discarded.

Line 18 makes a copy of the fields of the current node into the new variables σ', K', and Q'. The new state reached by the candidate system transition τ is updated appropriately at Lines 19–21. The observation index is incremented at Line 22.

At Line 23, the triggering events are dequeued from the relevant links (only if the latter are within the interface). The queues relevant to the output events of the candidate transition are updated at Lines 24–26.

At Line 27, the new node N' is created. If N' was not visited already (Line 34), *PStep* is recursively called at Line 35, and an edge from N to N' is possibly created (Line 41).

The loop terminates at Line 43. At this point, three cases are possible:

(1) N is final (Line 44): N is marked as consistent and returned (Lines 45–46);

(2) N is not final, but an edge from it to a new node was created (Line 47): N is simply returned (Line 48);

(3) N is not final and no edge was created from it (Line 50): N is marked as inconsistent and a null value is returned (Line 51).

Lines 58–61 in *Pjoin* remove from the graph $\tilde{\mathcal{J}}$ the possible dangling nodes and edges which are not part of any path from the root of $\tilde{\mathcal{J}}$ to a final node. These anomalous subgraphs of $\tilde{\mathcal{J}}$ correspond to cycles in which all the nodes are marked as unknown. However, consistent silent cycles possibly remain still in $\tilde{\mathcal{J}}$ after the pruning.

Lines 62–67 establish the isomorphism between the graph representation of the reconstruction $\mathcal{J}$ and its elements $\mathbb{S}$, $\mathbb{E}$, $\mathbb{T}$, S_0, and S_f. The resulting reconstruction is eventually returned at Line 68.

function *Pjoin*$(\mathbf{A}, \wp(\xi))$:
        The polymorphic join $\mathcal{J}_{\wp(\xi)}(\mathbf{A}) = (\mathbb{S}, \mathbb{E}, \mathbb{T}, S_0, S_f)$, $\mathbb{S} \subseteq \Sigma \times \mathbb{K} \times \mathbb{Q}$
    input
        $\mathbf{A} = \{Act(\wp(\xi_i)) \mid Act(\wp(\xi_i)) = (\mathbb{S}_i, \mathbb{E}_i, \mathbb{T}_i, S_{0_i}, S_{f_i}), i \in [1..m]\}$:
           The active spaces of a decomposition $\Xi(\xi) = \{\xi_1, \ldots, \xi_m\}$,
           $\wp(\xi) = (\Omega(\xi), OBS(\xi), \xi_0)$: A diagnostic problem for polymorphic cluster ξ;

function *PStep*$(\sigma, K, Q, \mathbf{out}\ \aleph)$: The root of a subgraph of $Act(\wp(\Theta))$
    input
        $\sigma = (S_1, \ldots, S_m) \in \Sigma$,
        $K \in \mathbb{K}$,
        $Q = Queue(\xi)$: a queue of ξ,
        $\aleph \subseteq \Sigma \times \mathbb{K} \times \mathbb{Q}$: the set of nodes visited up to the current call of *PStep*
    output
        $\aleph$: The updated set of visited nodes;
    begin {PStep}

```
1.      Insert N = (σ, K, Q) into ℵ and mark it as unknown;
2.      T_cand := Dom(N);
3.      for each τ ∈ T_cand, τ = {T_1, ..., T_p}, do
4.          L_α := {L | L = Link(T), T ∈ τ, L ∈ Interf(Ξ(ξ))};
5.          L_β := {L | L = Link(B), B ∈ β, S ⟶^{α|β} S' ∈ τ, L ∈ Interf(Ξ(ξ))};
6.          L_β^u := {L | L ∈ L_β, (either L is not saturated or L_β ∈ L_α)};
7.          L_β^s := L_β − L_β^u;
8.          L_β^{so} := {L_β | L_β ∈ L_β^s, the saturation policy of L_β is OVERRIDE};
9.          L_β^{sw} := {L_β | L_β ∈ L_β^s, the saturation policy of L_β is WAIT};

10.         if L ∈ L_α, L = Link(T), T = S ⟶^{α|β} S', α = (E, θ),
                              Head(D[L]) ≠ E  or  L_β^{sw} ≠ ∅ then
11.             goto 3
12.         end-if;
13.         for each j ∈ [1..n] do
14.             C_j^τ = C_j ∩ Comp(Obs(τ)) ∩ Vis(OBS(ξ));
```

15. $\quad\quad\quad \mu_j = obs(\mathbf{C}_j)[(K[j] + 1)..(K[j] + |\mathbf{C}_j^\tau|)];$
16. $\quad\quad\quad$ **if** $\mathbf{C}_j^\tau \neq \emptyset$ **and** $\mu_j \neq Obs_{\langle \mathbf{C}_j^\tau \rangle}(\tau))$ **then goto** 3
17. $\quad\quad$ **end-for**;
18. $\quad\quad \sigma' := \sigma;\ K' := K;\ Q' := Q;$
19. $\quad\quad$ **for each** $S_i \in \sigma'$ **do**
20. $\quad\quad\quad$ **if** $(\tau_i \in Group(\tau, \Xi(\xi))) \neq \emptyset, S \xrightarrow{\tau_i} S' \in \mathbb{T}_i$ **then** $S_i := S'$
21. $\quad\quad$ **end-for**;
22. $\quad\quad$ **for each** $j \in [1..m]$ **do** $K'[j] = K[j] + |\mathbf{C}_j^\tau|;$
23. $\quad\quad$ **if** $L_\alpha \in Interf(\Xi(\xi))$ **then** $Q'[L_\alpha] := Tail(Q[L_\alpha]);$
24. $\quad\quad$ **for each** event $(E, \vartheta) \in \beta, L_\beta = Link(\vartheta), L_\beta \in (\mathbb{L}_\beta^{u} \cup \mathbb{L}_\beta^{so})$ **do**
25. $\quad\quad\quad Q'[L_\beta] := Ins(Q[L_\beta], E)$
26. $\quad\quad$ **end-for**;
27. $\quad\quad N' := (\sigma', K', Q');$
28. $\quad\quad$ **if** $N' \in \aleph$ **then**
29. $\quad\quad\quad$ **if** N' is marked as *inconsistent* **then**
30. $\quad\quad\quad\quad N'_{\tilde{\mathcal{J}}} := \textbf{nil}$
31. $\quad\quad\quad$ **else**
32. $\quad\quad\quad\quad N'_{\tilde{\mathcal{J}}} := N'$
33. $\quad\quad\quad$ **end-if**
34. $\quad\quad$ **else**
35. $\quad\quad\quad N'_{\tilde{\mathcal{J}}} := PStep(\sigma', K', D', \aleph)$
36. $\quad\quad$ **end-if**;
37. $\quad\quad$ **if** $N'_{\tilde{\mathcal{J}}} \neq \textbf{nil}$ **then**
38. $\quad\quad\quad$ **if** N is marked as *unknown* and N' as *consistent* **then**
39. $\quad\quad\quad\quad$ Mark N as *consistent*
40. $\quad\quad\quad$ **end-if**;
41. $\quad\quad\quad$ Create an edge $N \xrightarrow{T} N'$
42. $\quad\quad$ **end-if**
43. $\quad$ **end-for**;
44. $\quad$ **if** $N = (\sigma, K, (\emptyset \cdots \emptyset))$ **and** $\forall S_i \in \sigma\ (S_i \in \mathbb{S}_{f_i})$ **and** $Complete(K)$ **then**
45. $\quad\quad$ Mark N as *consistent*;
46. $\quad\quad$ **return** N
47. $\quad$ **elsif** At least one edge was created at Line 41 **then**
48. $\quad\quad\quad$ **return** N
49. $\quad$ **else**
50. $\quad\quad$ Mark N as *inconsistent*;
51. $\quad\quad$ **return nil**
52. $\quad$ **end-if**
53. **end** {PStep};

54. **begin** {Pjoin}
55. $\quad \sigma_0 := (S_{01}, \ldots, S_{0m});\ K := (0 \cdots 0);\ Q := (\emptyset \cdots \emptyset);\ \aleph := \emptyset;$
56. $\quad N_0 := PStep(\sigma_0, K, D, \aleph);$
57. $\quad \tilde{\mathcal{J}} :=$ the graph rooted in $N_0;$
58. $\quad$ **for each** $N \in \aleph$ such that N is marked as *unknown* **do**
59. $\quad\quad$ **if** $\nexists$ a path $N \rightsquigarrow (\sigma_f, K_f, (\emptyset \cdots \emptyset))$ in $\tilde{\mathcal{J}}$ such that
$\quad\quad\quad\quad \forall S_i \in \sigma\ (S_i \in \mathbb{S}_{f_i})$ **and** $Complete(K_f)$ **then**
60. $\quad\quad\quad$ Remove N from $\tilde{\mathcal{J}};$
61. $\quad$ Remove from $\tilde{\mathcal{J}}$ all the dangling edges;

62. $\mathbb{S} :=$ the set of nodes in $\tilde{\mathcal{J}}$;
63. $\mathbb{E} :=$ the set of labels of edges in $\tilde{\mathcal{J}}$;
64. $\mathbb{T} :=$ the set of edges in $\tilde{\mathcal{J}}$;
65. $S_0 :=$ the root of $\tilde{\mathcal{J}}$;
66. $\mathbb{S}_{\mathrm{f}} \subseteq \mathbb{S} :=$ the set of nodes $N_f = (\sigma_{\mathrm{f}}, K_{\mathrm{f}}, (\emptyset \cdots \emptyset))$ such that
$$\forall S_i \in \sigma \ (S_i \in \mathbb{S}_{f_i}) \textbf{ and } \mathit{Complete}(K_{\mathrm{f}});$$
67. $\mathcal{J} := (\mathbb{S}, \mathbb{E}, \mathbb{T}, S_0, \mathbb{S}_{\mathrm{f}})$;
68. **return** $\mathcal{J}$
69. **end** {Pjoin}.

Appendix B
Proofs of Theorems

Theorem 6.1 *Let $\wp(\xi) = (\Omega(\xi), OBS(\xi), \xi_0)$ be a diagnostic problem for a polymorphic cluster ξ, $\Xi(\xi) = \{\xi_1, \ldots, \xi_m\}$ a decomposition of ξ,*

$$\mathbf{A}_\xi = \{Act(\wp(\xi_1)), \ldots, Act(\wp(\xi_m))\}$$

a set of relevant active spaces where

$$\forall i \in [1 .. m] \ (\wp(\xi_i) = (\Omega(\xi_i), OBS(\xi_i), \xi_{i_0})),$$

such that ξ_0 is the composition of $\xi_{1_0} \cdots \xi_{m_0}$ and

$$OBS(\xi) \trianglerighteq \{OBS(\xi_1), \ldots, OBS(\xi_m)\}.$$

Let $\mathbf{A}_C$ the set of atomic active spaces of components in ξ. Then,

$$\mathcal{J}_{\wp(\xi)}(\mathbf{A}_\xi) \equiv \Re(\wp(\xi)).$$

Proof. Let $R = (\mathbb{S}, \mathbb{E}, \mathbb{T}, S_0, \mathbb{S}_f)$ and $\hat{R} = (\hat{\mathbb{S}}, \hat{\mathbb{E}}, \hat{\mathbb{T}}, \hat{S}_0, \hat{\mathbb{S}}_f)$ denote the polymorphic join $\mathcal{J}_{\wp(\xi)}(\mathbf{A}_\xi)$ and the canonical reconstruction $\Re(\wp(\xi))$, respectively. Proving that $R = \hat{R}$ amounts to proving that $\mathbb{S} = \hat{\mathbb{S}}$, $\mathbb{E} = \hat{\mathbb{E}}$, $\mathbb{T} = \hat{\mathbb{T}}$, $S_0 = \hat{S}_0$, and $\mathbb{S}_f = \hat{\mathbb{S}}_f$. To this end, we have to show that the structure of R is isomorphic to the structure of $\hat{R}$, that is, the structure of each element in R is isomorphic to the structure of the respective element in $\hat{R}$. Hereafter, m, n, and p denote the number of clusters in $\Xi(\xi)$, the number of observation items in $OBS(\xi)$, and the number of components in ξ, respectively, while p_i and n_i denote the number of components in each $\xi_i \in \Xi(\xi)$ and the number of observation items in $OBS(\xi_i)$, respectively. Clearly,

$$p = \sum_{i=1}^m p_i, \qquad n = \sum_{j=1}^m n_j.$$

Lemma 6.1.1 $\mathbb{E} = \hat{\mathbb{E}}$.
    Proof. Since

$$\mathbb{E} = \bigcup_{i=1}^m \mathbb{E}_i, \text{ where } \forall i \in [1 .. m] \ (\mathbb{E}_i = \bigcup_{j=1}^{p_i} \hat{\mathbb{E}}_j), \text{ where } \hat{\mathbb{E}}_j = \mathbb{T}_{C_j}$$

being $\mathbb{T}_{C_j}$ the set of transitions in the atomic active space of the model of component $C_j \in \xi$, and

$$\hat{\mathbb{E}} = \bigcup_{c=1}^{p} \hat{\mathbb{E}}_c, \text{ where } \forall c \in [1\mathinner{.\,.}p] \ (\hat{\mathbb{E}}_c = \mathbb{T}_{C_c}, C_c \in \xi),$$

it follows that

$$\mathbb{E} = \bigcup_{i=1}^{m} \bigcup_{j=1}^{p_i} \hat{\mathbb{E}}_j = \bigcup_{c=1}^{p} \hat{\mathbb{E}}_c = \hat{\mathbb{E}}.$$

Lemma 6.1.2 $S_0 = \hat{S}_0$.

Proof. Since

$$
\begin{aligned}
\hat{S}_0 \;=\; & (\hat{\sigma}_0, \hat{K}_0, \hat{Q}_0), \text{ where} \\
& \hat{\sigma}_0 = (\hat{S}_{0_1}, \ldots, \hat{S}_{0_p}), \text{ where } \forall c \in [1\mathinner{.\,.}p] \ (\hat{S}_{0_c} \text{ is the initial state of } C_c \in \xi), \\
& \hat{K}_0 = (0, \ldots, 0), \text{ where } |\hat{K}_0| = n, \text{ and} \\
& \hat{Q}_0 = (\emptyset, \ldots, \emptyset), \text{ where } |\hat{Q}_0| = |\mathbb{L}(\xi)|,
\end{aligned}
$$

and

$$
\begin{aligned}
S_0 \;=\; & (\sigma_0, K_0, Q_0), \sigma_0 = (S_{0_1}, \ldots, S_{0_m}), \forall i \in [1\mathinner{.\,.}m] \ (S_{0_i} = (\hat{\sigma}_{0_i}, \hat{K}_{0_i}, \hat{Q}_{0_i})), \text{ where} \\
& \hat{\sigma}_{0_i} = (S_{0_1}, \ldots, S_{0_{p_i}}), \\
& \hat{K}_{0_i} = (0, \ldots, 0), \text{ where } |\hat{K}_{0_i}| = n_i, \text{ and} \\
& \hat{Q}_{0_i} = (\emptyset, \ldots, \emptyset), \text{ where } |\hat{Q}_{0_i}| = |\mathbb{L}(\xi_i)|,
\end{aligned}
$$

then, substituting and rearranging the association of the elements within the tuple, whilst disregarding the immaterial elements K_{0_i}, we obtain

$$
\begin{aligned}
S_0 \;=\; & (((\hat{\sigma}_{0_1}, \hat{K}_{0_1}, \hat{Q}_{0_1}), \ldots, (\hat{\sigma}_{0_m}, \hat{K}_{0_m}, \hat{Q}_{0_m})), K_0, Q_0) = \\
& ((\hat{\sigma}_{0_1}, \ldots, \hat{\sigma}_{0_m}), K_0, (Q_0, \hat{Q}_{0_1}, \ldots, \hat{Q}_{0_m})) = \\
& ((\hat{S}_{0_1}, \ldots, \hat{S}_{0_p}), \hat{K}_0, \hat{Q}_0) = (\hat{S}_0, \hat{K}_0, \hat{Q}_0) = \hat{S}_0.
\end{aligned}
$$

Lemma 6.1.3 $\mathbb{S} \subseteq \hat{\mathbb{S}}$, $\mathbb{S}_f \subseteq \hat{\mathbb{S}}_f$, and $\mathbb{T} \subseteq \hat{\mathbb{T}}$.

Proof. To prove $\mathbb{T} \subseteq \hat{\mathbb{T}}$, we have to show that

$$\forall N \xrightarrow{\tau} N' \in \mathbb{T}, N = \hat{N}, \hat{N} \in \hat{\mathbb{N}} \ (\hat{N} \xrightarrow{\tau} \hat{N}' \in \hat{\mathbb{T}}, \hat{N}' = N').$$

From the definition of polymorphic active space, if $N \xrightarrow{\tau} N' \in \mathbb{T}$ then the following conditions hold:

(1) $\tau \in Dom(N)$. According to the definition of $Dom(N)$, two cases are possible:

　(a) $\tau \in SDom(N)$, that is,

$$\tau \in (Sync(S_1) \boxtimes \cdots \boxtimes Sync(S_m))$$

　where for each link $L \in Kernel(N)$ there exists a transition $T \in \tau$ such that $Link(T) = L$. Since

$$\tau = \tau_1 \cup \cdots \cup \tau_m \text{ where } \forall i \in [1\mathinner{.\,.}m] \ (\tau_i \in Sync(S_i)),$$

by virtue of Proposition 6.1,

$$\forall i \in [1 \mathinner{..} m] \; (\tau_i \in (SDom(S_i))).$$

That is,

$$\forall i \in [1 \mathinner{..} m] \; (\tau_i \in (Sync(\hat{S}_{i_1}) \boxtimes \cdots \boxtimes Sync(\hat{S}_{i_{p_i}}))).$$

Since

$$\forall i \in [1 \mathinner{..} m] \; (\tau_i = \tau_{i_1} \cup \cdots \cup \tau_{i_{p_i}}),$$

we have

$$\forall i \in [1 \mathinner{..} m], \forall j \in [1 \mathinner{..} p_i] \; (\tau_{i_j} \in Sync(\hat{S}_{i_j})),$$

in other terms, combining i and j,

$$\forall c \in [1 \mathinner{..} p] \; (\tau_c \in Sync(\hat{S}_c)),$$

that is, $\tau \in SDom(\hat{N})$, therefore, $\tau \in Dom(\hat{N})$.

(b) $\tau \in ADom(N)$, that is,

$$\exists i \in [1 \mathinner{..} m] \; (\tau \in Async(S_i)).$$

In other words, by Proposition 6.1,

$$\tau \in (ADom(S_i) = \bigcup_{c=1}^{p_i} Async(\hat{S}_c)),$$

that is,

$$\exists c \in [1 \mathinner{..} p] \; (\tau \in Async(\hat{S}_c)),$$

in other words, $\tau \in ADom(\hat{N})$, therefore, $\tau \in Dom(\hat{N})$.

Thus, in either case, if $\tau \in Dom(N)$ then $\tau \in Dom(\hat{N})$.

(2) *Triggerable*(τ, N), that is,

$$\forall L_\alpha \in \mathbb{L}_\alpha, L_\alpha = Link(\alpha), \alpha = (E_\alpha, \theta_\alpha) \; (Head(Q[L_\alpha(Q)]) = E_\alpha).$$

Let

$$\hat{\mathbb{L}}_\alpha = \{\hat{L}_\alpha \mid \hat{L}_\alpha = Link(\alpha), T \in \tau, T = S \xrightarrow{\alpha|\beta} S', \alpha = (\hat{E}_\alpha, \hat{\theta}_\alpha), \hat{L}_\alpha \in \mathbb{L}(\xi)\}.$$

According to the decomposition $\Xi(\xi)$, $\hat{\mathbb{L}}_\alpha$ can be partitioned as

$$\hat{\mathbb{L}}_\alpha = \mathbb{L}_\alpha \cup (\mathbb{L}_{\alpha_1} \cup \cdots \cup \mathbb{L}_{\alpha_m}),$$

where $\forall i \in [1 \mathinner{..} m]$ the following condition holds:

$$\mathbb{L}_{\alpha_i} = \{\hat{L}_{\alpha_i} \mid \hat{L}_{\alpha_i} \in \mathbb{L}(\xi_i), \hat{L}_{\alpha_i} = Link(\alpha_i),$$
$$T = S \xrightarrow{\alpha_i|\beta_i} S', T \in \tau_i, \tau_i \in Group(\tau, \Xi(\xi))), \tau_i \neq \emptyset, \alpha_i = (E_{\alpha_i}, \theta_{\alpha_i})\}.$$

Furthermore, since $\hat{Q} = (Q, \hat{Q}_1, \ldots, \hat{Q}_m)$, being each $\hat{Q}_i$ the queue relevant to S_i, we may write

$$\forall i \in [1 \mathinner{..} m] \; (\forall \hat{L}_\alpha \in \mathbb{L}_{\alpha_i} (Head(\hat{L}_\alpha(\hat{Q}_i)) = E_{\alpha_i})),$$

that is

$$\forall \hat{L}_\alpha \in \hat{\mathbb{L}}_\alpha \ (Head(\hat{L}_\alpha(\hat{Q})) = \hat{E}_\alpha),$$

namely, $Triggerable(\tau, \hat{N})$.

(3) $Consistent(\tau, N)$, which amounts to the following two conditions:

(a) $\forall j \in [1 .. n], \mathbf{C}_j^\tau \neq \emptyset \ (\mu_j = Obs_{(\mathbf{C}_j^\tau)}(\tau))$, where

$$\mathbf{C}_j^\tau = \mathbf{C}_j \cap Comp(Obs(\tau)) \cap Vis(OBS(\xi)),$$
$$\mu_j = obs(\mathbf{C}_j)[(K[j] + 1) .. (K[j] + |\mathbf{C}_j^\tau|)].$$

Condition (a) is identical for N and $\hat{N}$, therefore it holds for $\hat{N}$ as well.

(b) $\mathbb{L}_\beta^{sw} = \emptyset$. Within the context of $\hat{N}$, we have:

$$\hat{\mathbb{L}}_\beta = \{L_\beta \mid L_\beta = Link(B), B \in \beta, T = S \xrightarrow{\alpha|\beta} S' \in \tau, L_\beta \in L(\xi)\},$$
$$\hat{\mathbb{L}}_\beta^u = \{L_\beta \mid L_\beta \in \hat{\mathbb{L}}_\beta, (\text{either } L_\beta \text{ is not saturated or } L_\beta \in \hat{\mathbb{L}}_\alpha)\},$$
$$\hat{\mathbb{L}}_\beta^s = \hat{\mathbb{L}}_\beta - \hat{\mathbb{L}}_\beta^u,$$
$$\hat{\mathbb{L}}_\beta^{so} = \{L_\beta \mid L_\beta \in \hat{\mathbb{L}}_\beta^s, \text{ the saturation policy of } L_\beta \text{ is } OVERRIDE\},$$
$$\hat{\mathbb{L}}_\beta^{sw} = \{L_\beta \mid L_\beta \in \hat{\mathbb{L}}_\beta^s, \text{ the saturation policy of } L_\beta \text{ is } WAIT\}.$$

such that

$$\hat{\mathbb{L}}_\beta^{sw} = \mathbb{L}_\beta^{sw} \cup (\mathbb{L}_{\beta_1}^{sw} \cup \cdots \cup \mathbb{L}_{\beta_m}^{sw})$$

where for each $i \in [1 .. m]$, $\mathbb{L}_{\beta_i}^{sw}$ is relevant to ξ_i. Since transitions τ_i in $Group(\tau, \Xi(\xi))$ are consistent themselves in the relevant active spaces, it follows that

$$\forall i \in [1 .. m] \ (\mathbb{L}_{\beta_i}^{sw} = \emptyset),$$

which brings to the conclusion $\hat{\mathbb{L}}_\beta^{sw} = \emptyset$.

The conjunction of Conditions 3(a) and 3(b) substantiate the claim $Triggerable(\tau, \hat{N})$.

Before showing the satisfaction of Conditions (4), (5), and (6), following a rewriting scheme similar to the one adopted in Lemma 6.1.2, we rewrite $N = (\sigma, K, Q)$ as follows:

$$
\begin{aligned}
N =& ((S_1, \ldots, S_m), K, Q) = (((\sigma_1, K_1, Q_1), \ldots, (\sigma_m, K_m, Q_m)), K, Q) \\
=& ((\hat{S}_1, \ldots, \hat{S}_p), K, Q) \\
=& ((((\hat{S}_{1_1}, \ldots, \hat{S}_{1_{p_1}}), K_1, Q_1), \ldots, ((\hat{S}_{1_m}, \ldots, \hat{S}_{m_{p_m}}), K_m, Q_m)), K, Q) \\
=& ((\hat{S}_1, \ldots, \hat{S}_p), (K_1, \ldots, K_m, K), (Q_1, \ldots, Q_m, Q)) = (\hat{\sigma}, \hat{K}, \hat{Q}, \mathbf{K}) \\
=& (\hat{N}, \mathbf{K}), \text{ where } \mathbf{K} = \{K_1, \ldots, K_m\}.
\end{aligned}
$$

Likewise,

$$
\begin{aligned}
N' =& (\sigma', K', Q') = ((\hat{S}'_1, \ldots, \hat{S}'_p), (K'_1, \ldots, K'_m, K'), (Q'_1, \ldots, Q'_m, Q')) \\
=& (\bar{\sigma}', K', \bar{Q}', \mathbf{K}'), \text{ where } \mathbf{K}' = \{K'_1, \ldots, K'_m\}.
\end{aligned}
$$

Note that $\mathbf{K}$ and $\mathbf{K}'$ are immaterial to the equivalence (isomorphism) because they are overwritten by $\hat{K}$ and K', respectively. Thus, we only have to prove that

$$(\bar{\sigma}', K', \bar{Q}') = (\hat{\sigma}', \hat{K}', \hat{Q}') = \hat{N}',$$

that is, $\bar{\sigma}' = \hat{\sigma}'$, $K' = \hat{K}'$, and $\bar{Q}' = \hat{Q}'$.

(4) $\bar{\sigma}' = \hat{\sigma}'$. According to the definition of polymorphic active space, on the one hand, $\hat{\sigma}'$ is such that the following condition holds ($\Xi_c(\xi)$ denotes the partition of components in ξ where each part is composed of a single component):

$$\forall c \in [1 \mathbin{..} p] \left(\hat{S}'_c = \left\{ \begin{array}{ll} \hat{S}_{c'} & \text{if } \tau_c \in Group(\tau, \Xi_c(\xi)), \tau_c \neq \emptyset, \hat{S}_c \xrightarrow{\tau_c} \hat{S}_{c'} \in \hat{\mathbb{T}}_c \\ \hat{S}_c & \text{otherwise} \end{array} \right. \right).$$

On the other hand, considering σ' we have:

$$\forall i \in [1 \mathbin{..} m] \left(S'_i = \left\{ \begin{array}{ll} S_{i'} & \text{if } \tau_i \in Group(\tau, \Xi(\xi)), \tau_i \neq \emptyset, S_i \xrightarrow{\tau_i} S_{i'} \in \mathbb{T}_i \\ S_i & \text{otherwise} \end{array} \right. \right),$$

where $S_{i'} = (\hat{S}'_1, \ldots, \hat{S}'_{p_i})$, $\tau_i = \{T_1, \cdots, T_\ell\}$, $\forall h \in [1 \mathbin{..} \ell](T_h = \hat{S}_{h'} \xrightarrow{\alpha|\beta} \hat{S}'_{h'})$. That is, since $\hat{\sigma} = (\hat{S}_1, \ldots, \hat{S}_p)$, the following condition holds:

$$\forall c \in [1 \mathbin{..} p] \left(\hat{S}'_c = \left\{ \begin{array}{ll} \hat{S}' & \text{if } T \in \tau, T = \hat{S}_c \xrightarrow{\alpha|\beta} \hat{S}' \\ \hat{S}_c & \text{otherwise} \end{array} \right. \right).$$

Comparing this result with the first formula above, we can conclude that the subset of elements in σ which have been updated in correspondence of τ are the same in both active spaces and, being both $\hat{\sigma}$ and the component transitions in τ the same, the new states of the involved components are the same as well, in other words, $\bar{\sigma}' = \hat{\sigma}'$.

(5) $K' = \hat{K}'$. According to the definition of polymorphic active space, K' is such that

$$\forall j \in [1 \mathbin{..} n] \, (K'[j] = K[j] + |\mathbf{C}_j^\tau|)$$

where

$$\forall j \in [1 \mathbin{..} n] \, (\mathbf{C}_j^\tau = \mathbf{C}_j \cap Comp(Obs(\tau)) \cap Vis(OBS(\xi))).$$

Since the above formula is valid for both N and $\hat{N}$, it follows that $K' = \hat{K}'$.

(6) $\bar{Q}' = \hat{Q}'$. We have to show that Conditions (a), (b), and (c) of Point (6) in the definition of polymorphic active space hold as well for $\hat{Q}'$. To this end, note that the counterparts of the sets of links $\mathbb{L}_\alpha$, $\mathbb{L}_\beta$, $\mathbb{L}_\beta^u$, $\mathbb{L}_\beta^s$, and $\mathbb{L}_\beta^{so}$ are defined for $\hat{Q}'$ as follows:

$$\hat{\mathbb{L}}_\alpha = \{L_\alpha \mid L_\alpha = Link(\alpha), T \in \tau, T = S \xrightarrow{\alpha|\beta} S', \alpha = (E_\alpha, \theta_\alpha), L_\alpha \in \mathbb{L}(\xi)\},$$
$$\hat{\mathbb{L}}_\beta = \{L_\beta \mid L_\beta = Link(B), B \in \beta, T = S \xrightarrow{\alpha|\beta} S' \in \tau, L_\beta \in \mathbb{L}(\xi)\},$$
$$\hat{\mathbb{L}}_\beta^u = \{L_\beta \mid L_\beta \in \hat{\mathbb{L}}_\beta, (\text{either } L_\beta \text{ is not saturated or } L_\beta \in \hat{\mathbb{L}}_\alpha)\},$$
$$\hat{\mathbb{L}}_\beta^s = \hat{\mathbb{L}}_\beta - \hat{\mathbb{L}}_\beta^u,$$
$$\hat{\mathbb{L}}_\beta^{so} = \{L_\beta \mid L_\beta \in \hat{\mathbb{L}}_\beta^s, \text{ the saturation policy of } L_\beta \text{ is } OVERRIDE\}.$$

Since $Group(\tau, \Xi(\xi)) = \{\tau_1, \ldots, \tau_m\}$ and $\hat{Q}' = (D'_1, \ldots, D'_m, D')$, where

$$\forall i \in [1 \mathbin{..} m] \, (Q'_i = Q'(S_i)),$$

we have

$$\hat{\mathbb{L}}_\alpha = \mathbb{L}_\alpha \cup \left(\bigcup_{i=1}^m \mathbb{L}_{\alpha_i} \right),$$

$$\hat{\mathbb{L}}_\beta = \mathbb{L}_\beta \cup \left(\bigcup_{i=1}^m \mathbb{L}_{\beta_i} \right),$$

$$\hat{\mathbb{L}}_\beta^{\mathrm{u}} = \mathbb{L}_\beta^{\mathrm{u}} \cup \left(\bigcup_{i=1}^m \mathbb{L}_{\beta_i}^{\mathrm{u}} \right),$$

$$\hat{\mathbb{L}}_\beta^{\mathrm{s}} = \hat{\mathbb{L}}_\beta - \hat{\mathbb{L}}_\beta^{\mathrm{u}} = \left(\mathbb{L}_\beta \cup \left(\bigcup_{i=1}^m \mathbb{L}_{\beta_i} \right) \right) - \left(\mathbb{L}_\beta^{\mathrm{u}} \cup \left(\bigcup_{i=1}^m \mathbb{L}_{\beta_i}^{\mathrm{u}} \right) \right)$$

$$= (\mathbb{L}_\beta - \mathbb{L}_\beta^{\mathrm{u}}) \cup \left(\bigcup_{i=1}^m (\mathbb{L}_{\beta_i} - \mathbb{L}_{\beta_i}^{\mathrm{u}}) \right),$$

$$\hat{\mathbb{L}}_\beta^{\mathrm{so}} = \mathbb{L}_\beta^{\mathrm{so}} \cup \left(\bigcup_{i=1}^m \mathbb{L}_{\beta_i}^{\mathrm{so}} \right),$$

where $\forall i \in [1 .. m]$, the sets of links relevant to ξ_i are defined as follows:

$$\mathbb{L}_{\alpha_i} = \{ L_\alpha \mid L_\alpha = Link(\alpha), T \in \tau_i, T = S \xrightarrow{\alpha|\beta} S', \alpha = (E_\alpha, \theta_\alpha), L_\alpha \in \mathbb{L}(\xi_i) \},$$

$$\mathbb{L}_{\beta_i} = \{ L_\beta \mid L_\beta = Link(B), B \in \beta, T = S \xrightarrow{\alpha|\beta} S' \in \tau_i, L_\beta \in \mathbb{L}(\xi_i) \},$$

$$\mathbb{L}_{\beta_i}^{\mathrm{u}} = \{ L_\beta \mid L_\beta \in \mathbb{L}_{\beta_i}, (\text{either } L_\beta \text{ is not saturated or } L_\beta \in \mathbb{L}_{\alpha_i}) \},$$

$$\mathbb{L}_{\beta_i}^{\mathrm{s}} = \mathbb{L}_{\beta_i} - \mathbb{L}_{\beta_i}^{\mathrm{u}},$$

$$\mathbb{L}_{\beta_i}^{\mathrm{so}} = \{ L_\beta \mid L_\beta \in \mathbb{L}_{\beta_i}^{\mathrm{s}}, \text{ the saturation policy of } L_\beta \text{ is } OVERRIDE \}.$$

Note that Conditions (a), (b), and (c) of Point (6), in the definition of polymorphic active space, hold as well for each active space $Act(\wp(\xi_i))$. This consideration along with the above equalities allow us to conclude that:

(a) $\forall L_\alpha \in \hat{\mathbb{L}}_\alpha \ (\hat{Q}'[L_\alpha] = Tail(\hat{Q}[L_\alpha]))$,

(b) $\forall (E, \vartheta) \in \beta, L_\beta = Link(\vartheta), L_\beta \in (\hat{\mathbb{L}}_\beta^{\mathrm{u}} \cup \hat{\mathbb{L}}_\beta^{\mathrm{so}}) \ (\hat{Q}'[L_\beta] = Ins(\hat{Q}[L_\beta], E))$,

(c) $\forall L \in (\mathbb{L}(\xi) - (\hat{\mathbb{L}}_\alpha \cup \hat{\mathbb{L}}_\beta^{\mathrm{u}} \cup \hat{\mathbb{L}}_\beta^{\mathrm{so}})) \ (\hat{Q}'[L] = \hat{Q}[L])$.

In other words, $\bar{Q}' = \hat{Q}'$.

So far, we proved that, disregarding the immaterial elements $\mathbf{K}$ and $\mathbf{K}'$ (which allows us to view the isomorphism as an equality), if $N = \hat{N}$, then

$$\forall N \xrightarrow{\tau} N' \in \mathbb{T} \ (N \xrightarrow{\tau} N' \in \hat{\tilde{\mathbb{T}}}),$$

that is, $\mathbb{T} \subseteq \hat{\tilde{\mathbb{T}}}$. This implies that, if $N = \hat{N}$, then $Succ(N) \subseteq Succ(\hat{N})$. From this consideration and Lemma 6.1.2, by induction, we may say that $\mathbb{S} \subseteq \hat{\tilde{\mathbb{S}}}$ and, specifically, $\mathbb{S}_{\mathrm{f}} \subseteq \hat{\tilde{\mathbb{S}}}_{\mathrm{f}}$. But, since each state in $N \in \mathbb{S}$ is convergent and each transition in $N \xrightarrow{\tau} N' \in \mathbb{T}$ is convergent, it follows that $\mathbb{S} \subseteq \hat{\mathbb{S}}$ and $\mathbb{T} \subseteq \hat{\mathbb{T}}$ as well, which concludes the proof of Lemma 6.1.3.

Lemma 6.1.4 $\forall i \in [1 \mathinner{\ldotp\ldotp} m]\ (\hat{\mathbb{T}}_{\langle \wp(\xi_i)\rangle} \subseteq \mathbb{T}_i,\ \hat{\mathbb{S}}_{\langle \wp(\xi_i)\rangle} \subseteq \mathbb{S}_i,\ \hat{\mathbb{S}}_{f_{\langle \wp(\xi_i)\rangle}} \subseteq \mathbb{S}_{f_i}).$

Proof. To prove $\forall i \in [1 \mathinner{\ldotp\ldotp} m](\hat{\mathbb{T}}_{\langle \wp(\xi_i)\rangle} \subseteq \mathbb{T}_i)$, we have to show that

$$\forall i \in [1 \mathinner{\ldotp\ldotp} m], \forall N_i \xrightarrow{\tau} N_i' \in \hat{\mathbb{T}}_{\langle \wp(\xi_i)\rangle}\ (N_i \xrightarrow{\tau} N_i' \in \mathbb{T}_i).$$

To this end, consider $\hat{N} \xrightarrow{\tau} \hat{N}' \in \hat{\mathbb{T}}$ such that

$$(\hat{N} \xrightarrow{\tau} \hat{N}')_{\langle \wp(\xi_i)\rangle} = \hat{N}_i \xrightarrow{\tau_i} \hat{N}_i'.$$

From the definition of polymorphic active space, if $\hat{N} \xrightarrow{\tau} \hat{N}' \in \hat{\mathbb{T}}$ then the following conditions hold:

(1) $\tau \in Dom(\hat{N})$. According to the definition of $Dom(\hat{N})$, two cases are possible:

(a) $\tau \in SDom(\hat{N})$, therefore, $\tau \in (Sync(\hat{N}_1) \boxtimes \cdots \boxtimes Sync(\hat{N}_p))$. According to the definition of polymorphic active space restriction,

$$(\tau_i = \tau_{\langle \xi_i \rangle} = \{T_{i_1}, \ldots, T_{i_k}\}) \in Group(\tau, \Xi(\xi)),$$

where for each $j \in [1 \mathinner{\ldotp\ldotp} k]$, T_{i_j} is a transition leaving node $\hat{N}_{i_j}$, therefore, $\tau_i \in SDom(N_i)$ and, thus, $\tau_i \in Dom(N_i)$.

(b) $\tau \in ADom(\hat{N})$, where

$$ADom(\hat{N}) = \bigcup_{c=1}^{p} Async(\hat{N}_c).$$

Thus, $\tau_i = \tau = \{T\}$, that is, $\tau_i \in ADom(N_i)$, therefore, $\tau_i \in Dom(N_i)$.

Thus, in either case, $\tau_i \in Dom(N_i)$.

(2) $Triggerable(\tau, \hat{N})$, that is, defining

$$\hat{\mathbb{L}}_\alpha = \{\hat{L}_\alpha \mid \hat{L}_\alpha = Link(\alpha), T \in \tau, T = S \xrightarrow{\alpha|\beta} S', \alpha = (\hat{E}_\alpha, \hat{\theta}_\alpha), \hat{L}_\alpha \in \mathbb{L}(\xi)\},$$

the following condition holds:

$$\forall \hat{L}_\alpha \in \hat{\mathbb{L}}_\alpha, Link(\alpha) = \hat{L}_\alpha, \alpha = (\hat{E}_\alpha, \hat{\theta}_\alpha)\ (Head(Q[\hat{L}_\alpha]) = \hat{E}_\alpha)$$

Since

$$\tau_i \in Group(\tau, \Xi(\xi))$$

and $\mathbb{L}_{\alpha_i}$ is a restriction of $\mathbb{L}(\xi)$, it follows that $Triggerable(\tau_i, N_i)$.

(3) $Consistent(\tau, \hat{N})$, which amounts to the following two conditions:

(a) $\forall j \in [1 \mathinner{\ldotp\ldotp} n], \mathbf{C}_j^\tau \neq \emptyset\ (\mu_j = Obs_{\langle \mathbf{C}_j^\tau \rangle}(\tau))$. Let

$$\forall i \in [1 \mathinner{\ldotp\ldotp} n]\ (\mathbf{C}_j^{\tau_i} = \mathbf{C}_j \cap Comp(Obs(\tau_i)) \cap Vis(OBS(\xi_i))).$$

Clearly,

$$\forall j \in [1 \mathinner{\ldotp\ldotp} n]\ (\mathbf{C}_j^{\tau_i} \subseteq \mathbf{C}_j^\tau).$$

Since, according to its definition,

$$\mu_j = obs(\mathbf{C}_j)[(K[j] + 1) .. (K[j] + |\mathbf{C}_j^\tau|)],$$

$K[j]$ denotes the position of the last message m of the prefix $\mathcal{O}_j$ of $obs(\mathbf{C}_j)$. Thus, μ_j represents the set of messages in the sequence $\mathcal{O}'_j$ which follow m in $obs(\mathbf{C}_j)$, where $|\mathcal{O}'_j| = |\mathbf{C}_j^\tau|$. Similarly, within the context of N_i, $K_i[j]$ identifies the last message m_i of a prefix $\mathcal{O}_{i_j}$ of $obs_{\langle\wp(\xi_i)\rangle}(\mathbf{C}_j)$, $j \in [1 .. n_i]$, while μ_{i_j} represents the set of messages in the sequence $\mathcal{O}'_{i_j}$ which follow m_i in $obs_{\langle\wp(\xi_i)\rangle}(\mathbf{C}_j)$, where $|\mathcal{O}'_{i_j}| = |\mathbf{C}_j^{\tau_i}|$. Since $\mathbf{C}_j^{\tau_i} \subseteq \mathbf{C}_j^\tau$, we have $\mathcal{O}_{j\langle\mathbf{C}^{\tau_i}\rangle} = \mathcal{O}_{i_j}$ and $\mathcal{O}'_{j\langle\mathbf{C}^{\tau_i}\rangle} = \mathcal{O}'_{i_j}$. But, the set of messages in $\mathcal{O}'_j$ are in fact $Obs_{\langle\mathbf{C}_j^\tau\rangle}(\tau)$, therefore, $\mathcal{O}'_{i_j} = Obs_{\langle\mathbf{C}^{\tau_i}\rangle}(\tau)$, which allows us to conclude

$$\forall j \in [1 .. n_i], \mathbf{C}_j^{\tau_i} \neq \emptyset \; (\mu_{i_j} = Obs_{\langle\mathbf{C}^{\tau_i}\rangle}(\tau)).$$

(b) $\hat{\mathbf{L}}_\beta^{sw} = \emptyset$. Since $\hat{\mathbf{L}}_{\beta_i}^{sw} \subseteq \hat{\mathbf{L}}_\beta^{sw}$, it follows that $\hat{\mathbf{L}}_{\beta_i}^{sw} = \emptyset$.

Since both conditions (a) and (b) hold, we may conclude $Consistent(\tau_i, N_i)$.

(4) Let $\hat{N} = (\hat{\sigma}, \hat{K}, \hat{Q})$, $\hat{\sigma} = (\hat{S}_1, \ldots, \hat{S}_p)$, $\hat{N}' = (\hat{\sigma}', \hat{K}', \hat{Q}')$, $\hat{\sigma}' = (\hat{S}'_1, \ldots, \hat{S}'_p)$. Then,

$$\forall c \in [1 .. p] \left(\hat{S}'_c = \left\{ \begin{array}{ll} \hat{S}_{c'} & \text{if } \hat{S}_c \xrightarrow{T} \hat{S}_{c'} \in \mathbb{T}_c, T \in \tau \\ \hat{S}_c & \text{otherwise} \end{array} \right. \right).$$

Since $N_i \xrightarrow{\tau_i} N'_i = (\hat{N} \xrightarrow{T} \hat{N}')_{\langle\wp(\xi_i)\rangle}$, $N_i = (\sigma_i, K_i, Q_i)$, $\sigma_i = (\hat{S}_{i_1}, \ldots, \hat{S}_{i_{p_i}})$, $N'_i = (\sigma'_i, K'_i, Q'_i)$, $\sigma'_i = (\hat{S}'_{i_1}, \ldots, \hat{S}'_{i_{p_i}})$, where, according to the definition of polymorphic active space restriction, σ_i and σ'_i are the projections of $\hat{\sigma}$ and $\hat{\sigma}'$, respectively, on components in ξ_i. Since $\sigma_i \subset \hat{\sigma}$ and $\sigma'_i \subset \hat{\sigma}'$, we have

$$\forall c \in [1 .. p_i] \left(\hat{S}'_{i_c} = \left\{ \begin{array}{ll} \hat{S}_{i'_c} & \text{if } \hat{S}_{i_c} \xrightarrow{T} \hat{S}_{i'_c} \in \mathbb{T}_c, T \in \tau_i \\ \hat{S}_{i_c} & \text{otherwise} \end{array} \right. \right),$$

which allows us to conclude that σ'_i satisfies Condition (4).

(5) $\forall j \in [1 .. n] \; (\hat{K}'[j] = \hat{K}[j] + |\mathbf{C}_j^\tau|)$. By definition of polymorphic active space restriction, the same property holds for the index of $OBS(\xi_i)$, that is,

$$\forall j \in [1 .. n_i] \; (K'_i[j] = K_i[j] + |\mathbf{C}_j^{\tau_i}|).$$

(6) Conditions (a), (b), (c), and (d) of Point (6) in the definition of polymorphic active space hold for $\hat{Q}'$. We have to show that the same conditions hold in the context of Q'_i too. To this end, it suffices noticing that:

(i) Links relevant to Q'_i are a subset of the links in $\hat{Q}'$, precisely, those in the restriction $\hat{Q}'_{\langle\wp(\xi_i)\rangle}$;

(ii) $Q_i = \hat{Q}_{\langle\wp(\xi_i)\rangle}$;

(iii) $\tau_i \subseteq \tau$.

Based on these relationships, we may conclude that $Q'_i = \hat{Q}'_{\langle\wp(\xi_i)\rangle}$, in other words, Conditions (a), (b), (c), and (d) of Point (6) hold as well within the context of Q'_i.

So far, we proved that $\hat{\mathbb{T}}_{\langle \wp(\xi_i) \rangle} \subseteq \tilde{\mathbb{T}}_i$. This brings us to the following entailment:

$$N_i = \hat{N}_{\langle \wp(\xi_i) \rangle} \Rightarrow Succ(\hat{N})_{\langle \wp(\xi_i) \rangle} \subseteq Succ(N_i).$$

Since, according to the definition of polymorphic active space restriction, $\hat{N}_{0_i} = N_{0_i}$, by induction, we may say that $\hat{\mathbb{S}}_i \subseteq \tilde{\mathbb{S}}_i$ and, specifically, $\hat{\mathbb{S}}_{f_i} \subseteq \mathbb{S}_{f_i}$. But, based on Proposition 6.2, since each state $\hat{N}_c \in \hat{\mathbb{S}}_{\langle \wp(\xi_i) \rangle}$ is convergent and each transition $\hat{T}_c \in \hat{\mathbb{T}}_{\langle \wp(\xi_i) \rangle}$ is convergent, it follows that $\hat{\mathbb{S}}_i \subseteq \mathbb{S}_i$ and $\hat{\mathbb{T}}_i \subseteq \mathbb{T}_i$, which concludes the proof of Lemma 6.1.4.

Lemma 6.1.5 $\hat{\mathbb{T}} \subseteq \mathbb{T}, \hat{\mathbb{S}} \subseteq \mathbb{S}$, and $\hat{\mathbb{S}}_f \subseteq \mathbb{S}_f$.

Proof. We have to show that

$$\forall \hat{N} \xrightarrow{\tau} \hat{N}' \in \hat{\mathbb{T}}, \hat{N} = N, N \in \mathbb{N} \ (N \xrightarrow{\tau} N' \in \mathbb{T}, N' = \hat{N}').$$

Based on the definition of polymorphic active space, if $\hat{N} \xrightarrow{\tau} \hat{N}' \in \hat{\mathbb{T}}$ then the following conditions hold:

(1) $\tau \in Dom(\hat{N})$. According to the definition of $Dom(\hat{N})$, two cases are possible:

    (a) $\tau \in SDom(\hat{N})$, that is, $\tau \in (Sync(\hat{S}_1) \boxtimes \cdots \boxtimes Sync(\hat{S}_m))$, where for each link $L \in Kernel(\hat{N})$ there exists a transition $T \in \tau$ such that $Link(T) = L$. Since

$$\forall i \in [1 \mathrel{..} m] \ (\tau_i = \tau_{\langle \wp(\xi_i) \rangle}, \tau_i \in Sync(N_i)),$$

        it follows that $\tau \in SDom(N)$, therefore, $\tau \in Dom(N)$.

    (b) $\tau \in ADom(N)$, thus,

$$\tau \in \bigcup_{c=1}^{p} Async(\hat{S}_c)).$$

    Since

$$\tau_{\langle \wp(\xi_i) \rangle} = \{\tau_i\}, S_i \xrightarrow{\tau_i} S_i' \in \mathbb{T}_i^a,$$

    it follows that $\tau \in ADom(N)$, therefore, $\tau \in Dom(N)$.

    Thus, in either case, if $\tau \in Dom(\hat{N})$ then $\tau \in Dom(N)$.

(2) $Triggerable(\tau, \hat{N})$, that is,

$$\forall \hat{L}_\alpha \in \hat{\mathbb{L}}_\alpha, \hat{L}_\alpha = Link(\alpha), \alpha = (\hat{E}_\alpha, \hat{\theta}_\alpha) \ (Head(Q[\hat{L}_\alpha]) = \hat{E}_\alpha),$$

where

$$\hat{\mathbb{L}}_\alpha = \{\hat{L}_\alpha \mid \hat{L}_\alpha = Link(\alpha), T \in \tau, T = S \xrightarrow{\alpha|\beta} S', \alpha = (\hat{E}_\alpha, \hat{\theta}_\alpha), \hat{L}_\alpha \in \mathbb{L}(\xi)\}.$$

We have to show that

$$\forall L_\alpha \in \mathbb{L}_\alpha, L_\alpha = Link(\alpha), \alpha = (E_\alpha, \theta_\alpha) \ (Head(Q[L_\alpha]) = E_\alpha),$$

where

$$\mathbb{L}_\alpha = \{L_\alpha \mid L_\alpha = Link(\alpha), T \in \tau, T = S \xrightarrow{\alpha|\beta} S', \alpha = (E_\alpha, \theta_\alpha), L_\alpha \in Interf(\Xi)\}.$$

The latter holds since $\mathbb{L}_\alpha \subseteq \hat{\mathbb{L}}_\alpha$, therefore, $Triggerable(\tau, N)$.

(3) $Consistent(\tau, \hat{N})$, which amounts to the following two conditions:

(a) $\forall j \in [1..n], \mathbf{C}_j^\tau \neq \emptyset \ (\mu_j = Obs_{\langle \mathbf{C}_j^\tau \rangle}(\tau))$. Once established the mapping

$$\forall i \in [1..m] \ (\tau_i = \tau_{\langle \wp(\xi_i) \rangle}),$$

we may prove that the above condition holds within the context of N too by following the same scheme given for the proof of Point (3) in Lemma 6.1.3.

(b) $\hat{\mathbb{L}}_\beta^{\mathrm{sw}} = \emptyset$. Since $\mathbb{L}_\beta^{\mathrm{sw}} \subseteq \hat{\mathbb{L}}_\beta^{\mathrm{sw}}$, it follows that $\mathbb{L}_\beta^{\mathrm{sw}} = \emptyset$.

The conjunction of Condition (a) and (b) within the context of N allows us to conclude $Consistent(\tau, N)$.

Before showing the satisfaction of Conditions (4), (5), and (6), we have to prove the isomorphism between $\hat{N}'$ an dN'. This can be accomplished by rewriting $\hat{N}'$ is a way similar to the one adopted in Lemma 6.1.2.

(4) Let $\hat{N} = (\hat{\sigma}, \hat{K}, \hat{Q})$, $\hat{\sigma} = (\hat{S}_1, \ldots, \hat{S}_p)$, $\hat{N}' = (\hat{\sigma}', \hat{K}', \hat{Q}')$, $\hat{\sigma}' = (\hat{S}_1', \ldots, \hat{S}_p')$. Then,

$$\forall c \in [1..p] \left(\hat{S}_c' = \left\{ \begin{array}{ll} \hat{S}_{c'} & \text{if } \hat{S}_c \xrightarrow{T} \hat{S}_{c'} \in \mathbb{T}_c, T \in \tau \\ \hat{S}_c & \text{otherwise} \end{array} \right. \right).$$

We have to show that σ' is such that:

$$\forall i \in [1..m] \left(S_i' = \left\{ \begin{array}{ll} S_{i'} & \text{if } \tau_i \in Group(\tau, \Xi(\xi)), \tau_i \neq \emptyset, S_i \xrightarrow{\tau_i} S_{i'} \in \mathbb{T}_i \\ S_i & \text{otherwise} \end{array} \right. \right).$$

In fact, on the one hand,

$$\tau_i = \tau_{\langle \wp(\xi_i) \rangle} \neq \emptyset \Rightarrow S_i' = \hat{S}'_{\langle \wp(\xi_i) \rangle} = S_{i'}.$$

On the other,

$$\tau_i = \tau_{\langle \wp(\xi_i) \rangle} = \emptyset \Rightarrow S_i' = \hat{S}'_{\langle \wp(\xi_i) \rangle} = \hat{S}_{\langle \wp(\xi_i) \rangle} = S_i.$$

(5) $\forall j \in [1..n] \ (\hat{K}'[j] = \hat{K}[j] + |\mathbf{C}_j^\tau|)$. To show that the same condition holds within the context of K' we may reuse the proof of Point (5) in Lemma 6.1.3.

(6) Conditions (a), (b), (c), and (d) of Point (6) in the definition of polymorphic active space hold for $\hat{Q}'$. We have to show that the same conditions hold in the context of Q' too. To this end, we may reuse the same scheme given in the proof of Point (6) of Lemma 6.1.3.

So far, we proved that $\hat{\mathbb{T}} \subseteq \tilde{\mathbb{T}}$. This implies that, if $N = \hat{N}$, then $Succ(\hat{N}) \subseteq Succ(N)$. Since, according to the definition of polymorphic active space restriction, $\hat{N}_{0_i} = N_{0_i}$, by induction, we may say that $\hat{\mathbb{S}} \subseteq \tilde{\mathbb{S}}$ and, specifically, $\hat{\mathbb{S}}_f \subseteq \mathbb{S}_f$. But, since each state in $\hat{N} \in \hat{\mathbb{S}}$ is convergent and each transition in $\hat{N} \xrightarrow{\tau} \hat{N}' \in \hat{\mathbb{T}}$ is convergent, it follows that $\hat{\mathbb{S}} \subseteq \mathbb{S}$ and $\hat{\mathbb{T}} \subseteq \mathbb{T}$ as well, which concludes the proof of Lemma 6.1.5.

Proof of Theorem 6.1. From Lemmas 6.1.3 and 6.1.5 it follows $\mathbb{T} = \hat{\mathbb{T}}$, $\mathbb{S} = \hat{\mathbb{S}}$, and $\mathbb{S}_f = \hat{\mathbb{S}}_f$. This result, along with Lemmas 6.1.1 and 6.1.2, substantiate the proof of Theorem 6.1. $\qquad\square$

Chapter 7

RULE-BASED DIAGNOSIS

Abstract   It is commonplace considering model-based and rule-based diagnosis as tasks with different ontological status, where the former is grounded on design knowledge and 'first principles', while the latter is merely based on empirical associations. In contrast with this view, this chapter presents a method that bridges model-based and rule-based diagnosis of polymorphic systems. Specifically, it focuses on off-line activities, by showing how system models can be automatically compiled into a set of diagnostic rules, these being associations between matching conditions for observable events and diagnoses. Given a diagnostic problem, diagnosis is generated at runtime by filtering out those rules matching the system observation. This way, both the accuracy of the model-based approach and the simplicity and efficiency of the reasoning mechanism of the rule-based counterpart are saved.

1.   Introduction

The diagnostic techniques presented in previous chapters share a common paradigm: in order to solve a diagnostic problem, a reconstruction of the system reaction is required, based on the topology of the system, the component models, the link models, and the system observation (as perceived by the observer). The relevant active space is made up by means of a simulation of the system behavior guided by the constraints imposed by the observation. As such, the diagnostic task is carried out on-line.

In the human world, diagnosis is normally performed by associating symptoms with possible causes. For example, in medicine, a disease is diagnosed based on empirical rules that depend on the experience and skill of the physician. This approach has both advantages and disadvantages. On the one hand, the diagnostic procedure is simple and efficient because no reasoning on deep knowledge is required on the (low-level) biological model of the human being. On the other, it is bound to fail when some rare disease is masked by common symptoms or, more generally, when the disease is beyond the domain of the specialist.

Automated diagnosis of industrial artefacts can be either rule-based or model-based, where the former is supported by the empirical knowledge of experts, while the latter is based on the model of the system to be diagnosed. Roughly, the accuracy (soundness and completeness) of model-based diagnosis is paid by the complexity of the reasoning on the system model. Dually, the simplicity (and efficiency) of rule-based diagnosis is negatively counterbalanced by inaccuracy (unsoundness and/or incompleteness).

It would be nice to define a diagnostic technique that is both accurate and efficient, thereby integrating the positive aspects of the model-based and rule-based approaches. To this end, we need to lighten the on-line reasoning process as much as possible by introducing additional off-line preprocessing activities.

Consider the mode in which simulation-based diagnosis of polymorphic systems is carried out. The construction of the active space is based both on the (compositional) model of the system, which does not depend on the diagnostic problem, and the system observation, which is the essential part of the diagnostic problem. Besides, the resulting active space has the following property: each history incorporated in it implies the system observation. Eventually, the set of candidate diagnoses is distilled from the active space.

The generation of the active space can be thought of as the selection, from the universal space of the system, of the histories that comply with the observation. Notice that the universal space does not depend on the system observation and can be virtually generated off-line. Therefore, a possible improvement might be as follows:

(1) Off-line generation of the universal space based on the (compositional) system model;

(2) On-line generation of the active space based on the universal space;

(3) On-line distillation of candidate diagnoses based on the active space.

According to the above scheme, the on-line reconstruction of the system behavior (active space) is lightened by the exploitation of the universal space, which is in fact the explicit representation of the behavioral model of the system[1].

---

[1] More precisely, the universal space is constrained by the initial state of the system, which is assumed to be the same of the diagnostic problem.

However, the distillation phase of candidate diagnoses is still confined on-line, after the generation of the active space. A further step can be made by processing the universal space off-line. The processing aims to make a partition of the system histories (implicitly incorporated within the universal space) where each part is relevant to a set of homogeneous histories that imply the same diagnosis. This way, we have a finite number of graphs, each of which embodies the whole set of histories relevant to a single diagnosis.

These graphs can be simplified by marking edges with relevant messages so that, in the end, each graph represents all possible strings of messages relevant to histories associated with a certain diagnosis.

On-line processing can be reduced to a pattern-matching activity between the system observation and the generated graphs: if the matching is successful for a certain graph, the relevant diagnosis will be part of the solution of the diagnostic problem. The whole diagnostic process can be summarized as follows:

(1) Off-line generation of the universal space based on the (compositional) system model;

(2) Off-line generation of a set of graphs, where each graph is relevant to a different diagnosis;

(3) On-line pattern-matching between the system observation and such graphs.

The technique is based on the compilation of the system model into diagnostic knowledge expressed by a series of rules. Each rule is the association between a matching condition for the system observation and a diagnosis, thereby confining on-line processing to mere pattern-matching.

In the remainder of the chapter, Section 2 deals with the problem of the compilation of diagnostic rules. The process of generating the graphs for pattern-matching are defined in Sections 3, 4, and 5. The notion of a diagnostic rule and relevant on-line exploitation are given in Sections 6 and 7, respectively. The introduced concepts are summarized in Section 8. Appendix A provides a pseudo-coded implementation of the matching algorithm. Proofs of relevant theorems, including the functional equivalence between simulation-based and rule-based diagnosis, are detailed in Appendix B.

2.   Rule Generation

Rule generation is the process of producing diagnostic rules based on the analysis of the universal space relevant to a blind diagnostic problem

$$\wp(\Pi) = (\Omega(\Pi), null, \Pi_0),$$

where $\Omega(\Pi) = (\emptyset, \mathbb{P})$, that is, where the visible set of the observer $\Omega(\Pi)$ is empty and (consequently) the observation is null (see Section 8 of Chapter 3). This is because, off-line, we know neither the actual observation nor the observer relevant to the on-line diagnostic problem. Thus, we are required to be as generic as possible about the domain of diagnostic problems to be solved.

To this end, we consider the universal space rooted in a given state of the polymorphic system Π and make some sort of graph-based reasoning so as to find out relationships between message patterns and diagnoses, namely, the diagnostic rules. Such rules are only applicable to diagnostic problems where the initial state is Π_0. Basically, the generation of the diagnostic rules requires the extraction of a set of graphs from the universal space, each graph incorporating the set of histories that entail one and only one diagnosis. In fact, a universal space $Usp(\Pi, \Pi_0)$ can be thought of as a container of system histories which are consistent with the blind diagnostic problem $\wp(\Pi)$. Therefore, the relevant (shallow) diagnostic set is expected to include a finite set of diagnoses, namely

$$\Delta(\wp(\Pi)) = \{\delta_1, \ldots, \delta_n\},$$

where each δ_i, $i \in [1 .. n]$, is implied by the set of histories $\mathbb{H}_i$ encompassed by the relevant graph. In other words, the main problem is to make up a set of graphs $\gamma_1, \ldots, \gamma_n$ such that:

(1) Each path in γ_i is a history in $\mathbb{H}(\wp(\Pi))$ entailing δ_i;

(2) The union of the history sets of each γ_i equals the history set relevant to $\wp(\Pi)$, namely

$$\bigcup_{i=1}^{n} \mathbb{H}(\gamma_i) = \mathbb{H}(\wp(\Pi)).$$

Based on the graphs γ_i, the next step is to derive the set of graphs $\mu_1, \ldots, \mu_n$, where each path in μ_i, $i \in [1 .. n]$, represents the sequence of sets of messages relevant to a history in $\mathbb{H}(\gamma_i)$. That is, each μ_i incorporates the whole set of modes in which Π may generate messages consistently with histories entailing δ_i. This allows us to make up a diagnostic rule that associates μ_i with δ_i to be exploited on-line, when the actual system observation is available in a given diagnostic problem $\bar{\wp}(\Pi)$. Intuitively, if the observation matches the graph μ_i, we may assert that δ_i is a diagnosis consistent with the actual diagnostic problem $\bar{\wp}(\Pi)$, namely

$$\delta_i \in \Delta(\bar{\wp}(\Pi)).$$

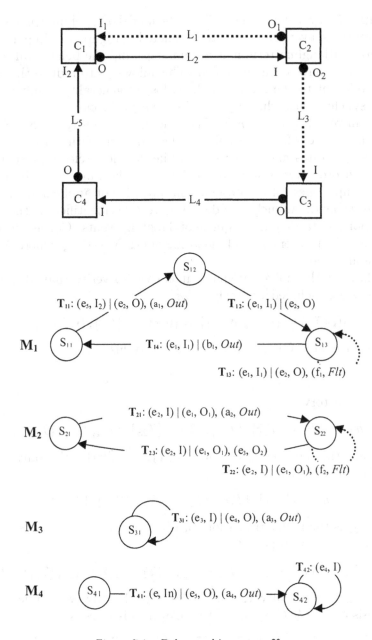

Figure 7.1. Polymorphic system Υ.

In order to present the technique for rule generation, we first introduce a reference polymorphic system and make up a relevant universal space.

Example 7.1. Shown in Figure 7.1 is the model of a polymorphic system Υ, obtained as a variation of system depicted in Figure 6.1. In particular, Υ is closed (dangling terminals are missing) and the behavioral models of C_3 and C_4 have been simplified. This allows us to reduce the search spaces relevant to the system without loss of generality. Links L_1 and L_3 are synchronous, while L_2, and L_4 are asynchronous.

The universal space $Usp(\Upsilon, \Upsilon_0)$, where $\Upsilon_0 = (S_{11}, S_{21}, S_{31}.S_{41})$, is shown in Figure 7.2. As usual, the dashed part of the graph denotes inconsistent states and transitions, while double circles represent final states. In order for a state to be final, all the links must be empty, as the completeness of the observation index is not applicable when the observer is blind. As such, a node in Figure 7.2 is identified by the record of component states and the queue of dangling events. Consistent nodes are labeled by integers $0 \cdots 14$, denoting nodes $N_0 \cdots N_{14}$, where N_0, N_6, and N_9 are final.

Looking at the universal space, it is easy to verify that all possible diagnoses are included[2], namely

$$\Delta(\wp(\Upsilon)) = \{\delta_1, \delta_2, \delta_3, \delta_4\} = \{\emptyset, \{C_1\}, \{C_2\}, \{C_1, C_2\}\}.$$

For instance, $\delta_1 = \emptyset$ is entailed by the empty history

$$h_1 = \langle \rangle,$$

as well as history

$$h_2 = \langle \{T_{41}\}, \{T_{11}\}, \{T_{21}\}, \{T_{12}\}, \{T_{23}\}, \{T_{14}, T_{31}\}, \{T_{42}\} \rangle.$$

These are the only histories entailing δ_1. Similarly, the only history entailing $\delta_2 = \{C_2\}$ is

$$h_3 = \langle \{T_{41}\}, \{T_{11}\}, \{T_{21}\}, \{T_{12}\}, \{T_{22}\}, \{T_{14}\} \rangle,$$

where T_{22} is the faulty transition for C_2. Considering $\delta_3 = \{C_3\}$, one of the relevant histories is

$$h_4 = \langle \{T_{41}\}, \{T_{11}\}, \{T_{21}\}, \{T_{12}\}, \{T_{23}\}, \{T_{13}, T_{31}\}, \{T_{42}\}, \{T_{21}\}, \{T_{14}\} \rangle,$$

where T_{13} is the faulty transition for C_3. Finally, with reference to diagnosis $\delta_4 = \{C_2, C_3\}$, one of the relevant histories is

$$h_5 = \langle \{T_{41}\}, \{T_{11}\}, \{T_{21}\}, \{T_{12}\}, \{T_{22}\}, \{T_{13}\}, \{T_{22}\}, \{T_{14}\} \rangle.$$

---

[2]This is only incidental: generally speaking, the diagnostic set of a universal space $Usp(\Theta, \Theta_0)$ is a subset of the powerset of components in Θ, namely

$$\Delta(\Theta, \Theta_0) \subseteq 2^{C(\Theta)}.$$

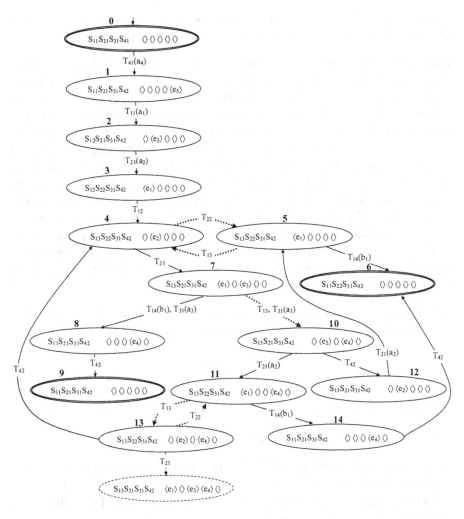

Figure 7.2. Universal space $Usp(\Upsilon, \Upsilon_0)$ (see Example 7.1).

Due to cycles within the universal space, both δ_3 and δ_4 are entailed by an infinite set of histories. $\diamond$

3. Route

Since the universal space $Usp(\Pi, \Pi_0)$ encompasses a history set that is heterogeneous from the diagnosis point of view, we need a technique for extracting parts of $Usp(\Pi, \Pi_0)$ so that each part is relevant to a homogeneous set of histories, specifically, those histories entailing just one diagnosis. However, this is not so straightforward as it might appear at a first glance. Generally speaking, it is not possible to extract subspaces of $Usp(\Pi, \Pi_0)$, as they are, to obtain homogeneous graphs.

Example 7.2. Consider the universal space $Usp(\Upsilon, \Upsilon_0)$ displayed in Figure 7.2. To isolate a part of the universal space that is relevant to diagnosis $\delta_4 = \{C_2, C_3\}$ we might extract the subgraph relevant to nodes N_0, N_1, N_2, N_3, N_4, N_5, and N_6, along with relevant edges. On the one hand, this subgraph involves histories entailing δ_4, such as h_5 of Example 7.1. On the other, it also incorporates a different kind of histories, such as h_1 and h_3 (see Example 7.1), which entail δ_1 and δ_2, respectively. Note that h_1 might be eliminated just by removing the qualification of final from node N_0. However, h_3 is still part of the history set of the subgraph. This is due to the cycle between nodes N_4 and N_5: the histories entailing δ_4 are required to traverse such a cycle so as to involve both faulty transitions. By contrast, this is not enforced by the graph itself, as from N_5 it is possible to go directly to N_6 without coming back to N_4, thereby involving just T_{22}. $\diamond$

The notion of a subspace can be formalized as follows. Let $Usp(\Pi, \Pi_0)$ be a universal space relevant to a diagnostic problem $\wp(\Pi)$, and

$$H \subseteq \mathbb{H}(\Pi, \Pi_0)$$

a set of histories pertinent to $Usp(\Pi, \Pi_0)$. The *subspace* of $Usp(\Pi, \Pi_0)$ relevant to H, namely

$$Subsp(Usp(\Pi, \Pi_0), H),$$

is the connected subgraph of $Usp(\Pi, \Pi_0)$ which consists of all the nodes and edges in $Usp(\Pi, \Pi_0)$ that are relevant to the histories in H.

   Let $\mathcal{E} = \{e_1, \ldots, e_k\}$ be a subset of the linearization of the diagnostic expression $Dex(\wp(\Pi))$, namely

$$\mathcal{E} \subseteq < Dex(\wp(\Pi)) > .$$

The subspace of $Usp(\Pi, \Pi_0)$ relevant to $\mathcal{E}$ is defined as follows:

$$Subsp(Usp(\Pi, \Pi_0), \mathcal{E}) = Subsp(Usp(\Pi, \Pi_0), H_{\mathcal{E}})$$

where

$$H_{\mathcal{E}} = \{h \mid h \in \mathbb{H}(\wp(\Pi)), e = \langle \tau \mid \tau \in h, \textit{Faulty}(\tau) \rangle, e \subseteq h, e \in \mathcal{E}\}.$$

Let δ be a diagnosis in $\Delta(\wp(\Pi))$. A *route* $\rho(\delta)$ is the smallest subspace of $Usp(\Pi, \Pi_0)$ such that each history in $Usp(\Pi, \Pi_0)$ entailing δ is also a history in $\rho(\delta)$.

Clearly, $Usp(\Pi, \Pi_0)$ includes as many (overlapping) routes as the cardinality of $\Delta(\wp(\Pi))$, that is,

$$Usp(\Pi, \Pi_0) = \bigcup_{\delta \in \Delta(\wp(\Pi))} \rho(\delta).$$

Each route is the graph representation of a (possibly infinite) set of histories, denoted by $\mathbb{H}(\rho(\delta))$. However, as discussed above, not all of the histories in $\mathbb{H}(\rho(\delta))$ entail the diagnosis δ. In particular, there exists a (possibly empty) subset

$$Spur(\rho(\delta)) \subset \mathbb{H}(\rho(\delta)),$$

called the *spurious set* of $\rho(\delta)$, such that each history in $Spur(\rho(\delta))$, called a *spurious history* of $\rho(\delta)$, entails a diagnosis $\widetilde{\delta} \subset \delta$.

Example 7.3. The routes relevant to the universal space displayed in Figure 7.2 are shown in Figure 7.3 ($\rho(\delta_1)$ and $\rho(\delta_2)$) and in Figure 7.4 ($\rho(\delta_3)$ and $\rho(\delta_4)$). Consider Figure 7.3. Note that $\rho(\delta_1)$ does not contain any spurious history, that is,

$$Spur(\rho(\delta)) = \emptyset,$$

since both histories included in $\mathbb{H}(\delta_1)$ entail the empty diagnosis δ_1. By contrast, the spurious set of $\rho(\delta_2)$ is not empty, as it incorporates the empty history (node N_0 is final). One might argue that, in order to remove the spurious history, it would suffice removing the qualification of final from node N_0. Even if this might be the solution for this particular route, such a transformation is not enough to remove the spurious set in the general case. To be convinced of this, consider the route $\rho(\delta_3)$ displayed on the left of Figure 7.4. Note that $\mathbb{H}(\delta_3)$ includes $\mathbb{H}(\delta_1)$. In other terms, $\mathbb{H}(\delta_1)$ is the spurious set of $\rho(\delta_3)$. However, although we may remove the empty spurious history by making node N_0 no longer final, we still have the spurious history

$$\langle \{T_{41}\}, \{T_{11}\}, \{T_{21}\}, \{T_{12}\}, \{T_{23}\}, \{T_{14}, T_{31}\}, \{T_{42}\} \rangle$$

that ends at final node N_9. The point is that N_9 cannot be transformed into a non-final node because this transformation is bound to remove relevant (non-spurious) histories from the route, specifically, a subset of the histories involving the cycles in $\rho(\delta_3)$. The same considerations apply to route $\rho(\delta_4)$, which is in fact the whole universal space $Usp(\Pi, \Pi_0)$ (see Figure 7.2), thereby involving all of the sets of histories relevant to the other routes, namely

$$Spur(\rho(\delta_4)) = \bigcup_{i=1}^{3} \mathbb{H}(\rho(\delta_i)).$$

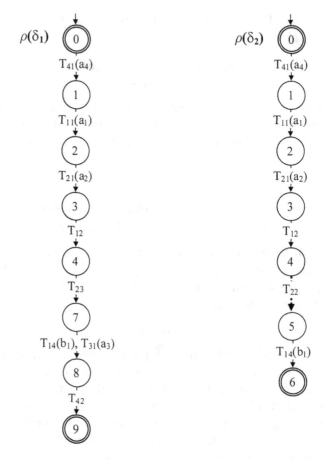

Figure 7.3. Routes $\rho(\delta_1)$ and $\rho(\delta_2)$, relevant to the universal space in Figure 7.2.

Even in this case, the only spurious history that can be removed, by simply requalifying the root as non-final, is the empty history. $\diamond$

Theorem 7.1. *Let* $Usp(\Pi, \Pi_0)$ *be a universal space relevant to a blind diagnostic problem* $\wp(\Pi)$, δ *a diagnosis in* $\Delta(\wp(\Pi))$,

$$\mathcal{E}(\delta) = \{e \mid e \in \langle Dex(\wp(\Pi)) \rangle, e \models \delta\}$$

the subset of the linearization of the corresponding diagnostic expression entailing δ, *and* $\rho(\delta)$ *a route in* $Usp(\Pi, \Pi_0)$. *Then*

$$Subsp(Usp(\Pi, \Pi_0), \mathcal{E}(\delta)) \equiv \rho(\delta).$$

4. Diagnostic Space

Once extracted the set of routes relevant to a universal space, we need to generate, for each route $\rho(\delta_i)$, the corresponding graph $\gamma_i(\delta)$, called

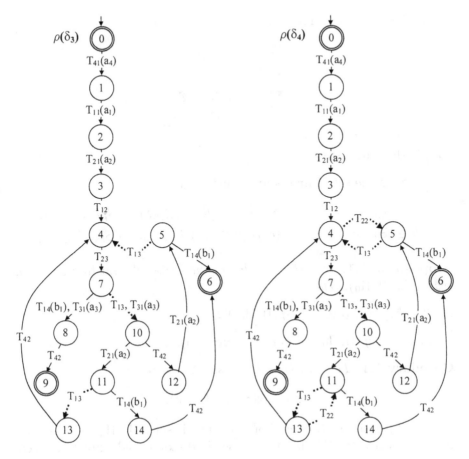

Figure 7.4. Routes $\rho(\delta_3)$ and $\rho(\delta_4)$, relevant to the universal space in Figure 7.2.

the *diagnostic space* of δ, where the set of paths from the root to final states corresponds to all the non-spurious histories of $\rho(\delta)$.

Let $\rho(\delta) = (\mathbb{S}, \mathbb{E}, \mathbb{T}, S_0, \mathbb{S}_f)$ be a route. A *diagnostic space* γ of δ is an automaton,

$$\gamma(\delta) = (\mathbb{S}_\gamma, \mathbb{E}_\gamma, \mathbb{T}_\gamma, S_{0\gamma}, \mathbb{S}_{f\gamma}),$$

where

$\mathbb{S}_\gamma$ is the set of states;

$\mathbb{E}_\gamma = \mathbb{E}$ is the set of events;

$\mathbb{T}_\gamma$ is the transition function, such that

$$\mathbb{T}_\gamma = \mathbb{T}^a_\gamma \cup \mathbb{T}^s_\gamma,$$
$$\mathbb{T}^a_\gamma \cap \mathbb{T}^s_\gamma = \emptyset,$$
$$\mathbb{T}^a_\gamma : \mathbb{S}_\gamma \times \mathbb{E}_\gamma \mapsto \mathbb{S}_\gamma,$$
$$\mathbb{T}^s_\gamma : \mathbb{S}_\gamma \times 2^{\mathbb{E}_\gamma} \mapsto \mathbb{S}_\gamma;$$

$S_{0\gamma}$ is the initial state;

$\mathbb{S}_{f\gamma} \subseteq \mathbb{S}_\gamma$ is the set of final states, such that

$$\forall h = S_{0\gamma} \leadsto S_{f\gamma} \in \gamma(\delta), S_{f\gamma} \in \mathbb{S}_{f\gamma} \; (h \in (\mathbb{H}(\rho(\delta)) - Spur(\rho(\delta)))),$$
$$\forall h \in (\mathbb{H}(\rho(\delta)) - Spur(\rho(\delta))) \; (h \in \gamma(\delta), h = S_{0\gamma} \leadsto S_{f\gamma}, S_{f\gamma} \in \mathbb{S}_{f\gamma}).$$

Proposition 7.1. *Let $\gamma(\delta)$ be a diagnostic space relevant to a universal space $Usp(\Pi, \Pi_0)$. Then:*

(1) $\forall h \in \mathbb{H}(\gamma(\delta)) \; (h \in \mathbb{H}(Usp(\Pi, \Pi_0)));$

(2) $\forall h \in \mathbb{H}(Usp(\Pi, \Pi_0)), h \models \delta \; (h \in \mathbb{H}(\gamma(\delta))).$

Corollary 7.1. *Let $\gamma(\delta)$ be a diagnostic space. Then,*

$$\forall h \in \mathbb{H}(\gamma(\delta)) \; (h \models \delta).$$

The *stratification hierarchy* ∇ of a diagnostic set $\Delta(\Pi, \Pi_0)$ relevant to a universal space $Usp(\Pi, \Pi_0)$, is a (possibly disconnected) directed acyclic graph,

$$\nabla(\Pi, \Pi_0) = (\mathbb{N}, \mathbb{E}, \mathbb{N}_0),$$

where

$\mathbb{N} = \Delta(\Pi, \Pi_0)$ is the set of nodes;

$\mathbb{E} : \mathbb{N} \mapsto 2^{\mathbb{N}}$ is the set of edges; $\mathbb{N}_0 \subseteq \mathbb{N}$ is the set of *roots*, such that, denoting with $Succ(N)$ the set of successive nodes of $N \in \mathbb{N}$, the following conditions hold:

  (1) $\forall N_0 \in \mathbb{N}_0 \; (\not\exists N(N \in \mathbb{N}, N_0 \in Succ(N)));$

  (2) $\forall N \in \mathbb{N} \; (Succ(N) = \{N' \mid N' \subset N\}, \forall N' \in Succ(N), \forall N'' \in Succ(N), N' \neq N'' \; (N' \not\subset N'')).$

The set of *descendants* of a node $N \in \mathbb{N}$, $Desc(N)$, is the set of nodes in $\mathbb{N}$ defined as follows:

(1) $N' \in Succ(N) \Rightarrow N' \in Desc(N);$

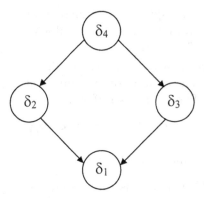

Figure 7.5. Stratification hierarchy (see Example 7.4).

(2) $N' \in Succ(N), N'' \in Desc(N') \Rightarrow N'' \in Desc(N)$.

Example 7.4. Consider the universal space $Usp(\Upsilon, \Upsilon_0)$ shown in Figure 7.2, where the relevant diagnostic set is

$$\Delta(\Upsilon, \Upsilon_0) = \{\delta_1, \delta_2, \delta_3, \delta_4\},$$

where $\delta_1 = \emptyset$, $\delta_2 = \{C_2\}$, $\delta_3 = \{C_3\}$, and $\delta_4 = \{C_2, C_3\}$. The corresponding stratification hierarchy is depicted in Figure 7.5. According to the definition we have:

$$\mathbb{N}_0 = \{\delta_4\},$$
$$Succ(\delta_4) = \{\delta_2, \delta_3\},$$
$$Desc(\delta_4) = \{\delta_1, \delta_2, \delta_3\},$$

where $\delta_2 \subset \delta_4$, $\delta_3 \subset \delta_4$, $\delta_2 \not\subset \delta_3$, $\delta_3 \not\subset \delta_2$, $\delta_1 \subset \delta_2$, and $\delta_1 \subset \delta_3$. ◇

Based on a stratification hierarchy $\nabla(\Pi, \Pi_0) = (\mathbb{N}, \mathbb{E}, \mathbb{N}_0)$, we can generate a diagnostic space as follows. If δ is a leaf node in $\mathbb{N}$, that is, a node such that $Succ(\delta) = \emptyset$, then the route relevant to δ is also the diagnostic space of δ, namely $\gamma(\delta) = \rho(\delta)$.

Example 7.5. With reference to the stratification hierarchy displayed in Figure 7.5, since $\delta_1 = \emptyset$ is a leaf node, the relevant diagnostic space will equal the route of δ_1 depicted on the left of Figure 7.3, in other words, $\gamma(\delta_1) = \rho(\delta_1)$. As a matter of fact, as pointed out in Example 7.3, all the (two) histories in $\rho(\delta_1)$ entail δ_1, that is, $Spur(\rho(\delta_1)) = \emptyset$. ◇

Things are different when considering internal nodes of the hierarchy, such as δ_2 or δ_3, which involve a non-empty set of spurious histories. In such cases, the diagnostic space can be made up by means of a difference operation between automata, defined as follows.

Let $A_1 = (\mathbb{S}_1, \mathbb{E}_1, \mathbb{T}_1, S_{0_1}, \mathbb{S}_{f_1})$ and $A_2 = (\mathbb{S}_2, \mathbb{E}_2, \mathbb{T}_2, S_{0_2}, \mathbb{S}_{f_2})$ be two deterministic automata, where the five fields are the set of states, the set of events, the transition function, the initial state, and the set of final states, respectively. The *spurious difference*

$$A \tilde{\boxminus} A' = (\tilde{\mathbb{S}}, \mathbb{E}, \tilde{\mathbb{T}}, S_0, \mathbb{S}_f)$$

is a deterministic automaton such that:

$\tilde{\mathbb{S}} \subseteq \mathbb{S}_1 \times (\mathbb{S}_2 \cup \{\bot\})$, where $\bot$ is the *null* state ($\bot \notin \mathbb{S}_2$);

$\mathbb{E} \subseteq \mathbb{E}_1$;

$\tilde{\mathbb{T}} : \tilde{\mathbb{S}} \times \mathbb{E} \mapsto \tilde{\mathbb{S}}$;

$S_0 = (S_{0_1}, S_{0_2})$;

$\mathbb{S}_f$ is defined by the following two rules:

$$S_0 \in \mathbb{S}_f \iff S_{0_1} \in \mathbb{S}_{f_1}, S_{0_2} \notin \mathbb{S}_{f_2};$$

$$S' \in \mathbb{S}_f, S' \neq S_0, S' = (S_1', S_2') \iff (S_1, S_2) \xrightarrow{E} (S_1', S_2') \in \tilde{\mathbb{T}},$$
$$S_1' \in \mathbb{S}_{f_1}, (S_2' \notin \mathbb{S}_{f_2} \text{ or } S_2' = \bot).$$

The transition function is defined by the following three rules:

$$(S_1, S_2) \xrightarrow{E} (S_1', S_2') \in \tilde{\mathbb{T}} \iff S_1 \xrightarrow{E} S_1' \in \mathbb{T}_1, S_2 \xrightarrow{E} S_2' \in \mathbb{T}_2;$$

$$(S_1, S_2) \xrightarrow{E} (S_1', \bot) \in \tilde{\mathbb{T}} \iff S_1 \xrightarrow{E} S_1' \in \mathbb{T}_1, S_2 \xrightarrow{E} S_2' \notin \mathbb{T}_2;$$

$$(S_1, \bot) \xrightarrow{E} (S_1', \bot) \in \tilde{\mathbb{T}} \iff S_1 \xrightarrow{E} S_1' \in \mathbb{T}_1, (S_1, \bot) \in \mathbb{S}.$$

The *difference*

$$A \boxminus A' = (\mathbb{S}, \mathbb{E}, \mathbb{T}, S_0, \mathbb{S}_f)$$

is the automaton obtained from $A \tilde{\boxminus} A'$ by selecting the states and transitions that converge to final states.

Example 7.6. Consider the routes $\rho(\delta_1)$ and $\rho(\delta_2)$ depicted on the right of Figure 7.3. The diagnostic space $\gamma(\delta_2)$ can be generated by the difference between $\rho(\delta_2)$ and $\rho(\delta_1)$, namely

$$\gamma(\delta_2) = \rho(\delta_2) \boxminus \rho(\delta_1).$$

Based on the definition above, the resulting automaton is shown in Figure 7.6. As expected, $\gamma(\delta_2)$ is isomorphic to $\rho(\delta_2)$, while the root is no

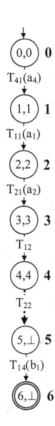

Figure 7.6. Diagnostic space $\gamma(\delta_2)$ (see Example 7.6).

longer final. This is due to the enforcement of the first rule on the set of final states $\mathbb{S}_f$, that requires that only the initial state of $\rho(\delta_2)$ be final, which is not the case. Intuitively, if both the initial states of the routes were final as well as the initial state of the diagnostic space, the latter would include the empty history, which is also included in $\rho(\delta_1)$, thereby violating the definition of a diagnostic space. Starting from the root $N_0 = (0,0)$ of $\gamma(\delta_2)$, the following four transitions are determined based on the first rule for the transition function $\mathbb{T}$ (see above). When in state $N_4 = (4,4)$, since T_{22} leaves state 4 of $\rho(\delta_2)$ but not state 4 of $\rho(\delta_1)$, the applied rule on the transition function is the second one, where $E = \{T_{22}\}$, making up the state $N_5 = (5, \perp)$. The last transition is determined based on the third rule. Informally, this rule is based on the fact that, once reached a state $(S_1, \perp)$, it means that the relevant history is not in $\mathbb{H}(\rho(\delta_1))$, thereby allowing all the successive transitions in $\rho(\delta_2)$ to be part of $\gamma(\delta_2)$. Accordingly with the second rule on final states, node $(6, \perp)$ is final. ◇

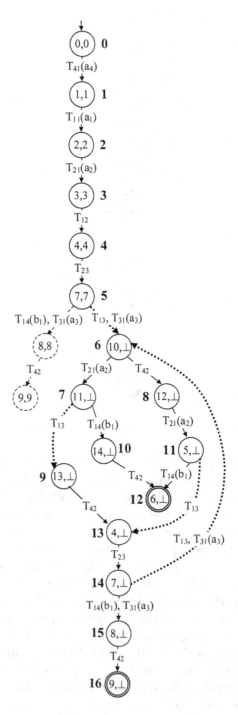

Figure 7.7. Diagnostic space $\gamma(\delta_3)$ (see Example 7.7).

Example 7.7. The computation of the diagnostic space $\gamma(\delta_3)$ is based on the route $\rho(\delta_3)$ depicted on the left of Figure 7.4, from which histories in $\rho(\delta_1)$ must be removed. The resulting graph $\rho(\delta_3) \boxminus \rho(\delta_1)$ is displayed in Figure 7.7. As for $\gamma(\delta_2)$, the root $(0,0)$ is not final. Two different transitions leave node $(7,7)$, marked by events $\{T_{14}, T_{31}\}$ and $\{T_{13}, T_{31}\}$, respectively. The former is part of the spurious difference $\rho(\delta_3) \boxminus \rho(\delta_1)$, the latter is relevant to $\rho(\delta_3)$ only, thereby moving to state $(10, \perp)$. Note that node $(9,9)$ is not final insofar as the second rule for final states is not met (both states in $\rho(\delta_1)$ and $\rho(\delta_3)$ are final). This is correct as, if not so, the path $(0,0) \rightsquigarrow (9,9)$ would be a history in $\mathbb{H}(\gamma(\delta_3))$ entailing $\delta_1 = \emptyset$ instead of $\delta_3 = \{C_3\}$, as expected. Besides, no transition can leave node $(9,9)$ because no transition leaves nodes 9 in the respective routes $\rho(\delta_1)$ and $\rho(\delta_3)$. This is why the dashed (spurious) part of the graph is not encompassed by $\gamma(\delta_3)$. On the other hand, since event $\{T_{13}, T_{31}\}$ is only relevant to $\rho(\delta_3)$, the reached node is $(10, \perp)$, from which on, all paths in $\rho(\delta_3)$ will be part of $\gamma(\delta_3)$. As expected, each path in the resulting diagnostic space of Figure 7.7 is a history of $\rho(\delta_3)$ entailing δ_3, and vice versa. $\diamond$

In the previous two examples the computation of the diagnostic space is performed by a difference between two routes, namely $\rho(\delta) \boxminus \rho(\delta')$, which is the case when the descendants of δ in the stratification hierarchy includes just a node, in other words,

$$Desc(\delta) = \{\delta'\}.$$

In the general case, however, the second term of the difference is not simply a route, but rather the automaton-based union of several routes, specifically, the union of all the routes relevant to the descendants of δ. The union of two automata can be formalized as follows.

Let $A_1 = (\mathbb{S}_1, \mathbb{E}_1, \mathbb{T}_1, S_{0_1}, \mathbb{S}_{f_1})$ and $A_2 = (\mathbb{S}_2, \mathbb{E}_2, \mathbb{T}_2, S_{0_2}, \mathbb{S}_{f_2})$ be two deterministic automata, with the usual meaning. The *automaton-based union*

$$A \sqcup A' = (\mathbb{S}, \mathbb{E}, \mathbb{T}, S_0, \mathbb{S}_f)$$

is a deterministic automaton such that:

$\mathbb{S} \subseteq ((\mathbb{S}_1 \cup \{\perp\}) \times (\mathbb{S}_2 \cup \{\perp\})) - \{(\perp, \perp)\};$

$\mathbb{E} \subseteq (\mathbb{E}_1 \cup \mathbb{E}_2);$

$\mathbb{T} : \mathbb{S} \times \mathbb{E} \mapsto \mathbb{S};$

$S_0 = (S_{0_1}, S_{0_2});$

$\mathbb{S}_f = \{(S_1, S_2) \in \mathbb{S} \mid (S_1 \in \mathbb{S}_{f_1} \text{ or } S_1 = \perp), (S_2 \in \mathbb{S}_{f_2} \text{ or } S_2 = \perp)\}.$

The transition function is defined by the following five rules:

$$(S_1, S_2) \xrightarrow{E} (S_1', S_2') \in \mathbb{T} \iff S_1 \xrightarrow{E} S_1' \in \mathbb{T}_1, S_2 \xrightarrow{E} S_2' \in \mathbb{T}_2;$$

$$(S_1, S_2) \xrightarrow{E} (S_1', \perp) \in \mathbb{T} \iff S_1 \xrightarrow{E} S_1' \in \mathbb{T}_1, S_2 \xrightarrow{E} S_2' \notin \mathbb{T}_2;$$

$$(S_1, S_2) \xrightarrow{E} (\perp, S_2') \in \mathbb{T} \iff S_2 \xrightarrow{E} S_2' \in \mathbb{T}_2, S_1 \xrightarrow{E} S_1' \notin \mathbb{T}_1;$$

$$(S_1, \perp) \xrightarrow{E} (S_1', \perp) \in \mathbb{T} \iff S_1 \xrightarrow{E} S_1' \in \mathbb{T}_1, (S_1, \perp) \in \mathbb{S};$$

$$(\perp, S_2) \xrightarrow{E} (\perp, S_2') \in \mathbb{T} \iff S_2 \xrightarrow{E} S_2' \in \mathbb{T}_2, (\perp, S_2) \in \mathbb{S}.$$

Proposition 7.2. *Let $\rho(\delta_1)$ and $\rho(\delta_2)$ be two routes relevant to the same universal space. Then,*

$$\mathbb{H}(\rho(\delta_1) \sqcup \rho(\delta_2)) \equiv \mathbb{H}(\rho(\delta_1)) \cup \mathbb{H}(\rho(\delta_2))$$

Example 7.8. With reference to the routes $\rho(\delta_1)$ and $\rho(\delta_2)$ displayed in figure 7.3, the union $\rho(\delta_1) \sqcup \rho(\delta_2)$ is shown in Figure 7.8. Accordingly with Proposition 7.2, the set of histories embodied in the union equals the union of the set of histories of the operands $\rho(\delta_1)$ and $\rho(\delta_1)$. ◇

The notion of an automaton-based union can be easily extended to $n \geq 2$ operands, as informally shown in the following example.

Example 7.9. Consider routes $\rho(\delta_1)$ and $\rho(\delta_2)$ displayed in Figure 7.3, and route $\rho(\delta_3)$ depicted on the left of Figure 7.4. The extended union of $\rho(\delta_1)$, $\rho(\delta_2)$, and $\rho(\delta_3)$ is shown in Figure 7.9. ◇

The generalized union of routes allows us to easily compute the diagnostic space of any internal node of the stratification hierarchy as formalized below.

Theorem 7.2. *Let $\nabla(\Pi, \Pi_0) = (\mathbb{N}, \mathbb{E}, \mathbb{N}_0)$ be a stratification hierarchy, $\delta \in \mathbb{N}$ a diagnosis, and $\rho(\delta)$ the relevant route in $Usp(\Pi, \Pi_0)$. Then*

$$\mathbb{H}\left(\rho(\delta) \boxminus \left(\bigsqcup_{\delta' \in Desc(\delta)} \rho(\delta')\right)\right) \equiv \mathbb{H}(\rho(\delta)) - Spur(\rho(\delta)).$$

Corollary 7.2. *Let $\nabla(\Pi, \Pi_0) = (\mathbb{N}, \mathbb{E}, \mathbb{N}_0)$ be a stratification hierarchy, $\delta \in \mathbb{N}$ a diagnosis, and $\rho(\delta)$ the relevant route in $Usp(\Pi, \Pi_0)$. Then, a diagnostic space $\gamma(\delta)$ can be made up as follows:*

$$\gamma(\delta) = \rho(\delta) \boxminus \left(\bigsqcup_{\delta' \in Desc(\delta)} \rho(\delta')\right).$$

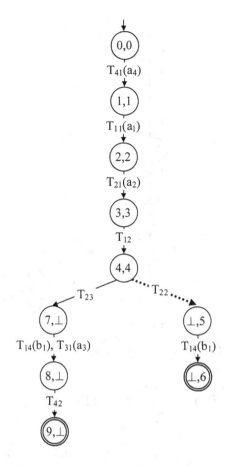

Figure 7.8. Automaton-based union $\rho(\delta_1) \sqcup \rho(\delta_2)$.

Corollary 7.3. *Let* $Usp(\Pi, \Pi_0)$ *be a universal space. It is possible to generate a set*

$$\Gamma(\Pi, \Pi_0) = \{\gamma(\delta_1), \ldots, \gamma(\delta_n)\}$$

of diagnostic spaces which is isomorphic to the diagnostic set

$$\Delta(\Pi, \Pi_0) = \{\delta_1, \ldots, \delta_n\}.$$

The set $\Gamma(\Pi, \Pi_0)$ referenced in Corollary 7.3 is called the *diagnostic partition* of $Usp(\Pi, \Pi_0)$.

Example 7.10. Based on Corollary 7.2, the diagnostic space relevant to δ_4 can be obtained by the automaton-based difference between route $\rho(\delta_4)$ (displayed on the right of Figure 7.4) and the automaton-based union of $\rho(\delta_1)$, $\rho(\delta_2)$, and $\rho(\delta_3)$ (depicted in Figure 7.9), namely

$$\gamma(\delta_4) = \rho(\delta_4) \boxminus (\rho(\delta_1) \sqcup \rho(\delta_2) \sqcup \rho(\delta_3)).$$

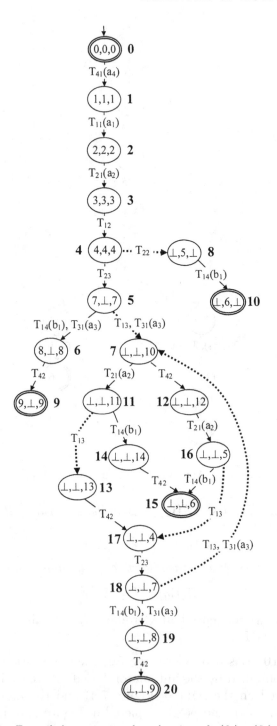

Figure 7.9. Extended, automaton-based union of $\rho(\delta_1)$, $\rho(\delta_2)$, and $\rho(\delta_3)$.

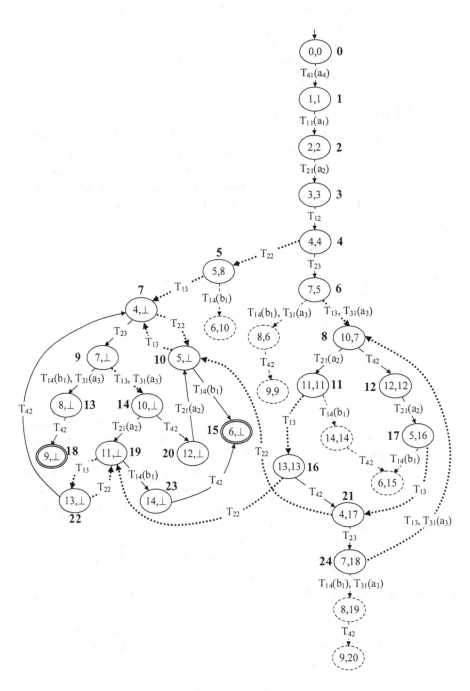

Figure 7.10. Diagnostic space $\gamma(\delta_4)$ (see Example 7.10).

The resulting graph is displayed in Figure 7.10. Note that the second filed within the pair identifying a state in $\gamma(\delta_4)$ is the (renamed) label of a node in Figure 7.9. As usual, the dashed elements denote the spurious part of the graph, which is not encompassed by any history (due to lack of convergence to final states).   ◇

5.   Matching Graph

Once generated the set of diagnostic spaces based on the stratification hierarchy relevant to a universal space $Usp(\Pi, \Pi_0)$, we are expected to make a further transformation of such graphs, where the focus is on messages rather than component transitions. A diagnostic space $\gamma(\delta)$ is in fact the graph representation of the subset of histories of $Usp(\Pi, \Pi_0)$ that entail δ. Formally, if

$$\Gamma(\Pi, \Pi_0) = \{\gamma(\delta_1), \ldots, \gamma(\delta_n)\}$$

is the diagnostic partition of $Usp(\Pi, \Pi_0)$ (see Corollary 7.3), the following properties hold:

$$\bigcup_{i \in [1 \, .. \, n]} \mathbb{H}(\gamma(\delta_i)) = \mathbb{H}(Usp(\Pi, \Pi_0)),$$

$$\bigcap_{i \in [1 \, .. \, n]} \mathbb{H}(\gamma(\delta_i)) = \emptyset.$$

That is, the diagnostic partition is an intensional (graph-based) partition of the history set of the universal space, where the i-th part (the diagnostic space $\gamma(\delta_i)$) refers to all and only histories entailing δ_i.

As a consequence, a specific $\gamma(\delta_i)$ implicitly embodies the constraints that govern the mode in which messages are generated by the histories entailing δ_i. We may therefore highlight messages as labels of the edges in $\gamma(\delta_i)$ by simply isolating the observable transitions and removing the non-observable ones. The resulting automaton is in general nondeterministic, owing to possible empty edges (when the relevant transitions are not observable), which can be possibly transformed into an equivalent deterministic one. All this is formalized by the notion of a matching graph given below.

Let $\gamma(\delta) = (\mathbb{S}, \mathbb{E}, \mathbb{T}, S_0, \mathbb{S}_f)$, $\mathbb{T} = \mathbb{T}^a \cup \mathbb{T}^s$, be a diagnostic space relevant to a diagnosis δ in the universal diagnostic set $\Delta(\Pi, \Pi_0)$. The *matching graph* μ for δ is the finite automaton

$$\mu(\delta) = (\mathbb{S}, \mathbb{E}_\mu, \mathbb{T}_\mu, S_0, \mathbb{S}_f),$$

such that

$\mathbb{E}_\mu = 2^{(\mathbb{E}_{obs} \times \mathbb{C}(\Pi))}$ is the set of events, where $\mathbb{E}_{obs}$ is the domain of messages involved in the models of components in Π;

$\mathbb{T}_\mu : \mathbb{S}_\mu \times \mathbb{E}_\mu \mapsto 2^{\mathbb{S}}$ is the transition function obtained from $\mathbb{T}$ as follows:

(1) $\forall S \xrightarrow{E} S' \in \mathbb{T}^a$ of component C, $E = S_1 \xrightarrow{\alpha|\beta} S_2, (m, Out) \in \beta$
$(S \xrightarrow{\{(m,C)\}} S' \in \mathbb{T}_\mu)$;

(2) $\forall S \xrightarrow{E} S' \in \mathbb{T}^s$, $E = \{T_1, \ldots, T_k\}$, $E_\mu = \{T'_1, \ldots, T'_{k'}\}$ is the subset of observable transitions of E, $E_\mu \neq \emptyset$,

$$\forall i \in [1 .. k'] \, (T'_i = S_1 \xrightarrow{\alpha|\beta} S_2 \text{ of component } C_i, (m'_i, Out) \in \beta)),$$

the following condition holds:

$$S \xrightarrow{\{(m'_1,C_1),\ldots,(m'_{n'},C_{n'})\}} S' \in \mathbb{T}_\mu;$$

(3) $\forall S \xrightarrow{E} S' \in \mathbb{T}, \neg Observable(E) \, (S \xrightarrow{\emptyset} S' \in \mathbb{T}_\mu)$.

The *matching expression* of δ, $Mex(\delta)$, is the regular expression corresponding to $\mu(\delta)$. A string $\mathbf{O} = \langle \mathcal{O}_1, \ldots, \mathcal{O}_r \rangle$ in $Mex(\delta)$ is a *matching string* of $Mex(\delta)$. The *extension* of $\mathbf{O}$ is defined as follows:

$$\|\mathbf{O}\| = \begin{cases} \{\langle\rangle\} & \text{if } \mathbf{O} = \langle\rangle \\ \{Q\} & \text{otherwise,} \end{cases}$$

where Q is a concatenation of the elements in $\mathcal{O}_1, \ldots, \mathcal{O}_r$ in the given order. Owing to synchronism, the symbols of each string in $Mex(\delta)$ consist in general of a set of messages. That is, a string in $Mex(\delta)$ is a sequence of sets of messages relevant to components in Π.

The whole set of matching strings in $Mex(\delta)$ is denoted either as $\|Mex(\delta)\|$ or $\|\mu(\delta)\|$.

Example 7.11. Consider the diagnostic spaces $\gamma(\delta_1) = \rho(\delta_1)$ and $\gamma(\delta_2)$ displayed on the right of Figure 7.3 and in Figure 7.6, respectively. The corresponding matching graphs are shown in Figure 7.11. Each matching graph is displayed both as a nondeterministic (left) and deterministic (right) automaton. Draw the attention to the deterministic $\mu(\delta_1)$: it embodies two matching strings, namely

$$\mathbf{O}_1 = \langle\rangle,$$
$$\mathbf{O}_2 = \langle \{a_4\}, \{a_1\}, \{a_2\}, \{a_3, b_1\} \rangle,$$

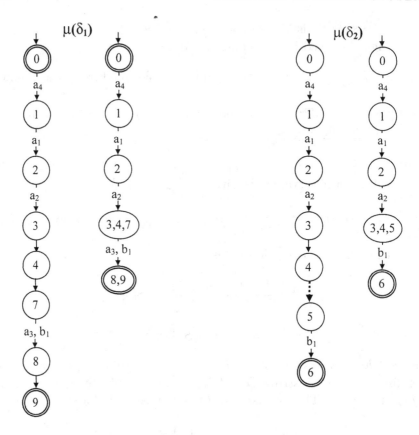

Figure 7.11.   Matching graphs $\mu(\delta_1)$ (left) and $\mu(\delta_2)$ (right) (see Example 7.11).

with corresponding extensions

$$\|\mathbf{O}_1\| = \{\langle\rangle\},$$
$$\|\mathbf{O}_2\| = \{\langle a_4, a_1, a_2, a_3, b_1\rangle, \langle a_4, a_1, a_2, b_1, a_3\rangle\}.$$

In other words, since $\delta_1 = \emptyset$, the union $\|\mathbf{O}_1\|\cup\|\mathbf{O}_2\|$ represents the whole set of signatures (see Section 5.2 of Chapter 3) relevant to nominal (non-faulty) reactions of the system Υ outlined in Figure 7.1.

Considering $\mu(\delta_2)$, where $\delta_2 = \{C_2\}$, the only possible signature is

$$\langle a_4, a_1, a_2, b_1\rangle,$$

corresponding to the system history

$$\langle T_{41}, T_{11}, T_{21}, T_{12}, T_{22}, T_{14}\rangle$$

incorporated in the diagnostic space $\gamma(\delta_2)$ of Figure 7.6. $\diamond$

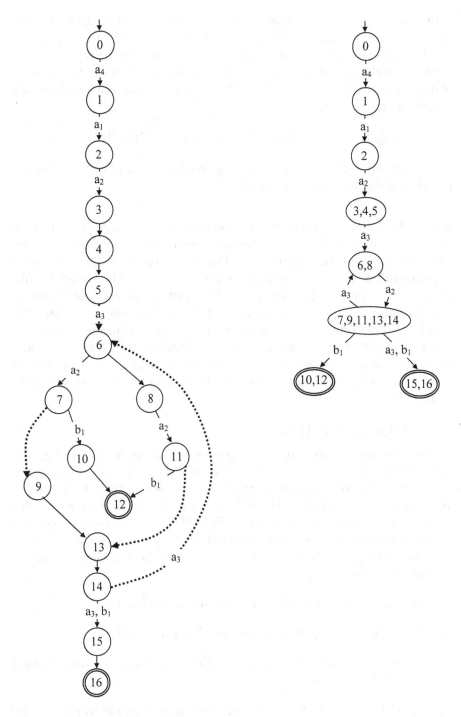

Figure 7.12. Matching graph $\mu(\delta_3)$ (see Example 7.12).

Example 7.12. The matching graph $\mu(\delta_2)$ obtained from the diagnostic space depicted in Figure 7.7 is displayed in Figure 7.12, specifically, the nondeterministic automaton on the left and the deterministic one on the right. Considering the latter, note that, owing to the cycle, the number of matching strings is unbounded. They are represented by the following matching expression:

$$Mex(\delta_3) = \{a_4\}\{a_1\}\{a_2\}\{a_3\}\{a_2\}(\{a_3\}\{a_2\})^\star(\{b_1\}|\{a_3, b_1\}),$$

where, accordingly with the definition, each symbol of the regular expression alphabet is a set of messages.   ◇

Example 7.13. The nondeterministic matching graph relevant to $\delta_4 = \{C_2, C_3\}$, obtained from the diagnostic space $\gamma(\delta_4)$ displayed in Figure 7.10, is shown in Figure 7.13. The equivalent deterministic one is depicted in Figure 7.14. A comparison of the latter with the deterministic matching graphs $\mu(\delta_2)$ and $\mu(\delta_3)$ displayed on the right of Figure 11 and Figure 12, respectively, shows that both $\mu(\delta_2)$ and $\mu(\delta_3)$ are subgraphs of $\mu(\delta_4)$ (identifiers of nodes are irrelevant to the comparison). In other words, all the matching strings in $Mex(\delta_2)$ and $Mex(\delta_3)$ are also matching strings in $Mex(\delta_4)$. This inclusion property has interesting consequences on the diagnosability of the system, as pointed out in the next section.   ◇

6.   Diagnostic Rule

The matching graphs relevant to a universal space $Usp(\Pi, \Pi_0)$ can be conveniently exploited on-line when an actual diagnostic problem is given. Specifically, a matching graph $\mu(\delta)$ allows us to test whether or not the given observation $OBS(\Pi)$ matches the diagnostic expression $Mex(\delta)$. However, since in general $Ukn(OBS(\Pi)) \neq \emptyset$, this matching is required to account for the unavailability of messages for components in $Ukn(OBS(\xi))$. The actual exploitation of the matching graphs is formalized as follows. Let

$\wp(\Pi) = (\Omega(\Pi), OBS(\Pi), \Pi_0)$ be a diagnostic problem,

$\mu(\delta) = (\mathbb{S}, \mathbb{E}, \mathbb{T}, S_0, \mathbb{S}_f)$ a matching graph for $\delta \in \Delta(\Pi, \Pi_0)$,

$\mathbb{E}_{Ukn}$ the subset of events in $(\mathbb{E}_{obs} \times \mathbb{C}(\Pi))$ relevant to components in $Ukn(OBS(\Pi))$, and

$\bar{\mu}(\delta, \wp(\Pi)) = (\mathbb{S}, \mathbb{E}, \bar{\mathbb{T}}, S_0, \mathbb{S}_f)$ an automaton isomorphic to $\mu(\delta)$, called the *actual matching graph* of δ, where each transition $S \xrightarrow{E} S' \in \mathbb{T}$ is

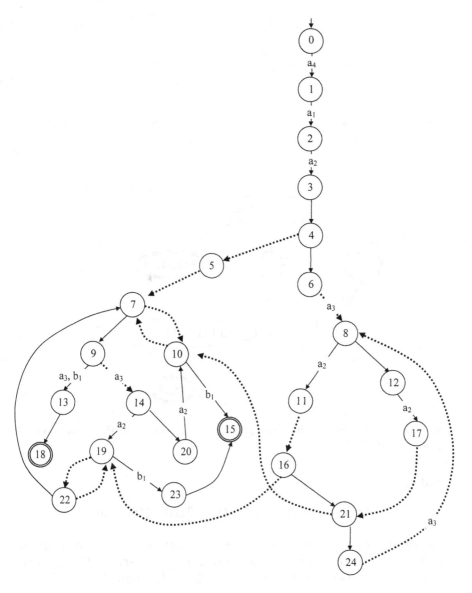

Figure 7.13. Nondeterministic matching graph $\mu(\delta_4)$ (see Example 7.13).

replaced by a transition

$$S \xrightarrow{E - \mathbb{E}_{Ukn}} S' \in \bar{\mathbb{T}}.$$

The observation $OBS(\Pi)$ is said to *match* $\mu(\delta)$, denoted by

$$OBS(\Pi) \asymp \mu(\delta),$$

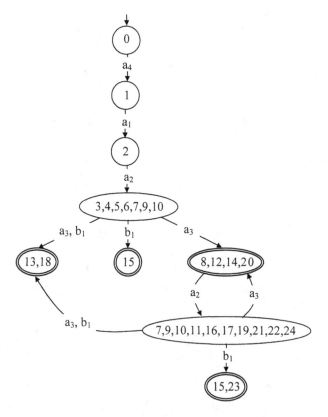

Figure 7.14. Deterministic matching graph $\mu(\delta_4)$ equivalent to the automaton displayed in Figure 7.13.

if and only if

$$\exists \mathbf{O} \ (\mathbf{O} \in \|\bar{\mu}(\delta, \wp(\Pi))\|, \|\mathbf{O}\| \cap \|OBS(\Pi)\| \neq \emptyset).$$

A *diagnostic rule*, $Rule(\delta, \mu(\delta))$, is an association between a diagnosis $\delta \in \Delta(\Pi, \Pi_0)$ and a matching graph $\mu(\delta)$. The *rule set* of (Π, Π_0) is the whole set of diagnostic rules relevant to $\Delta(\Pi, \Pi_0)$, namely

$$Rules(\Pi, \Pi_0) = \{Rule(\delta, \mu(\delta)) \mid \delta \in \Delta(\Pi, \Pi_0)\}.$$

The *matching rule set* of the diagnostic problem $\wp(\Pi)$ is the subset of $Rules(\Pi, \Pi_0)$, denoted by $Matching(\wp(\Pi))$, defined as follows:

$$\{R \mid R \in Rules(\Pi, \Pi_0), R = Rule(\delta, \mu(\delta)), OBS(\Pi) \asymp \mu(\delta)\}.$$

Theorem 7.3. Let $Usp(\Pi, \Pi_0)$ be a universal space, $\mu(\delta)$ the matching graph for a diagnosis $\delta \in \Delta(\Pi, \Pi_0)$, and $\wp(\Pi) = (\Omega(\Pi), OBS(\Pi), \Pi_0)$ a diagnostic problem. Then,

$$OBS(\Pi) \asymp \mu(\delta) \iff \delta \in \Delta(\wp(\Pi)).$$

Corollary 7.4. The diagnostic set of a diagnostic problem $\wp(\Pi)$ equals the set of diagnoses relevant to the matching rule set of $\wp(\Pi)$, that is,

$$\Delta(\wp(\Pi)) = \{\delta \mid Rule(\delta, \mu(\delta)) \in Matching(\wp(\Pi))\}.$$

7. Rule Exploitation

The formal result of the previous section is that the diagnostic set relevant to an actual diagnostic problem $\wp(\Pi)$ can be determined by matching the relevant observation with each diagnostic rule. If the match succeeds, the corresponding diagnosis will be part of the diagnostic set of $\wp(\Pi)$. In other words, rule based diagnosis allows us to obtain the same result as simulation-based diagnosis.

Example 7.14. In order to verify the equivalence between simulation-based and rule-based diagnosis, we solve a diagnostic problem for the system Υ displayed in Figure 7.1 by means of both techniques. Consider the actual diagnostic problem

$$\bar{\wp}(\Upsilon) = (\Omega(\Upsilon), OBS(\Upsilon), \Upsilon_0),$$
$$\Omega(\Upsilon) = (\mathbb{C}(\Upsilon), \{\{C_1, C_2\}, \{C_3, C_4\}\}),$$
$$OBS(\Upsilon) = (\langle a_1, a_2, b_1\rangle, \langle a_4, a_3\rangle),$$
$$\Upsilon_0 = (S_{11}, S_{21}, S_{31}, S_{41}).$$

The simulation-based diagnostic technique is expected to generate the active space $Act(\bar{\wp}(\Upsilon))$ displayed in Figure 15. The relevant diagnostic set is $\Delta(\bar{\wp}(\Upsilon)) = \{\delta_1, \delta_4\}$, where $\delta_1 = \emptyset$ and $\delta_4 = \{C_1, C_2\}$.

Now, based on the (deterministic) matching graphs generated from the universal space $Usp(\Upsilon, \Upsilon_0)$ and displayed in Figures 11, 12, and 14, we solve the same problem by means of the corresponding matching rules. Note that the actual matching graphs $\bar{\mu}(\delta_i, \bar{\wp}(\Upsilon))$, $i \in [1 .. 4]$, coincide with the matching graphs $\mu(\delta_i)$, respectively, as $Ukn(OBS(\Upsilon)) = \emptyset$. Testing the matching of $OBS(\Upsilon)$ with each matching graph we obtain the results enumerated below.

(1) $OBS(\Upsilon) \asymp \mu(\delta_1)$, with matching string $\langle a_4, a_1, a_2, a_3, b_1\rangle$, therefore $\delta_1 \in \Delta(\bar{\wp}(\Upsilon))$;

(2) $OBS(\Upsilon) \not\asymp \mu(\delta_2)$, therefore $\delta_2 \notin \Delta(\bar{\wp}(\Upsilon))$;

(3) $OBS(\Upsilon) \not\asymp \mu(\delta_3)$, therefore $\delta_3 \notin \Delta(\bar{\wp}(\Upsilon))$

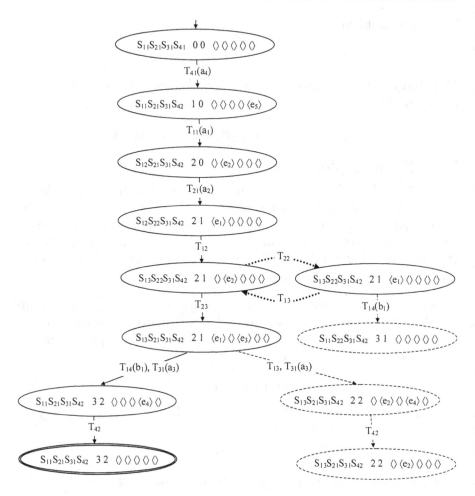

Figure 7.15. Active space $Act(\bar{\wp}(\Upsilon))$ (see Example 7.14).

(4) $OBS(\Upsilon) \asymp \mu(\delta_4)$, with matching string $\langle a_4, a_1, a_2, a_3, b_1 \rangle$, therefore $\delta_4 \in \Delta(\bar{\wp}(\Upsilon))$.

The above results bring us to the conclusion that $\Delta(\bar{\wp}(\Upsilon)) = \{\delta_1, \delta_4\}$, where $\delta_1 = \emptyset$ and $\delta_4 = \{C_1, C_2\}$, which is the same diagnostic set obtained with the simulation-based technique.

It is worthwhile noticing that the two matching strings for δ_1 and δ_4 are in fact the same. This is not incidental insofar as $\mu(\delta_1)$ is a subgraph of $\mu(\delta_4)$, with the exception of the initial state, which is final in $\mu(\delta_1)$, while not so in $\mu(\delta_4)$. This means that whenever a non-empty observation matches $\mu(\delta_1)$, it matches $\mu(\delta_4)$ too. This fact can be easily interpreted by looking at the universal space depicted in Figure 2. As a matter of fact, it is possible to reach the final node N_9 either straightly, by means of the non-faulty transition, or encompassing the cycle involving the faulty transitions T_{22} and T_{13}. In any case, the list of (groups of) messages is the same, owing to the fact that the faulty cycle is not observable. $\diamond$

The following examples apply the rule-based diagnostic technique only, without any generation of the active space. The reader is invited to solve the same problem by means of the simulation-based technique too. This can be easily accomplished based on the universal space outlined in Figure 7.2.

Example 7.15. Consider a variation $\bar{\wp}'(\Upsilon)$ of the actual diagnostic problem given in Example 7.14, where the observer is the same, while the observation is

$$OBS'(\Upsilon) = (\langle a_1, a_2, a_2, b_1 \rangle, \langle a_4, a_3 \rangle).$$

Matching the observation with the graphs yields the following results.

(1) $OBS'(\Upsilon) \not\asymp \mu(\delta_1)$, therefore $\delta_1 \notin \Delta(\bar{\wp}'(\Upsilon))$;

(2) $OBS'(\Upsilon) \not\asymp \mu(\delta_2)$, therefore $\delta_2 \notin \Delta(\bar{\wp}'(\Upsilon))$;

(3) $OBS'(\Upsilon) \asymp \mu(\delta_3)$, with matching string $\langle a_4, a_1, a_2, a_3, a_2, a_3, b_1 \rangle$, therefore $\delta_3 \in \Delta(\bar{\wp}'(\Upsilon))$

(4) $OBS'(\Upsilon) \asymp \mu(\delta_4)$, with matching string $\langle a_4, a_1, a_2, a_3, a_2, a_3, b_1 \rangle$, therefore $\delta_4 \in \Delta(\bar{\wp}'(\Upsilon))$.

Thus, $\Delta(\bar{\wp}'(\Upsilon)) = \{\delta_3, \delta_4\}$, where $\delta_3 = \{C_3\}$ and $\delta_4 = \{C_1, C_2\}$. Even in this case, the two matching strings coincide for a specific reason: the matching graph relevant to δ_3 is a subgraph of $\mu(\delta_4)$, with the same set of final states. Consequently, each matching string in $\mu(\delta_3)$ is also a matching string in $\mu(\delta_4)$, namely

$$\|\mu(\delta_3)\| \subset \|\mu(\delta_4)\|.$$

In other words, given an actual diagnostic problem $\wp(\Upsilon)$, the following entailment holds:

$$\delta_3 \in \Delta(\wp(\Upsilon)) \Rightarrow \delta_4 \in \Delta(\wp(\Upsilon)).$$

The inverse does not hold, as the extension of $\mu(\delta_4)$ incorporates strings beyond the extension of $\mu(\delta_3)$. $\diamond$

Example 7.16. Consider another variation $\bar{\wp}''(\Upsilon)$ of the actual diagnostic problem given in Example 7.14, with the same observer and observation

$$OBS''(\Upsilon) = (\langle a_1, a_2, b_1 \rangle, \langle a_4 \rangle).$$

Matching the observation with the graphs generates the following results.

(1) $OBS''(\Upsilon) \not\asymp \mu(\delta_1)$, therefore $\delta_1 \notin \Delta(\bar{\wp}''(\Upsilon))$;

(2) $OBS''(\Upsilon) \asymp \mu(\delta_2)$, with matching string $\langle a_4, a_1, a_2, b_1 \rangle$, therefore $\delta_2 \in \Delta(\bar{\wp}''(\Upsilon))$;

(3) $OBS''(\Upsilon) \not\asymp \mu(\delta_3)$, therefore $\delta_3 \notin \Delta(\bar{\wp}''(\Upsilon))$

(4) $OBS''(\Upsilon) \asymp \mu(\delta_4)$, with matching string $\langle a_4, a_1, a_2, b_1 \rangle$, therefore $\delta_4 \in \Delta(\bar{\wp}''(\Upsilon))$.

Thus, $\Delta(\bar{\wp}''(\Upsilon)) = \{\delta_2, \delta_4\}$, where $\delta_2 = \{C_2\}$ and $\delta_4 = \{C_1, C_2\}$. As in the previous example, the two matching strings coincide because $\mu(\delta_3)$ is a subgraph of $\mu(\delta_4)$, with the same final state. Consequently,

$$\delta_2 \in \Delta(\wp(\Upsilon)) \Rightarrow \delta_4 \in \Delta(\wp(\Upsilon)).$$

Still, the inverse does not hold, as the extension of $\mu(\delta_4)$ is a strict superset of the extension of $\mu(\delta_3)$. $\diamond$

Example 7.17. A further variation $\bar{\wp}'''(\Upsilon)$ of the actual diagnostic problem given in Example 7.14 involves the same observer and the empty observation

$$OBS'''(\Upsilon) = (\langle \rangle, \langle \rangle).$$

Rule-based diagnosis leads to the following results.

(1) $OBS'''(\Upsilon) \asymp \mu(\delta_1)$, with matching string $\langle \rangle$, therefore $\delta_1 \in \Delta(\bar{\wp}'''(\Upsilon))$;

(2) $OBS'''(\Upsilon) \not\asymp \mu(\delta_2)$, therefore $\delta_2 \notin \Delta(\bar{\wp}'''(\Upsilon))$;

(3) $OBS'''(\Upsilon) \not\asymp \mu(\delta_3)$, therefore $\delta_3 \notin \Delta(\bar{\wp}'''(\Upsilon))$

(4) $OBS'''(\Upsilon) \not\asymp \mu(\delta_4)$, therefore $\delta_4 \notin \Delta(\bar{\wp}'''(\Upsilon))$.

Thus, $\Delta(\bar{\wp}'''(\Upsilon)) = \{\delta_1\} = \{\emptyset\}$. The result is consistent with the universal space $Usp(\Upsilon)$ outlined in Figure 7.2, where the only history entailing a null observation is the empty history. $\diamond$

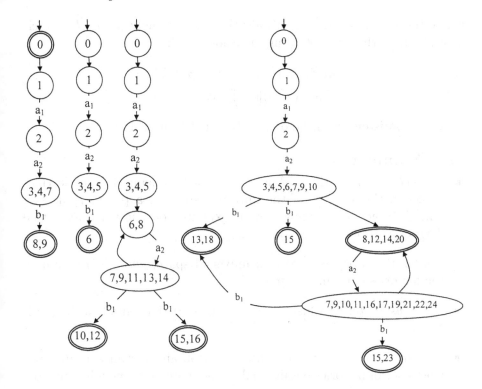

Figure 7.16. Actual matching graphs $\bar{\mu}(\delta_i, \hat{\wp}(\Upsilon))$, $i \in [1..4]$ (from left to right) (see Example 7.18).

Example 7.18. As a final example, we consider an actual diagnostic problem $\hat{\wp}(\Upsilon)$ where the unknown set of the observation is not empty, specifically

$$\hat{\wp}(\Upsilon) = (\Omega(\Upsilon), OBS(\Upsilon), \Upsilon_0),$$
$$\Omega(\Upsilon) = (\{C_1, C_2\}, \{\{C_1, C_4\}, \{C_2, C_3\}\}),$$
$$OBS(\Upsilon) = (\langle a_1, b_1 \rangle, \langle a_2 \rangle),$$
$$\Upsilon_0 = (S_{11}, S_{21}, S_{31}, S_{41}).$$

Note that $Ukn(\Omega(\Upsilon)) = \{C_3, C_4\}$. Therefore, the actual matching graphs $\bar{\mu}(\delta_i, \tilde{\wp}(\Upsilon))$, $i \in [1..4]$, differ from the matching graphs $\mu_i(\delta_i)$, respectively. The relevant (nondeterministic) actual matching graphs are displayed in Figure 7.16. Matching the actual observation with these actual matching graphs leads to the following conclusion:

$$\Delta(\hat{p}(\Upsilon)) = \{\delta_1, \delta_2, \delta_4\} = \{\emptyset, \{C_2\}, \{C_2, C_3\}\},$$

where the matching string is always the same, namely $\langle a_1, a_2, b_1 \rangle$. Incidentally, note that the following inclusions hold:

$$\|\bar{\mu}(\delta_2, \tilde{\wp}(\Upsilon))\| \subset \|\bar{\mu}(\delta_4, \tilde{\wp}(\Upsilon))\|,$$
$$\|\bar{\mu}(\delta_3, \tilde{\wp}(\Upsilon))\| \subset \|\bar{\mu}(\delta_4, \tilde{\wp}(\Upsilon))\|,$$

which is consistent with the diagnostic set determined above. $\diamond$

8.   Summary

- *Rule Generation*: the off-line process of producing diagnostic rules based on the analysis of the universal space of a polymorphic system. The generated knowledge is exploited on-line when an actual diagnostic problem must be solved.

- *Route*: the subspace $\rho(\delta)$ of a universal space, which is relevant to the histories entailing diagnosis δ.

- *Spurious History*: a history in a route $\rho(\delta)$ that entails a diagnosis $\delta' \subset \delta$.

- *Diagnostic Space*: an automaton $\gamma(\delta)$ obtained from a route $\rho(\delta)$, whose set of histories entail δ only. Therefore, $\gamma(\delta)$ includes the non-spurious histories of $\rho(\delta)$ only.

- *Stratification Hierarchy*: an acyclic graph relevant to the diagnostic set Δ of a universal space. Each node of the graph is a diagnosis in Δ. The edges of the graph specify the inclusion relationship among diagnoses in Δ.

- *Matching Graph*: the automaton $\mu(\delta)$ obtained from a diagnostic space $\gamma(\delta)$ by replacing each label, which is a system transition τ, with the set of messages generated by τ. It represents all possible modes in which messages can be generated by the system histories entailing δ.

- *Matching Expression*: the regular expression $Mex(\delta)$ relevant to a matching graph $\mu(\delta)$. The alphabet of $Mex(\delta)$ is a subset of the powerset relevant to the set of messages encompassed by system histories entailing δ.

- *Actual Matching Graph*: the automaton $\bar{\mu}(\delta, \tilde{\wp}(\Pi))$ obtained from $\mu(\delta)$ by removing from the labels of edges the set of messages that are not visible by the observer of a given (actual) diagnostic problem $\tilde{\wp}(\Pi)$.

- *Matching Observation*: an observation $OBS(\Pi)$ relevant to an actual diagnostic problem $\bar{\wp}(\Pi)$, which matches a string in $Mex(\delta)$.

- *Diagnostic Rule*: an association between a diagnosis δ and the relevant matching graph $\mu(\delta)$. Given an actual diagnostic problem $\bar{\wp}(\Pi) = (\Omega, OBS(\Pi), \Pi_0)$, $Rule(\delta, \mu(\delta)$ is successful when $OBS(\Pi)$ matches $\mu(\delta)$, denoted $OBS(\Pi) \asymp \mu(\delta)$.

- *Rule Set*: the whole set of diagnostic rules isomorphic to a diagnostic set of a universal space $Usp(\Pi, \Pi_0)$, namely $Rules(\Pi, \Pi_0)$.

- *Rule Exploitation*: the on-line process of generating the diagnostic set relevant to an actual diagnostic problem based on the rule set generated off-line, rather than performing simulation-based diagnosis.

Appendix A
Algorithms

This appendix provides a pseudo-coded implementation of the matching operator introduced in Section 6. Remember that, given an actual diagnostic problem

$$\wp(\Pi) = (\Omega(\Pi), OBS(\Pi), \Pi_0),$$

the matching

$$OBS(\Pi) \asymp \mu(\delta)$$

is performed on the actual matching graph $\bar{\mu}(\delta, \wp(\Pi))$, which is in general nondeterministic. Therefore, backtracking may be required by the matching algorithm.

Algorithm A.1. (*Match*) The *Match* function implements the matching operator. It takes as input the observation $OBS(\Pi)$ of an actual diagnostic problem $\wp(\Pi)$ and a matching graph $\mu(\delta)$, while returning the boolean value corresponding to the matching $OBS(\Pi) \asymp \mu(\delta)$. To do so, it makes use of the recursive function *Recurse*, whose definition is embedded within *Match*, between the parameter specification and the body of *Match*, the latter starting at Line 27.

At Line 28, the actual matching graph relevant to $\wp(\Pi)$ is determined. Parameter K_0 of the first call to *Recurse* is instantiated at Line 29. *Recurse* is first called at Line 30.

Considering the body of *Recurse* (Lines 1–26), at Line 1 the function returns, provided that N is final in the actual matching graph and K is complete.

If the return condition is not met, a loop on the set of edges leaving N is carried out (Lines 2–24). At Line 3, if the current edge is empty (owing to the nondeterminism of the automaton), *Recurse* is recursively called on the node N' entered by the current edge and, if successful, *true* is returned by the current function also (Line 5), otherwise (Line 7) a new edge is tried (Line 7).

If, by contrast, the current edge is not empty, a copy of $\mathcal{O}$ and K is assigned to $\mathcal{O}'$ and K', respectively (Line 10). Then, a nested loop is carried out (Lines 11–23) on the observation items of $OBS(\Pi)$. Specifically, at Line 12, if the relevant item is not consumed and incorporates a message within $\mathcal{O}'$, then such a message is removed from $\mathcal{O}'$ (Line 13), while the index is incremented (Line 14).

At this point (Line 14), if the set of messages in $\mathcal{O}'$ has been consumed (Line 15), *Recurse* is called on the node reached by the current edge (Line 16) and, if successful, a successful return is made by the current function also (Line 17), otherwise a new edge is tried (Line 7).

When the computation reaches Line 25, it means that all of the calls to *Recurse*, relevant to the set of edges leaving the current node, have failed, therefore the current call to *Recurse* has failed too (Line 24). Clearly, if such failure is concerned with the initial node S_0 (Line 30), the matching fails, otherwise it succeeds.

function *Match*($OBS(\Pi), \mu(\delta)$): **boolean**
    input
        $OBS(\Pi)$: an observation of the polymorphic system Π,
        $\mu(\delta)$: a matching graph;

    function *Recurse*($N, OBS(\Pi), K$): **boolean**
        input
            N: a node of the actual matching graph $\bar{\mu}(\delta, \wp(\Pi))$,
            $OBS(\Pi) = (obs(\mathbf{C}_1), \ldots, obs(\mathbf{C}_n))$: the observation relevant to $\wp(\Pi)$,
            K: the index of $OBS(\Pi)$;
        begin {Recurse}

```
1.      if N ∈ 𝕊_f(μ̄(δ, ℘(Π))) and K is complete then return true;
2.      for each edge N ⎯O→ N' in μ̄(δ, ℘(Π)) do
3.         if O = ∅ then
4.            if Recurse(N', OBS(Π), K) = true then
5.               return true
6.            else
7.               goto 2
8.            end-if
9.         end-if;
10.        O' := O; K' := K;
11.        for i := 1 to n do
12.           if K'[i] < |obs(C_i)|, obs(C_i)[K'[i]] = m, (m, C) ∈ O', C ∈ C_i then
13.              O' := O' − {(C, m)};
14.              K'[i] := K'[i] + 1;
15.              if O' = ∅ then
16.                 if Recurse(N', OBS(Π), K') = true then
17.                    return true
18.                 else
19.                    goto 2
20.                 end-if
21.              end-if
22.           end-if
23.        end-for;
24.     end-for;
25.     return false
26.  end {Recurse};

27. begin {Match}
28.    μ̄(δ, ℘(Π)) = (𝕊, 𝔼, 𝕋̄, S_0, 𝕊_f) := the actual matching graph of δ;
29.    K_0 := (0, 0, ..., 0);
30.    return Recurse(S_0, OBS(Π), K_0)
31. end {Match}.
```

Appendix B
Proofs of Theorems

Theorem 7.1 *Let $Usp(\Pi, \Pi_0)$ be a universal space relevant to a blind diagnostic problem $\wp(\Pi)$, δ a diagnosis in $\Delta(\wp(\Pi))$,*

$$\mathcal{E}(\delta) = \{e \mid e \in< Dex(\wp(\Pi)) >, e \models \delta\}$$

the subset of the linearization of the corresponding diagnostic expression entailing δ, and $\rho(\delta)$ a route in $Usp(\Pi, \Pi_0)$. Then

$$Subsp(Usp(\Pi, \Pi_0), \mathcal{E}(\delta)) \equiv \rho(\delta).$$

The proof of the theorem makes use of the following lemma (the proof of which is omitted):

Lemma 7.1.1 *Let $H_\delta = \{h \mid h \in \mathbb{H}(\wp(\Pi)), h \models \delta\}$. Then,*

$$\exists h \in H_\delta \; (e \in< Dex(\wp(\Pi)) >, e \models \delta, e = \langle \tau \mid \tau \in h, Faulty(\tau) \rangle, e \subseteq h).$$

Proof of Theorem 7.1.   Let

$$(\mathbb{N}, \mathbb{E}) = Subsp(Usp(\Pi, \Pi_0), \mathcal{E}(\delta)),$$
$$(\mathbb{N}', \mathbb{E}') = \rho(\delta),$$

where $\mathbb{N}$, $\mathbb{N}'$, and $\mathbb{E}$, $\mathbb{E}'$ are the set of nodes and the set of edges of the corresponding graphs. We have to show that $\mathbb{N} = \mathbb{N}'$ and $\mathbb{E} = \mathbb{E}'$, that is,

(1) If $N \in \mathbb{N}$ then $N \in \mathbb{N}'$;

(2) If $N \in \mathbb{N}'$ then $N \in \mathbb{N}$;

(3) If $E \in \mathbb{E}$ then $E \in \mathbb{E}'$;

(4) If $E \in \mathbb{E}'$ then $E \in \mathbb{E}$.

Assuming $N \in \mathbb{N}$ implies that N is a node relevant to a history $h \in \mathbb{H}(\wp(\Pi))$ such that $\langle \tau \mid \tau \in h, Faulty(\tau) \rangle = e$, $e \subseteq h$, $e \in \mathcal{E}(\delta)$, $e \models \delta$. Therefore, $h \models \delta$, that is,

231

$h \in \mathbb{H}(\rho(\delta))$, hence, $N \in \mathbb{N}'$. Following the same scheme we can prove the third rule, namely, if $E \in \mathbb{E}$ then $E \in \mathbb{E}'$.

Assume now $N \in \mathbb{N}'$. This means that N is a node relevant to a set H of histories in $\mathbb{H}(\rho(\delta))$ such that $\forall h \in H$ $(h \models \delta)$. Based on Lemma 7.1.1, there exists $h \in H$ such that $\langle \tau \mid \tau \in h, Faulty(\tau) \rangle = e$, $e \subseteq h$, $e \in \mathcal{E}(\delta)$. Therefore, $h \in \mathbb{H}(Subsp(Usp(\Pi, \Pi_0)))$, hence, $N \in \mathbb{N}$. Following the same scheme we can prove the fourth rule, namely, if $E \in \mathbb{E}'$ then $E \in \mathbb{E}$, which brings use to the conclusion of the proof of Theorem 7.1. $\qquad \square$

Theorem 7.2 *Let* $\nabla(\Pi, \Pi_0) = (\mathbb{N}, \mathbb{E}, \mathbb{N}_0)$ *be a stratification hierarchy,* $\delta \in \mathbb{N}$ *a diagnosis, and* $\rho(\delta)$ *the relevant route in* $Usp(\Pi, \Pi_0)$. *Then*

$$\mathbb{H}\left(\rho(\delta) \boxminus \left(\bigsqcup_{\delta' \in Desc(\delta)} \rho(\delta')\right)\right) \equiv \mathbb{H}(\rho(\delta)) - Spur(\rho(\delta)).$$

Proof. According to the definition of stratification hierarchy, the union

$$A = \bigcup_{\delta' \in Desc(\delta)} \rho(\delta')$$

represents an automaton A whose paths from initial to final states are spurious histories in $\rho(\delta)$. Based on the definition of descendant nodes of δ in the stratification hierarchy, all of the spurious histories relevant to $\rho(\delta)$, namely $Spur(\rho(\delta))$, are included in A. Therefore, the difference

$$\rho(\delta) \boxminus A$$

represents an automaton whose set of histories corresponds to all and only the set of histories entailing δ, in other terms,

$$\mathbb{H}(\rho(\delta) \boxminus A) \equiv \mathbb{H}(\rho(\delta)) - Spur(\rho(\delta)).$$

$\qquad \square$

Theorem 7.3 *Let* $Usp(\Pi, \Pi_0)$ *be a universal space,* $\mu(\delta)$ *the matching graph for a diagnosis* $\delta \in \Delta(\Pi, \Pi_0)$, *and* $\wp(\Pi) = (\Omega(\Pi), OBS(\Pi), \Pi_0)$ *a diagnostic problem. Then,*

$$OBS(\Pi) \asymp \mu(\delta) \iff \delta \in \Delta(\wp(\Pi)).$$

We limit the proof to the case in which $Ukn(OBS(\Pi)) = \emptyset$, so that $\mu(\delta) = \bar{\mu}(\delta, \wp(\Pi))$. The proof of the theorem is based on the following lemmas.

Lemma 7.3.1 $OBS(\Pi) \asymp \mu(\delta) \Rightarrow \delta \in \Delta(\wp(\Pi))$.

Proof. According to the definition of match, $OBS(\Pi) \asymp \mu(\delta)$ means that $\exists \mathbf{O} \in \|\mu(\delta)\|$ such that $\|\mathbf{O}\| \cap \|OBS(\Pi)\| \neq \emptyset$. Based on the mode in which $\mu(\delta)$ is derived from $\gamma(\delta)$, we can say that $\exists h \in \|\gamma(\delta)\|$ such that $\|Sign(h)\| \cap \|OBS(\xi)\| \neq \emptyset$. By virtue of Proposition 7.1 and Corollary 7.2 we can assert that $\exists h \in \mathbb{H}(\wp(\Pi))$, such that $h \models \delta$, therefore, $\delta \in \Delta(\wp(\Pi))$, which brings us to the conclusion of the proof of Lemma 7.3.1.

Lemma 7.3.2 $\delta \in \Delta(\wp(\Pi)) \Rightarrow OBS(\Pi) \asymp \mu(\delta)$.

Proof. By definition of diagnosis, if $\delta \in \Delta(\wp(\Pi))$, then $\exists h \in \mathbb{H}(\wp(\Pi))$ such that $h \models \delta$. By virtue of Proposition 7.1 and the definition of system signature, we may assert that $\exists h \in \mathbb{H}(\gamma(\delta))$ such that $h \implies \delta$, in other words, such that $\|Sign(h)\| \cap \|OBS(\Pi)\| \neq \emptyset$. According to the mode in which $\mu(\delta)$ is derived from $\gamma(\delta)$, we can say that $\exists \mathbf{O} \in \|\mu(\delta)\|$, $\mathbf{O} = Sign(h)$, such that $\|\mathbf{O}\| \cap \|OBS(\Pi)\| \neq \emptyset$, therefore, $OBS(\Pi) \asymp \mu(\delta)$, which brings us to the conclusion of the proof of Lemma 7.3.2.

Proof of Theorem 7.3. The proof follows directly Lemmas 7.3.1 and 7.3.2. $\square$

Corollary 7.4 *The diagnostic set of a diagnostic problem $\wp(\Pi)$ equals the set of diagnoses relevant to the matching rule set of $\wp(\Pi)$, that is,*

$$\Delta(\wp(\Pi)) = \{\delta \mid Rule(\delta, \mu(\delta)) \in Matching(\wp(\Pi))\}.$$

Proof. On the one hand we prove that if $\delta \in \Delta(\wp(\Pi))$ then $Rule(\delta, \mu(\delta)) \in Matching(\wp(\Pi))$. According to Lemma 7.3.2, if $\delta \in \Delta(\wp(\Pi))$ then $OBS(\xi) \asymp \mu(\delta)$, that is, by definition of diagnostic rule, $Rule(\delta, \mu(\delta)) \in Matching(\wp(\Pi))$.

On the other, assuming $Rule(\delta, \mu(\delta)) \in Matching(\wp(\Pi))$, we have, according to the definition of diagnostic rule, $OBS(\Pi) \asymp \mu(\delta)$, that is, by virtue of Lemma 7.3.1, $\delta \in \Delta(\wp(\Pi))$, which concludes the proof of Corollary 7.4. $\square$

Chapter 8

MONITORING-BASED DIAGNOSIS

Abstract The diagnostic techniques for diagnosis of polymorphic systems presented in the previous chapters share a common assumption: the diagnostic process starts once the system has become quiescent after a reaction. Since all such techniques are concerned with a posteriori diagnosis, they do not support the monitoring of the system, specifically, the continuous diagnosis of the reacting system. Monitoring-based diagnosis requires the generation of diagnostic information at the occurrence of each system message. Consequently, candidate diagnoses are to be produced while reconstructing the system behavior, rather than after the complete generation of the relevant active space. Continuous diagnosis can be carried out in two different ways, either based on off-line preprocessing or not. The former requires the generation of a graph that is even wider than the universal space, thereby making the approach prohibitive in real contexts. The latter, instead, combines behavior reconstruction and diagnosis generation without any off-line preprocessing.

1. Introduction

A basic assumption of the diagnostic techniques introduced in Chapters 4–7 (including rule-based diagnosis) is that the diagnosis task starts after the complete reaction of the system, based on the system observation specified in the diagnostic problem. Even though this assumption complies with a variety of application domains[1], it becomes unacceptable whenever the system requires supervision (monitoring) capabilities.

For example, if the system is a chemical plant, the occurrence of an observable event may be the signal of abnormal behavior, which is bound to deteriorate and become harmful to the plant and/or the environment.

---

[1]Chapter 13 presents the application of such diagnostic techniques in the domain of power transmission networks.

Therefore, such a message should be processed by the diagnostic task in real-time or at least before possible negative consequences.

Besides, a wide range of real systems are permanently in a reactive mode and never quiescent, so that it does not make sense to wait for the completion of the reaction before starting diagnosis.

These considerations offer evidence for the claim that a posteriori diagnosis is inappropriate either when supervisioning a critical system or when the system is permanently in the reacting mode: in either case, diagnosis must be continuous.

Both notions of a diagnostic problem and a diagnosis are to be changed appropriately and, consequently, the diagnostic technique too. In a posteriori diagnosis, the diagnostic problem is a triple involving the observer, the observation, and the initial state of the system. In continuous diagnosis things are different. The occurrence of each newly generated system message is associated with a new diagnosis.

Specifically, two kinds of diagnoses can be envisaged: snapshot diagnosis and historic diagnosis. Snapshot diagnosis encompasses the faults that comply with the new message, disregarding the previous behavior of the system. Historic diagnosis, instead, embraces the whole set of faults that comply with the behavior of the system since its initial state[2].

A continuous, or monitoring-based, diagnostic problem can therefore be defined as a triple involving the observer, the sequence of system messages[3] generated by the system during operation, and the initial state of the system. The diagnostic output is a sequence of diagnostic pairs, each pair involving both the snapshot and historic diagnoses relevant to each newly generated system message. In practice, monitoring-based diagnosis is required to continuously generate diagnostic pairs. The implicit assumption is that the diagnostic pair relevant to the occurrence of a system message be computed (virtually) before the occurrence of the successive system message.

Several alternative techniques can be envisaged in order to solve a monitoring-based diagnostic problem, among which are the following.

(1) A graph similar to the active space, called abductive space, is incrementally made up at each newly generated system message. Each node of this graph is extended with an additional field that is a (either shallow or deep) diagnosis. For instance, in case of shallow diagnosis, the field contains the set of components that have been faulty for each path from the root to that node. Intuitively, when generating a new

---

[2]Historic diagnosis is very similar to a posteriori diagnosis.
[3]A system message is the set of observable events generated by a transition of the polymorphic system, that is, a group of component messages.

node of the graph, if the system transition is abnormal, the relevant faulty components are added to the current value of the diagnosis field. Both snapshot and historic diagnoses can be yielded based on the diagnosis field relevant to the nodes of the graph corresponding to the newly generated system message.

(2) A graph isomorphic to the active space, called diagnostic space, is incrementally made up at each newly generated system message. Unlike the abductive space, each node of the diagnostic space is extended with a new field that represents the set of diagnoses (rather than a single diagnosis) that are implied by each path from the root to that node. In this case, the number of nodes (and edges) equals those of the active space[4]. Owing to possible cycles in the graph, the generation of the diagnoses field is not as straightforward as in the abductive space and requires some propagation techniques in order to be consistent for each newly generated node. As in the abductive space, the computation of snapshot and historic diagnostic sets is based on the diagnoses field.

(3) A universal abductive space might be generated off-line. Nodes of the resulting graph associate a node of the universal space with a diagnosis. Such a graph can be transformed into a deterministic automaton, called monitoring space, where each edge is marked by a system messages and each node is associated with a diagnostic set, specifically, the set of candidate diagnoses relevant to the sequence of system messages corresponding to each path from the root to that node. A diagnostic pair is straightforwardly generated based on the diagnostic set of the node reached by the edge marked by the current system message.

(4) At each newly generated system message, the diagnostic subspace (see point (2)) reachable by means of such a message is made up dynamically, while monitoring the system. The new diagnostic pair is generated based on the diagnoses fields of such subspace and the old diagnostic pair. The set of generated nodes (diagnostic subspaces) and edges (marked by system messages) makes up a so-called monitoring graph.

In the remainder of the chapter, Sections 2, 3, 4, and 5 introduce the notions of abductive space, diagnostic space, monitoring space, and monitoring graph, respectively. Section 6 copes with continuous diagnosis.

---

[4]The abductive space is in general wider than the active space, as different diagnosis fields may be associated with the same active space node.

A summary is provided in Section 7. The appendix details the pseudo-coded algorithms relevant to the generation of the diagnostic space and the continuous diagnosis based on the dynamic construction of the monitoring graph.

2.   Abductive Space

A major limitation of the diagnostic approach presented in previous chapters consists of the fact that the reconstruction of the system behavior is disjoint from the distillation of the diagnostic set. Besides, the latter requires either the generation of the diagnostic graph or the multiple traversing of the active space in order to find out all significant paths that connect the root with a final state, which is quite inefficient in nature. To overcome this shortcoming, we introduce a technique which integrates the reconstruction of the system behavior with the generation of the candidate diagnoses. This can be achieved by extending the content of a node in the active space with a new field that represents a candidate diagnosis relevant to the current state. This way, we can generate a candidate diagnosis at the completion of each node.

Example 8.1. To this end, we consider a variation of the polymorphic system Υ introduced in Chapter 7 and displayed in Figure 7.1. The new system, called Λ, is outlined in Figure 8.1. The only difference of Λ with respect to Υ is the qualification of transition T_{42}, which is faulty in Λ, while the rest is kept unchanged.

The corresponding universal space $Usp(\Lambda, \Lambda_0)$ is depicted in Figure 8.2. This is topologically isomorphic to the active space $Usp(\Upsilon, \Upsilon_0)$ of Figure 7.2, with the exception of the qualification of the faulty edges marked by transition T_{42}.

Then, we consider the diagnostic problem $\wp(\Lambda)$ defined as follows:

$$\wp(\Lambda) = (\Omega(\Lambda), OBS(\Lambda), \Lambda_0),$$
$$\Omega(\Lambda) = (\mathbb{C}(\Lambda), \{\{C_1, C_2\}, \{C_3, C_4\}\}),$$
$$OBS(\Lambda) = (\langle a_1, a_2, b_1 \rangle, \langle a_4, a_3 \rangle),$$
$$\Lambda_0 = (S_{11}, S_{21}, S_{31}, S_{41}).$$

The corresponding active space is displayed in Figure 8.3. Each node in $Act(\wp(\Lambda))$ is identified by a pair (N, K), where N is the identifier of a node in the universal space $Usp(\Lambda, \Lambda_0)$, while $K = (k_1, k_2)$ is the index of the relevant observation $OBS(\Lambda)$. As a matter of fact, a state of the active space is the association of a system state and the prefix of the observation generated so far, namely, an observation index. $\diamond$

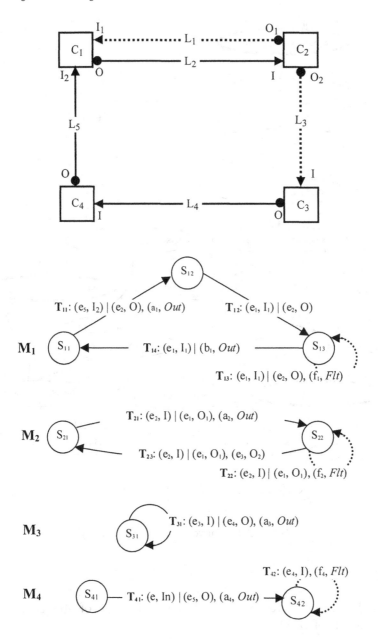

Figure 8.1. Polymorphic system Λ.

We now introduce the notion of an abductive space. The *abductive space* of a diagnostic problem $\wp(\Lambda)$ is an automaton

$$Abd(\wp(\Lambda)) = (\mathbb{S}, \mathbb{E}, \mathbb{T}, S_0, \mathbb{S}_f)$$

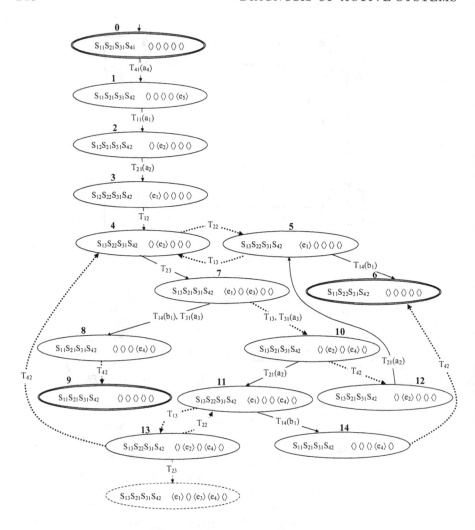

Figure 8.2.   Universal space $Usp(\Lambda, \Lambda_0)$.

where

$\mathbb{S}$ is the set of states $S = (N, \delta)$, N is a state of $Act(\wp(\Lambda))$, δ is a diagnosis of $\wp(\Lambda)$;

$\mathbb{E}$ is the same set of events of $Act(\wp(\Lambda))$;

$\mathbb{S}_f$ is the set of final states defined as follows:

$$\mathbb{S}_f = \{S \mid S \in \mathbb{S}, S = (N, \delta), N \text{ is a final state of } Act(\wp(\Lambda))\};$$

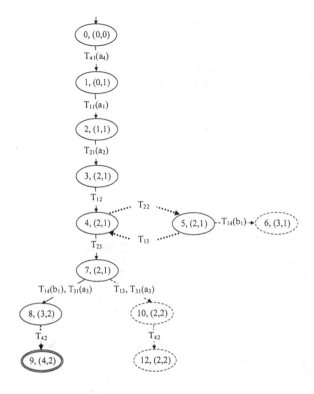

Figure 8.3. Active space $Act(\wp(\Lambda))$.

$\mathbb{T}$ is the transition function, $\mathbb{T} : \mathbb{S} \times \mathbb{E} \mapsto \mathbb{S}$ such that

$$(N, \delta) \xrightarrow{\tau} (N', \delta') \in \mathbb{T}$$

if and only if the following conditions hold:

(1) $N \xrightarrow{\tau} N'$ is a transition of $Act(\wp(\Lambda))$;

(2) $\delta' = \delta \cup \{C \mid C \text{ is a faulty component in } \tau\}$.

Example 8.2. Shown in Figure 8.4 is the abductive space for the diagnostic problem given in Example 8.1. Each node is identified by three fields, namely (N, K, δ), where (N, K) is a node of the active space displayed in Figure 8.3, while δ is a diagnosis of $\wp(\Lambda)$.

A comparison with the active space of Figure 8.3 shows that $Abd(\wp(\Lambda))$ embodies more states than $Act(\wp(\Lambda))$. Specifically, the same node (N, K) of the active space can be duplicated in the abductive space, with different values for δ.

For instance, node $(4, (2, 1))$ in $Act(\Lambda)$ is duplicated in $Abd(\Lambda)$ as $(4, (2, 1), \emptyset)$ and $(4, (2, 1), \{C_1, C_2\})$, where the former is qualified by the

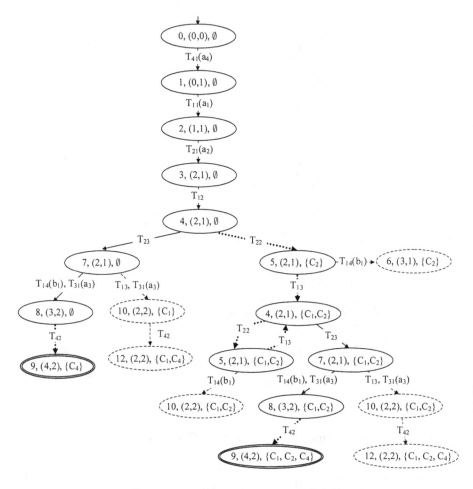

Figure 8.4. Abductive space $Abd(\wp(\Lambda))$.

empty diagnosis, while the latter is characterized by faulty components $\{C_1, C_2\}$.

The set of candidate diagnoses for the given diagnostic problem corresponds to the δ field associated with the final states, specifically, $(9, (4, 2), \{C_4\})$ and $(9, (4, 2), \{C_1, C_2, C_4\})$, which, consistently, yields the same diagnostic set relevant to the active space in Figure 8.3, namely $\Delta(\wp(\Lambda)) = \{\{C_4\}, \{C_1, C_2, C_4\}\}$. $\Diamond$

Generally speaking, the abductive space relevant to a diagnostic problem may be considerably larger than the active space relevant to the same diagnostic problem. If $\mathcal{F}$ is the number of possible faulty components in the system, and $\mathcal{N}$ the number of states in the active space, the number

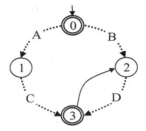

Figure 8.5. Abstract diagnostic space (see Example 8.3).

$\mathcal{N}'$ of nodes in the abductive space are, in the worst case,

$$\mathcal{N}' = \mathcal{N} * 2^{\mathcal{F}},$$

that is, the size of the abductive space grows by a factor which is exponential in the cardinality of the domain of possible faulty components. In order not to enlarge the size of the active space while retaining the advantages of the abductive approach, we introduce a technique that allows us to dynamically 'decorate' the nodes of the active space with a *diagnostic attribute*, that is, the set of candidate diagnoses relevant to the current state.

3.   Diagnostic Space

The *diagnostic space* of a diagnostic problem $\wp(\Lambda)$ is an automaton

$$Dgn(p(\Lambda)) = (\mathbb{S}, \mathbb{E}, \mathbb{T}, S_0, \mathbb{S}_f)$$

isomorphic to the active space $Act(\wp(\Lambda))$, where each node N of the active space is marked (extended) with a diagnostic attribute

$$\Delta = \{\delta \mid \exists \text{ a path } S_0 \rightsquigarrow N \text{ in } Act(\wp(\Lambda)) \text{ that implies } \delta\}.$$

Compared with the abductive space, the generation of the diagnostic space is complicated by the fact that the decoration (instantiation) of a node with the relevant diagnostic attribute is incremental in nature. Such incrementality stems from the existence of cycles, that makes it possible to connect the root with a node in several different ways.

Example 8.3. Consider the abstract active space displayed in Figure 8.5, where edges are marked by faulty components A, B, C, and D, while details on system transitions are omitted. To obtain a diagnostic space we need to instantiate for each node N the relevant diagnostic attribute $\Delta(N)$. However, such a decoration is supposed to be done when building the nodes of the diagnostic space, rather than after the complete generation of it. Thus, we start from the initial node 0, which is decorated with the empty diagnosis, that is,

$$\Delta(0) = \{\emptyset\}.$$

Then we create node 1 , for which the diagnostic attribute becomes

$$\Delta(1) = \{\{A\}\}.$$

Successively, the creation of node 3 brings us to the decoration

$$\Delta(3) = \{\{A, C\}\}.$$

The next node to be created is 2, which (owing to the empty edge from 3) inherits the diagnostic attribute of 3, that is,

$$\Delta(2) = \Delta(3) = \{\{A, C\}\}.$$

Up to now, the creation of the nodes of the diagnostic space mirrors that of the abductive space. However, at this point, the edge from 2 to 3 yields a new candidate diagnosis for node 3, namely $\{\{A, C, D\}\}$, which is to be added to the current set of diagnoses. Thus, the updated diagnostic set for node 3 becomes

$$\Delta(3) = \{\{A, C\}, \{A, C, D\}\}.$$

Since this node was already created, we need to propagate the newly added diagnosis to the nodes currently reachable from 3, namely node 2. Consequently, the new diagnostic attribute for node 2 becomes

$$\Delta(2) = \{\{A, C\}, \{A, C, D\}\}.$$

Such a propagation is recursive in nature: since node 2 was itself previously created, we need to propagate the new added diagnosis to the nodes reachable from it, namely node 3. However, such a propagation does not produce any change in the current diagnostic attribute of node 3 (diagnosis $\{A, C, D\}$ is already included), so that the propagation terminates. At this point, the edge from 0 to 2 is created, which causes $\Delta(2)$ to become

$$\Delta(2) = \{\{A, C\}, \{A, C, D\}, \{B\}\}.$$

Since node 2 was already created, a further propagation is required, which leads to the new diagnostic attribute values

$$\Delta(3) = \{\{A, C\}, \{A, C, D\}, \{B, D\}\}$$

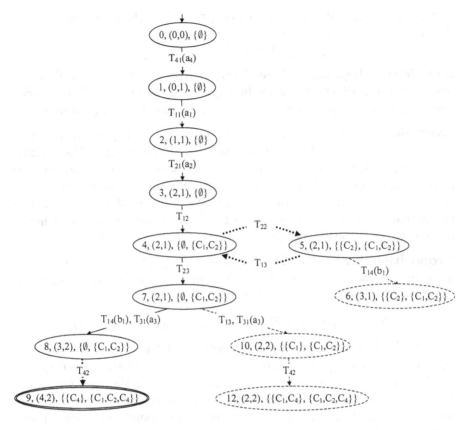

Figure 8.6. Diagnostic space $Dgn(\wp(\Lambda))$.

and

$$\Delta(2) = \{\{A,C\},\{A,C,D\},\{B\},\{B,D\}\}.$$

The final set of candidate diagnoses will be the union of the diagnostic attributes associated with the final nodes (0 and 3):

$$\Delta(\wp(\Lambda)) = \{\emptyset,\{A,C\},\{A,C,D\},\{B,D\}\}.$$

$\diamond$

Essentially, the generation of the diagnostic space for the diagnostic problem $\wp(\Lambda)$ mirrors the generation of the abductive space for $\wp(\Lambda)$, where the replication of nodes of the active space are replaced by the propagation mechanism, thereby allowing us to maintain the exact topology (nodes and edges) of the active space with each node marked by the additional diagnostic attribute.

A pseudo-coded algorithm for the generation of the spurious diagnostic space

$$\tilde{D}gn(\wp(\Lambda)) = (\tilde{\mathbb{N}}, \mathbb{E}, \tilde{\mathbb{T}}, N_0, N_f)$$

is provided in Appendix A. The diagnostic space $Dgn(\wp(\Lambda))$ is obtained from a spurious diagnostic space by eliminating from $\tilde{\mathbb{N}}$ and $\tilde{\mathbb{T}}$ the nodes and transitions, respectively, that are not connected to a final node.

Example 8.4. Shown in Figure 8.6 is the spurious diagnostic space for the diagnostic problem $\wp(\Lambda)$ given in Example 8.1. The dashed part of the graph denotes the spurious part of the diagnostic space, the latter being depicted in plain lines. Once eliminated the spurious part of the graph, the topology of the diagnostic space is identical to that of the active space depicted in Figure 8.3 (not including the extra nodes incorporated in the abductive space shown in Figure 8.4).   ◇

Proposition 8.1. *Let $\wp(\Lambda)$ be a diagnostic problem, $\mathbb{S}_{abd}$ the set of states of the abductive space $Abd(\wp(\Lambda))$, and $\mathbb{S}_{dgn}$ the set of states of the diagnostic space $Dgn(\wp(\Lambda))$. Then,*

$$\forall S \in \mathbb{S}_{dgn}, S = (\lambda, K, \Delta) \ (\Delta = \{\delta \mid (\lambda, K, \delta) \in \mathbb{S}_{abd}\}).$$

Essentially, the above proposition asserts that the diagnostic attribute Δ associated with a node (λ, K, Δ) of the diagnostic space corresponds to the union of all the diagnoses δ associated with the set of nodes of the abductive space sharing the same active-space node (λ, K). In other words, each node of the diagnostic space corresponds to several nodes of the abductive space, specifically, those nodes which share the same system state and observation index, namely λ and K.

4.   Monitoring Space

The final goal of monitoring-based diagnosis is to define a diagnostic technique which is capable of monitoring a polymorphic system. In contrast with the one-shot diagnostic methods based on the abductive space and diagnostic space, continuous diagnosis requires that the diagnostic information be produced at the occurrence of each new system message. To this end, we introduce the notion of a monitoring space, which is essentially an automaton where each node is associated with a diagnostic set and transitions are marked by system messages. Thus, starting from the initial state of the system, the arrival of a new system message causes a transition from the current state to a new one, that is associated with a different set of candidate diagnoses.

First we have to define the notion of a universal abductive space, which is the abductive counterpart of the universal space defined in Section 2.2 of Chapter 4.

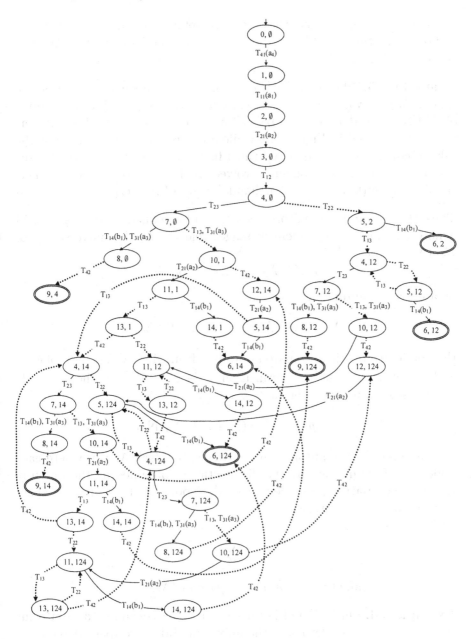

Figure 8.7. Universal abductive space $Abd(\Lambda, \Lambda_0)$.

A *universal abductive space* $Abd(\Lambda, \Lambda_0)$ is the abductive space relevant to the diagnostic problem where no observation is given, in other words, where the system transitions are not constrained by any observation (the observer is blind). As such, the diagnostic set associated with $Abd(\Lambda, \Lambda_0)$ embodies all the possible diagnoses that are entailed by the universal space $Usp(\Lambda, \Lambda_0)$.

Example 8.5. Shown in Figure 8.7 is the universal abductive space $Abd(\Lambda, \Lambda_0)$ relevant to the system Λ depicted in Figure 8.1. Each node of $Abd(\Lambda, \Lambda_0)$ is denoted by a pair (λ, δ), where λ is a node of the universal space outlined in Figure 8.2, while δ is a relevant diagnosis. For the sake of conciseness, we denoted the δ field with the string of digits corresponding to the components involved in the diagnosis. For instance, the final node $(6, 124)$ is a shorthand for $(6, \{C_1, C_2, C_4\})$.   ◇

Thus, the notion of a monitoring space can be defined as follows. Let $\mathbf{M}$ denote the domain of system messages relevant to the universal space $Usp(\Lambda, \Lambda_0)$. Let

$$Abd(\Lambda, \Lambda_0) = (\mathbb{S}, \mathbb{E}, \mathbb{T}, S_0, \mathbb{S}_f)$$

be a universal abductive space, and

$$Msp^n(\Lambda, \Lambda_0) = (\mathbb{S}^n, \mathbf{M}, \mathbb{T}^n, S_0^n, \mathbb{S}_f^n),$$

the nondeterministic automaton obtained from $Abd(\Lambda, \Lambda_0)$ by replacing each event τ relevant to a transition $S \xrightarrow{\tau} S' \in \mathbb{T}$ with the (possibly empty) system message relevant to τ. Let

$$Msp(\Lambda, \Lambda_0) = (\mathbf{N}, \mathbf{M}, \mathbf{T}, N_0, \mathbf{N}_f)$$

be the deterministic automaton equivalent to $Msp^n(\Lambda, \Lambda_0)$, where each node $N \in \mathbf{N}$ is marked by the diagnostic set obtained as the union of the diagnostic attributes relevant to the nodes of the abductive space incorporated[5] in N, namely

$$\Delta(N) = \bigcup_{S \in N} \Delta(S).$$

$Msp(\Lambda, \Lambda_0)$ is called the *monitoring space* of (Λ, Λ_0).

Example 8.6. Outlined in Figure 8.8 is the nondeterministic monitoring space $Msp^n(\Lambda, \Lambda_0)$ relevant to the universal abductive space depicted in Figure 8.7. As expected, it is isomorphic to $Abd(\Lambda, \Lambda_0)$, with the edges possibly marked by system transitions only.

---

[5]This is in accordance with the *subset construction algorithm* [Aho et al., 1986], which generates the equivalent deterministic automaton, where each state is identified by a subset of the states in the nondeterministic automaton.

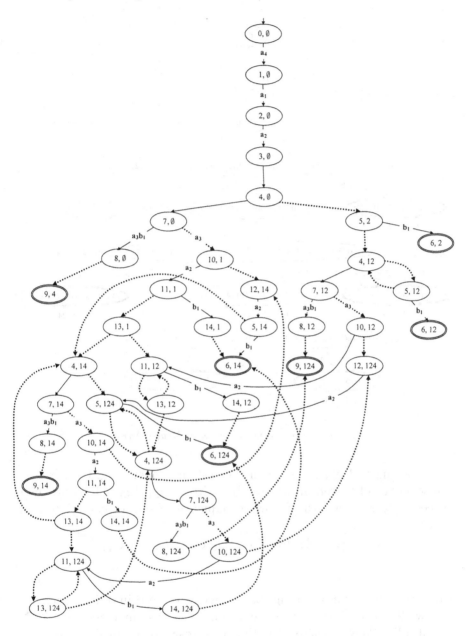

Figure 8.8. Nondeterministic monitoring space $Msp^n(\Lambda, \Lambda_0)$.

The equivalent deterministic automaton, namely the monitoring space $Msp(\Lambda, \Lambda_0)$ is shown in Figure 8.9. Each node of the latter is identified by a set of nodes of $Msp^n(\Lambda, \Lambda_0)$ and is marked by the union of the diagnoses associated with such nodes.

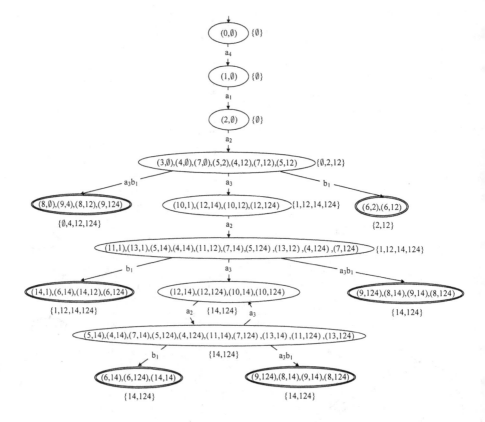

Figure 8.9. Monitoring space $Msp(\Lambda, \Lambda_0)$.

According to Figure 8.9, when the system is operating, the arrival of the sequence of system messages $\langle \{a_4\}, \{a_1\}, \{a_2\}, \{a_3, b_1\} \rangle$ allows us to successively generate the following sequence of sets of candidate diagnoses:

$$\langle \{\emptyset\}, \{\emptyset\}, \{\emptyset\}, \{\emptyset, \{C_4\}\}, \{C_1, C_2\}, \{C_1, C_2, C_4\}\} \rangle.$$

◇

Conceptually, the monitoring space is what we need in order to generate a new diagnostic set at the occurrence of a system message $\mu \in \mathbf{M}$, specifically, $\Delta(N)$, where N is the state reached by means of the edge marked by μ.

Interestingly enough, $Msp(\Lambda, \Lambda_0)$ can be made up dynamically without generating the universal abductive space (and therefore, without the generation of the deterministic equivalent automaton either).

However, even generated 'on demand', that is, at each newly produced system message, the monitoring space may involve a huge number of nodes. This might be a problem if we are required to maintain the image of the monitoring space visited during the monitoring of Λ. A more efficient monitoring technique involves the notion of monitoring graph, as introduced in the next section.

5. Monitoring Graph

Let Λ be a polymorphic system with initial state Λ_0. A *monitoring graph* of (Λ, Λ_0) is an automaton

$$Mgr(\Lambda, \Lambda_0) = (\mathbb{N}, \mathbb{E}, \mathbb{T}, N_0)$$

where $\mathbb{N}$ is the set of nodes. Each node $N \in \mathbb{N}$ is in its turn an automaton

$$N = (\mathbf{S}, \mathbf{E}, \mathbf{T}, S_0, \mathbf{S}_{\text{out}})$$

called the *silent closure* of S_0, namely $Sclosure(S_0)$, defined as follows.

Each state $S \in \mathbf{S}$ is a pair (λ, Δ) where λ is a state of $Usp(\Lambda, \Lambda_0)$ and Δ a set of diagnoses. $\mathbf{E}$ is the set of transitions of Λ. $\mathbf{T} : \mathbf{S} \times \mathbf{E} \mapsto \mathbf{S}$ is the transition function such that

$$(\lambda, \Delta) \xrightarrow{\tau} (\lambda', \Delta') \in \mathbf{T}$$

if and only if τ is a silent transition of Λ and $\lambda \xrightarrow{\tau} \lambda'$ is a transition in $Usp(\Lambda, \Lambda_0)$. $S_0 = (\lambda_0, \Delta_0)$ is the *root*. $\mathbf{S}_{\text{out}}$ is the *leaving set*, defined as follows:

$$\mathbf{S}_{\text{out}} = \{S = (\lambda, \Delta), S \in \mathbf{S}, \lambda \xrightarrow{\tau} \lambda' \text{ is an observable trans. in } Usp(\Lambda, \Lambda_0)\}.$$

Furthermore, the following condition holds:

$$\forall (\lambda, \Delta) \in \mathbf{S} \; (\Delta = \{\delta \mid \lambda_0 \rightsquigarrow \lambda \text{ is a path in } Usp(\Lambda, \Lambda_0) \text{ that implies } \delta\}).$$

Let $\mathbf{M}$ be the domain of system messages in $Usp(\Lambda, \Lambda_0)$ and $\mathbb{F}$ the domain of faulty components in $Usp(\Lambda, \Lambda_0)$. $\mathbb{E} \subseteq \mathbf{M} \times 2^{\mathbb{F}}$ is the set of events. Let

$$\mathbb{S} = \bigcup_{N \in \mathbb{N}} \mathbf{S}(N).$$

$\mathbb{T} : \mathbb{S} \times \mathbf{M} \times 2^{\mathbb{F}} \mapsto 2^{\mathbb{S}}$ is the nondeterministic transition function of $Mgr(\Lambda, \Lambda_0)$, such that

$$S \xrightarrow{(\mu, \mathcal{F})} S' \in \mathbb{T}$$

(where $S = (\lambda, \Delta)$ and $S' = (\lambda', \Delta')$ are internal nodes of N and N', respectively), if and only if the following conditions hold:

(1) $S' = S_0(N')$;

(2) $\lambda \xrightarrow{\tau} \lambda'$ is a transition in $Usp(\Lambda, \Lambda_0)$;

(3) μ is the set of messages generated by τ;

(4) $\mathcal{F}$ is the set of faulty components associated with τ.

Finally, N_0 is the initial node, such that its root is $S_0 = (\Lambda_0, \Delta_0)$.

Example 8.7. Shown in Figure 8.10 is the monitoring graph $Mgr(\Lambda, \Lambda_0)$ relevant to the universal space displayed in Figure 8.2. Such a graph is composed of ten (shaded) nodes. The initial node N_0 involves the only internal state $(0, \{\emptyset\})$, the root of N_0, which is the leaving set of N_0 as well. Plain arrows within nodes denote transitions between internal states (identifiers of transitions are omitted). By definition of a monitoring graph, all such transitions are silent. If an internal transition is faulty, it is marked by the relevant set of faulty components.

Consider node N_3, rooted in $(3, \{\emptyset\})$. In accordance with Figure 8.2, there exists a system transition $4 \xrightarrow{\tau} 5$ in $Usp(\Lambda, \Lambda_0)$, where $\tau = \{T_{22}\}$. Consequently, N_3 incorporates the internal transition

$$(4, \{\emptyset, 12\}) \xrightarrow{C_2} (5, \{2, 12\}).$$

All the internal nodes are qualified with the relevant diagnostic attribute Δ. Clearly, Δ is local to the node and, therefore, does not represent the actual diagnostic set during the monitoring, whose value depends on the mode in which the monitoring graph has been traversed from the beginning of the system operation.

Instances of the transition function $\mathbb{T}$ of the monitoring graph are depicted by either dashed or dotted arrows, the latter involving faulty components. According to the definition, each 'external' transition is marked by the pair $(\mu, \mathcal{F})$, that is, a system message and a set of faulty components. When empty, $\mathcal{F}$ is omitted, as is in all dashed arrows. For example, the label marking the arrow from node N_3 to node N_4 stands for $(\mu = \{a_3, b_1\}, \mathcal{F} = \emptyset)$. Instead the label marking the arrow from N_3 to N_5 stands for $(\mu = \{a_3\}, \mathcal{F} = \{C_1\})$. □

6. Continuous Diagnosis

Suppose to install a polymorphic system Λ, for example, a protection apparatus (see Chapter 13). During its operation, Λ is expected to react to external events and to generate a series of system messages. The goal of the monitoring is to compute the set of candidate diagnoses at the occurrence of each newly generated system message.

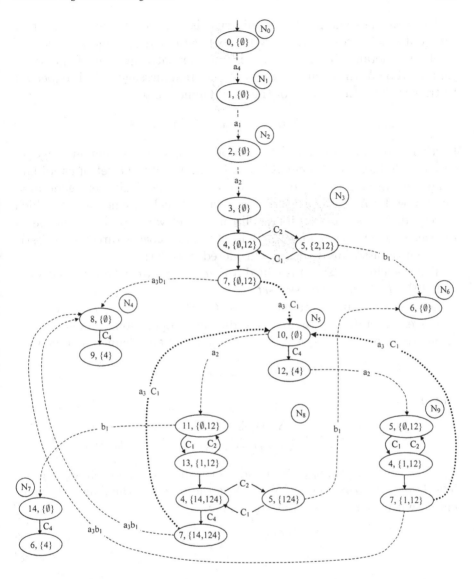

Figure 8.10. Monitoring graph $Mgr(\Lambda, \Lambda_0)$.

At the beginning, no system message is available, that is,

$$OBS(\Lambda) = \langle\rangle.$$

However, in a strict sense, a first diagnostic set should be provided, as Λ might have a silent reaction involving faulty transitions. Thus, $\Delta(\langle\rangle)$ is the initial set of candidate diagnoses. At the occurrence of the first system message μ_1, the monitoring is required to yield the set of candidate diagnoses relevant to the observation $\langle \mu_1 \rangle$, namely $\Delta(\langle \mu_1 \rangle)$.

More generally, if $\langle \mu_1, \ldots, \mu_k \rangle$ is the current sequence of messages generated by Λ, the occurrence of a new system message μ_{k+1} is expected to trigger the computation of the new diagnostic set

$$\Delta(\langle \mu_1, \ldots, \mu_k, \mu_{k+1} \rangle).$$

Furthermore, we would like to compute, for each system message μ, the so-called *snapshot diagnostic set* $\mathfrak{D}$, informally, the set of candidate diagnoses implied by the occurrence of μ as if no fault were generated up to now by Λ. Intuitively, $\mathfrak{D}$ is a sort of 'local' diagnostic set, which disregards the set of candidate diagnoses relevant to the sequence of messages generated before μ. This way, we have a direct diagnostic information associated with μ considered in isolation.

The association $(\mathfrak{D}, \Delta)$ is called a *diagnostic pair*, where the latter is called the *historic diagnostic set*.

Let Λ_0 be the initial state of Λ, and $OBS(\Lambda) = \langle \mu_1, \ldots, \mu_n \rangle$ the list of messages generated up to now by Λ during its operation. The *diagnostic sequence* $\boldsymbol{\Delta}$ of $OBS(\Lambda)$, is a list of diagnostic pairs defined as follows:

$$\boldsymbol{\Delta}(OBS(\Lambda)) = \langle \Delta_0, \Delta_1, \ldots, \Delta_n \rangle,$$

where

$$\forall i \in [0 .. n] \left(\Delta_i = \left\{ \begin{array}{ll} (\Delta(\langle \rangle), \Delta(\langle \rangle)) & \text{if } i = 0 \\ (\mathfrak{D}(\mu_i), \Delta(\langle \mu_1, \ldots, \mu_i \rangle)) & \text{otherwise} \end{array} \right. \right).$$

Let $N = (\mathbf{S}, \mathbf{E}, \mathbf{T}, S_0, \mathbf{S}_{\text{out}})$ be a node of the monitoring graph $Mgr(\Lambda, \Lambda_0)$. The *local diagnostic set* $\Delta^\ell(N)$ of N is the union of the diagnoses associated with each internal state of N, namely

$$\Delta^\ell(N) = \bigcup_{(\lambda, \Delta) \in \mathbf{S}} \Delta.$$

Let Δ_1 and Δ_2 be two diagnostic sets. The *composition* of Δ_1 and Δ_2 is defined as follows:

$$\Delta_1 \otimes \Delta_2 = \{ \delta \mid \delta = \delta_1 \cup \delta_2, \delta_1 \in \Delta_1, \delta_2 \in \Delta_2 \}.$$

Proposition 8.2. *The following equivalences hold:*

$$\Delta_1 \otimes \Delta_2 \equiv \Delta_2 \otimes \Delta_1$$
$$\Delta_1 \otimes (\Delta_2 \otimes \Delta_3) \equiv (\Delta_1 \otimes \Delta_2) \otimes \Delta_3$$
$$\Delta_1 \otimes (\Delta_2 \cup \Delta_3) \equiv (\Delta_1 \otimes \Delta_2) \cup (\Delta_1 \otimes \Delta_3).$$

Example 8.8. With reference to the monitoring graph displayed in Figure 8.10, consider the evolution of system Λ relevant to the following sequence of system messages:

$$\langle \{a_4\}, \{a_1\}, \{a_2\}, \{a_3\}, \{a_2\}, \{a_3\}\rangle.$$

When no messages are considered, the initial set of candidate diagnoses (historic diagnostic set) coincides with the local diagnostic set of the root N_0, namely

$$\Delta(\langle\rangle) = \Delta^\ell(N_0) = \{\emptyset\}.$$

At the occurrence of message $\{a_4\}$, a (non-faulty) transition from the root of N_0 to the root of N_1 happens, which gives rise to a snapshot diagnosis that equals the local diagnostic set, namely $\mathfrak{D} = \{\emptyset\}$. Since the diagnostic attribute associated with the root of N_0 is the singleton $\{\emptyset\}$, the historic diagnostic set equals the snapshot diagnostic set, namely

$$\Delta = \mathfrak{D} = \{\emptyset\}.$$

The occurrence of $\{a_1\}$ moves the monitoring state from N_1 to N_2, by means of a normal (non-faulty) transition. Thus, both the snapshot and the historic diagnostic sets are still $\{\emptyset\}$.

The occurrence of the next message, $\{a_2\}$, brings the monitoring state from N_2 to N_3 by means of a normal transition. Thus, both the snapshot and the historic diagnostic sets corresponds to the local diagnostic set of N_3, namely

$$\mathfrak{D} = \Delta = \Delta^\ell(N_3) = \{\emptyset, \{C_2\}, \{C_1, C_2\}\}.$$

The occurrence of message $\{a_3\}$ moves the monitoring state from N_3 to N_5 by means of a faulty transition marked by "a_3, C_1". Thus, the snapshot diagnostic set is given by the composition

$$\Delta^\ell(N_5) \otimes \{\{C_1\}\} = \{\emptyset, \{C_4\}\} \otimes \{\{C_1\}\} = \{\{C_1\}, \{C_1, C_4\}\}.$$

Instead, the historic diagnostic set is generated by composing the snapshot diagnostic set above with the diagnostic attribute relevant to the node of the leaving set of N_3, namely $(7, \{\emptyset, \{C_1, C_2\}\})$, that is,

$$\Delta = \{\{C_1\}, \{C_1, C_4\}\} \otimes \{\emptyset, \{C_1, C_2\}\}$$
$$= \{\{C_1\}, \{C_1, C_4\}, \{C_1, C_2\}, \{C_1, C_2, C_4\}\}.$$

The occurrence of the next message $\{a_2\}$ introduces a nondeterministic pattern, as N_5 involves two leaving edges marked by a_2, one to N_8 and the other to N_9. To be consistent with the monitoring graph, we have to consider both possibilities in the transition to the new monitoring state, which in fact includes both N_8 and N_9. This means that the current monitoring state is composed of two nodes of the monitoring graph.

Since both edges are not marked by any faulty components, the snapshot diagnostic set corresponds to the union of the local diagnostic sets relevant to nodes N_8 and N_9, namely

$$\mathfrak{D} = \Delta^\ell(N_8) \cup \Delta^\ell(N_9)$$
$$= \{\emptyset, \{C_1\}, \{C_1, C_2\}, \{C_1, C_4\}, \{C_1, C_2, C_4\}\} \cup \{\{C_1, C_4\}, \{C_1, C_2, C_4\}\}$$
$$= \{\emptyset, \{C_1\}, \{C_1, C_2\}, \{C_1, C_4\}, \{C_1, C_2, C_4\}\}.$$

In order to compute the new historic set, we have to make two compositions, corresponding to the two edges leaving N_5 and marked by a_2, and then merge the results. However, this requires a previous relocation of the diagnostic attributes of nodes in N_5 so as to keep the historic diagnostic information up to date. The relocation $\rho(10)$ of the diagnostic attribute of the root of N_5 is computed as follows:

$$\rho(10) = \{\emptyset\} \otimes \{\{C_1\}\} \otimes \{\emptyset, \{C_1, C_2\}\} = \{\{C_1\}, \{C_1, C_2\}\},$$

where $\{\emptyset, \{C_1, C_2\}\}$ is the (relocated) diagnostic attribute of the leaving node of N_3. Similarly, the relocation $\rho(12)$ of the diagnostic attribute of the non-root node of N_5 is determined by the following composition:

$$\rho(12) = \{\{C_4\}\} \otimes \{\{C_1\}\} \otimes \{\emptyset, \{C_1, C_2\}\} = \{\{C_1, C_4\}, \{C_1, C_2, C_4\}\}.$$

The new historic diagnostic set is generated based on the following formula:

$$\Delta = (\Delta^\ell(N_8) \otimes \rho(10)) \cup (\Delta^\ell(N_9) \otimes \rho(12))$$
$$= \{\emptyset, \{C_1\}, \{C_1, C_4\}, \{C_1, C_2, C_4\}\} \otimes \{\{C_1\}, \{C_1, C_2\}\} \cup$$
$$\{\emptyset, \{C_1\}, \{C_1, C_2\}\} \otimes \{\{C_1, C_4\}, \{C_1, C_2, C_4\}\}$$
$$= \{\{C_1\}, \{C_1, C_2\}, \{C_1, C_4\}, \{C_1, C_2, C_4\}\} \cup \{\{C_1, C_4\}, \{C_1, C_2, C_4\}\}$$
$$= \{\{C_1\}, \{C_1, C_2\}, \{C_1, C_4\}, \{C_1, C_2, C_4\}\}.$$

Finally, the occurrence of message $\{a_3\}$ matches two edges, one form N_8 to N_5 and the other from N_9 to N_5, both being marked by the faulty component C_1. Therefore, the snapshot diagnostic set is determined by composing the local diagnostic set of N_5 with C_1, namely

$$\mathfrak{D} = \Delta^\ell(N_5) \otimes \{\{C_1\}\} = \{\emptyset, \{C_4\}\} \otimes \{\{C_1\}\} = \{\{C_1\}, \{C_1, C_4\}\}.$$

As above, to determine the new historic diagnostic set, we need to relocate the diagnostic attributes relevant to the leaving states in N_8 and

N_9. The relocation of the leaving state in N_8 is computed as follows:

$$\rho(7) = \{\{C_1, C_4\}, \{C_1, C_2, C_4\}\} \otimes \rho(10)$$
$$= \{\{C_1, C_4\}, \{C_1, C_2, C_4\}\} \otimes \{\{C_1\}, \{C_1, C_2\}\}$$
$$= \{\{C_1, C_4\}, \{C_1, C_2, C_4\}\}.$$

Incidentally, the relocation leaves the diagnostic attribute unchanged.

The relocation of the leaving state in N_9 is as follows:

$$\rho(7') = \{\{C_1\}, \{C_1, C_2\}\} \otimes \rho(12)$$
$$= \{\{C_1\}, \{C_1, C_2\}\} \otimes \{\{C_1, C_4\}, \{C_1, C_2, C_4\}\}$$
$$= \{\{C_1, C_4\}, \{C_1, C_2, C_4\}\}.$$

The new historic diagnostic set is generated based on the following formula:

$$\Delta = (\Delta^\ell(N_5) \otimes \rho(7)) \cup (\Delta^\ell(N_5) \otimes \rho(7'))$$
$$= \Delta^\ell(N_5) \otimes (\rho(7) \cup \rho(7'))$$
$$= \{\emptyset, \{C_4\}\} \otimes \{\{C_1, C_4\}, \{C_1, C_2, C_4\}\}$$
$$= \{\{C_1, C_4\}, \{C_1, C_2, C_4\}\}.$$

In the above transformation we exploited the distributive property of composition on the union asserted in Proposition 8.2.

As expected, the diagnostic set

$$\Delta(\langle \{a_4\}, \{a_1\}, \{a_2\}, \{a_3\}, \{a_2\}, \{a_3\}\rangle) = \{\{C_1, C_4\}, \{C_1, C_2, C_4\}\}$$

equals the set of diagnoses associated with the node of the monitoring space displayed in Figure 8.9 that is reached by the above sequence of system messages. $\square$

As pointed out in the previous example, the monitoring context may involve several monitoring nodes, which is the case when several transitions marked by the same system message move to different nodes. Therefore, a new message moves from a *monitoring hyperstate*

$$\mathcal{M} = \{N_1, \ldots, N_k\}$$

to another hyperstate

$$\mathcal{M}' = \{N_1', \ldots, N_{k'}'\}.$$

Besides, during the monitoring, some additional information must be kept in order to correctly compute the diagnostic pair $(\mathfrak{D}, \Delta)$. In particular, $\mathfrak{D}$ requires the knowledge of the set of faulty components marking the (external) transitions of the monitoring graph.

Specifically, the snapshot diagnostic set associated with a transition from $\mathcal{M}$ to $\mathcal{M}'$ triggered by a system message μ will be

$$\mathfrak{D}(\mathcal{M} \xrightarrow{\mu} \mathcal{M}') = \bigcup_{N' \in \mathcal{M}'} \Delta^{\ell}(N') \otimes \mathbb{F}_{\mu},$$

where

$$\mathbb{F}_{\mu} = \{\mathcal{F} \mid S \xrightarrow{(\mu, \mathcal{F})} S' \in \mathbb{T}, S \in N, S' \in N', N \in \mathcal{M}\}.$$

Instead, the computation of the historic diagnostic set Δ requires the additional knowledge of the relocated diagnostic attributes relevant to the leaving states of nodes in $\mathcal{M}$, namely

$$\Delta(\mathcal{M} \xrightarrow{\mu} \mathcal{M}') = \bigcup_{N' \in \mathcal{M}'} \Delta^{\ell}(N') \otimes \mathbb{F}_{\rho},$$

where

$$\mathbb{F}_{\rho} = \{\Delta^{*}(S) \otimes \mathcal{F} \mid S \xrightarrow{(\mu, \mathcal{F})} S' \in \mathbb{T}, S \in N, S' \in N', N \in \mathcal{M}\},$$

where $\Delta^{*}(S)$ is the relocated diagnostic attribute associated with S (see Example 8.8).

Thus, more precisely, a monitoring hyperstate $\mathcal{M}$ is a set of triples,

$$\mathcal{M} = \{M \mid M = (N, \mathbb{F}_{\mu}, \mathbb{F}_{\rho})\},$$

where $\mathbb{F}_{\mu}$ is the *transitional context* and $\mathbb{F}_{\rho}$ is the *historical context*. An algorithm for the monitoring of a polymorphic system is given in Appendix A.

7. Summary

- *Abductive Space*: It is an automaton where each node is an association between a node of the active space relevant to the same diagnostic problem and a diagnosis.

- *Diagnostic Space*: It is an automaton isomorphic to the active space relevant to the same diagnostic problem, where each node is decorated with a set of candidate diagnoses.

- *Universal Abductive Space*: the abductive space relevant to a blind diagnostic problem.

- *Monitoring Space*: an automaton where each node is marked by a set of candidate diagnoses and each edge is labeled by a system message.

- *Monitoring Graph*: a directed graph where each node is associated with a set of candidate diagnoses and each edge is labeled by the association between a system message and a set of faulty components.

- *Snapshot Diagnostic Set*: a set of candidate diagnoses relevant to the occurrence of a system message considered in isolation.

- *Historic Diagnostic Set*: a set of candidate diagnoses relevant to the sequence of system messages generated so far.

- *Local Diagnostic Set*: a set of candidate diagnoses relevant to a node of the monitoring graph.

- *Continuous Diagnosis*: the process of generating a new diagnostic pair at each system message occurrence.

- *Composition*: an algebraic operation that combines two sets of diagnoses into a new set.

Appendix A
Algorithms

This appendix provides two algorithms, expressed in pseudo-code, relevant to monitoring-based diagnosis. Based on a given diagnostic problem, Algorithm A.1 implements the generation of the spurious diagnostic space $\tilde{D}gn(\wp(\Lambda))$.

Algorithm A.2 dynamically generates the monitoring graph $Mgr(\Lambda, \Lambda_0)$, yielding the diagnostic pair relevant to each occurrence of a system message.

Algorithm A.1. (*DiagSpace*) The *DiagSpace* function generates the spurious diagnostic space $\tilde{D}gn(\wp(\Lambda))$, along with the relevant diagnostic set, that is, the solution of the diagnostic problem $\wp(\Lambda)$ (assuming, without loss of generality, linear observation).

To this end, it makes use of two auxiliary subprograms, namely function *Candidates* (Lines 7-20), to compute the set of candidate system transitions, and procedure *Propagate* (Lines 21-39), to recursively update the diagnostic attributes associated with the nodes created so far. The actual body of *DiagSpace* starts at Line 40. After the initialization of the fields of $\tilde{D}gn(\wp(\Lambda))$ (Lines 41-44) and the diagnostic set Δ_Λ (line 45), a loop is iterated on all the nodes dynamically inserted into $\tilde{\mathbb{N}}$ (Lines 46-72). In the beginning, the only included node is N_0, which is unmarked. Thus, a node N is picked (Line 47) and the set of candidate system transitions leaving N are computed by means of the auxiliary function *Candidates*.

To make up the candidates, the set $\mathbb{T}_{cand}^a$ of triggerable asynchronous system transitions and the set $\mathbb{T}_{cand}^s$ of triggerable synchronous system transitions are generated (Lines 13-17), where the former includes only singletons, while the latter incorporates all the sets of synchronous (component) transitions that can be triggered in parallel. The actual set of candidate diagnoses is determined at Line 19, based on the fact that synchronous transitions have higher priority than asynchronous ones.

For each system transition within the candidates, a new node N' is generated (Lines 50-58). If N' was not generated already, it is inserted (as an unmarked node) into the set of created nodes (Line 62) and, if final, it is added to the set of final nodes (Line 64), while the set of candidate diagnoses Δ_Λ is extended with the diagnostic attribute Δ' (Line 65). Instead, if N was previously generated (Line 67), the additional set of candidate diagnoses Δ^+ is propagated to the relevant nodes (Line 69) by means of the auxiliary procedure *Propagate*.

Such a procedure recursively updates both the diagnostic attributes associated with the nodes reachable from input node N and the set of candidate diagnoses

Δ_Λ. To do so, the diagnostic attribute Δ' of each neighboring node N' is possibly extended with the additional set of diagnoses Δ^+ (Line 34) and, if N' is final, Δ_Λ is updated too. The propagation is recursive in nature (Line 36), as the extra diagnoses ($\Delta' - \Delta_{\text{old}}$) are propagated from N' as well. The propagation mechanism terminates when no extra diagnoses are generated.

```
1.  procedure DiagSpace(℘(Λ))
2.     input
3.         ℘(Λ) = (Ω(Λ), OBS(λ), Λ₀): a diagnostic problem for the system Λ
4.     output
5.         D̃gn(℘(Λ)) = (Ñ, E, T̃, N₀, Nf): the spurious diagnostic space,
6.         Δ_Λ: the diagnostic set Δ(℘(Λ));

7.  function Candidates(N)
8.     input
9.         N = (λ, K, Δ): a node of D̃gn(℘(Λ))
10.    output
11.        T_cand: the set of candidate system transitions leaving N;
12.    begin{Candidates}
13.        T_a := {T₁, ..., T_k} := the set of asynchronous system transitions
                     leaving λ which are consistent with OBS(Λ);
14.        T^a_cand := {{T₁}, ..., {T_k}};
15.        T_s := the set of triggerable synchronous system transitions leaving λ;
16.        T̂_s := {T₁, ..., T_p} where ∀i ∈ [1..p]
                     (T_i = {T | T ∈ T_s, T is a transition relevant to component C_i});
17.        T^s_cand := {τ₁, ..., τ_h} := T₁ × ··· × T_p;
18.        Remove from T^s_cand the τ which are not consistent with OBS(Λ);
19.        T_cand := if T^s_cand ≠ ∅ then T^s_cand else T^a_cand
20.    end{Candidates};

21. procedure Propagate(N, Δ⁺, T̃, Δ_Λ)
22.    input
23.        N: a node of D̃gn(℘(Λ)),
24.        Δ⁺: the additional set of faulty components attached to N,
25.        T̃: the current set of edges in D̃gn(℘(Λ)),
26.        Δ_Λ: the current set of candidate diagnoses
27.    side effects
28.        Recursively updating of the diagnostic attributes in the
                 current set of nodes of D̃gn(℘(Λ)),
29.        Recursively updating of Δ_Λ;
30.    begin{Propagate}
31.        for each N →ᵀ N' ∈ T̃, where N' = (λ', K', Δ'), do
32.            if Δ⁺ ⊄ Δ' then
33.                Δ_old := Δ';
34.                Δ' := Δ' ∪ Δ⁺;
35.                if N' is final then Δ_Λ := Δ_Λ ∪ Δ⁺ end-if;
36.                Propagate(N', Δ' - Δ_old, T̃, Δ_Λ)
37.            end-if
38.        end-for
```

39.   **end**{Propagate};

40. **begin** {DiagSpace}
41. $\lambda_0 := \Lambda_0;\ K_0 := 0;\ \Delta_0 := \{\emptyset\};$
42. $N_0 := (\lambda_0, K_0, \Delta_0);$
43. $\tilde{\mathbb{N}} := \{N_0\};\ \mathbb{E} := \emptyset;\ \tilde{\mathbb{T}} := \emptyset;$
44. $\mathbb{N}_f :=$ **if** N_0 is complete **then** $\{N_0\}$ **else** $\emptyset;$
45. $\Delta_\Lambda := \emptyset;$
46.   **repeat**
47.   Pick an unmarked node $N = (\lambda, K, \Delta)$ in $\tilde{\mathbb{N}};$
48. $\mathbb{T} := Candidates(N);$
49.   **for each** $\tau \in \mathbb{T}$ **do**
50.   Create a copy $N' = (\lambda', K', \Delta')$ of $N;$
51.   Update λ' with the new system state reached by $\tau;$
52.   Increment K by the number of observable transitions in $\tau;$
53. $\mathbb{F} :=$ the set of faulty components relevant to transitions in $\tau;$
54.   **if** $\mathbb{F} \neq \emptyset$ **then**
55.   **for each** $\delta' \in \Delta'$ **do**
56. $\delta' := \delta' \cup \mathbb{F}$
57.   **end-for**
58.   **end-if**;
59.   Insert edge $N \overset{\tau}{\to} N'$ into $\tilde{\mathbb{T}};$
60.   Insert event τ into $\mathbb{E};$
61.   **if** $N' \notin \tilde{\mathbb{N}}$ **then**
62.   Insert N' into $\tilde{\mathbf{N}};$
63.   **if** N' is final **then**
64.   Insert N' into $\mathbb{N}_f;$
65. $\Delta_\Lambda := \Delta_\Lambda \cup \Delta'$
66.   **end-if**
67.   **else**
68. $\Delta^+ := \Delta' - \Delta;$
69. $Propagate(N', \Delta^+, \tilde{\mathbb{T}}, \Delta_\Lambda)$
70.   **end-if**
71.   **end-for**
72.   **until** all nodes in $\tilde{\mathbb{N}}$ are marked
73. **end** {DiagSpace}.

Algorithm A.2. (*Monitor*) The *Monitor* procedure generates the sequence of diagnostic pairs relevant to the monitoring of a polymorphic system Λ with initial state Λ_0. In doing so, starting from scratch, it incrementally makes up the part of the monitoring graph $Mgr(\Lambda, \Lambda_0)$ relevant to the actual sequence of messages.

At Line 10, the initial node N_0 is generated as the silent closure of Λ_0, which, essentially, mirrors the construction of a diagnostic space for a diagnostic problem in which the observation is empty (see Algorithm A.1). Then, the initial monitoring hyperstate is defined (Line 11), and the diagnostic sequence initialized (Line 12). The actual body of the procedure is the (unconstrained) loop enclosed by Lines 13 and 41. After the fetching of the new system message μ, the new monitoring hyperstate $\mathcal{M}'$ is yielded (Lines 15-33). In particular, if the involving internal states have not yet been generated, they are created and inserted into $\mathbb{N}$ (Lines 19-20) . The same applies to the new events and transitions (Lines 23-26). If the new internal state

N' was already created, the transitional and historical contexts ($\mathbb{F}_\mu$ and $\mathbb{F}_\rho$) will be updated (Lines 26-28), otherwise a new triple will be inserted into $\mathcal{M}'$ (Line 30). The new diagnostic pair $(\mathfrak{D}', \Delta')$ is computed after the completion of $\mathcal{M}'$ (Lines 34-39), and appended to the diagnostic sequence $\boldsymbol{\Delta}$ (Line 40).

1. **procedure** *Monitor*(Λ, Λ_0)
2.   **input**
3. Λ: a polymorphic system to be monitored,
4. Λ_0: the initial state of Λ
5.   **side effects**
6.   Dynamic construction of the monitoring graph $Mgr(\Lambda, \Lambda_0) = (\mathbb{N}, \mathbb{E}, \mathbb{T}, N_0)$,
7.   Computation of a diagnostic pair at each new message generated by Λ;
8.   **begin** {Monitor}
9. $\mathbb{N} := \emptyset$; $\mathbb{E} := \emptyset$; $\mathbb{T} := \emptyset$;
10. $S_0 := (\Lambda_0, \emptyset)$; $N_0 := Sclosure(\Lambda_0)$;
11. $\mathcal{M} := \{(N_0, \emptyset, \emptyset)\}$;
12. $\boldsymbol{\Delta} := \langle (\Delta^\ell(N_0), \Delta^\ell(N_0)) \rangle$;
13.   **loop**
14. $\mu :=$ the new system message generated by Λ;
15. $\mathcal{M}' := \emptyset$;
16.   **for each** $(N, \mathbb{F}_\mu, \mathbb{F}_\rho) \in \mathcal{M}$ **do**
17.   **for each** $\lambda \xrightarrow{\tau} \lambda'$ relevant to Λ,
               where $S = (\lambda, \Delta)$, $S \in \mathbf{S}_{\text{out}}(N)$, τ generates μ, $\tau \models \delta$ **do**
18.   **if** $\nexists N'$ $(N' \in \mathbb{N}, (\lambda', \Delta') = S_0(N'))$ **then**
19. $N' := Sclosure(\lambda')$;
20. $\mathbb{N} := \mathbb{N} \cup \{N'\}$
21.   **end-if**;
22.   **if** $(\mu, \delta) \notin \mathbb{E}$ **then** $\mathbb{E} := \mathbb{E} \cup \{(\mu, \delta)\}$ **end-if**;
23.   **if** $(\lambda, \Delta) \xrightarrow{(\mu, \delta)} (\lambda', \Delta') \notin \mathbb{T}$ **then**
24. $\mathbb{T} := \mathbb{T} \cup \{(\lambda, \Delta) \xrightarrow{(\mu, \delta)} (\lambda', \Delta')\}$
25.   **end-if**;
26.   **if** $(N', \mathbb{F}'_\mu, \mathbb{F}'_\rho) \in \mathcal{M}'$, $S_0(N') = (\lambda', \Delta')$ **then**
27. $\mathbb{F}'_\mu := \mathbb{F}'_\mu \cup \{\delta\}$;
28. $\mathbb{F}'_\rho := \mathbb{F}'_\rho \cup (\Delta(S) \otimes \mathbb{F}_\rho(N) \otimes \{\delta\})$
29.   **else**
30. $\mathcal{M}' := \mathcal{M}' \cup \{(N', \{\delta\}, \Delta(S) \otimes \mathbb{F}_\rho(N) \otimes \{\delta\})\}$
31.   **end-if**
32.   **end-for**
33.   **end-for**;
34. $\mathfrak{D}' := \emptyset$; $\Delta' := \emptyset$;
35.   **for each** $(N', \mathbb{F}'_\mu, \mathbb{F}'_\rho) \in \mathcal{M}'$ **do**
36. $\Delta'_\ell := \Delta^\ell(N')$;
37. $\mathfrak{D}' := \mathfrak{D}' \cup (\Delta'_\ell \otimes \mathbb{F}'_\mu)$;
38. $\Delta' := \Delta' \cup (\Delta'_\ell \otimes \mathbb{F}'_\rho)$
39.   **end-for**;
40. $\boldsymbol{\Delta} := \boldsymbol{\Delta} \cup \langle (\mathfrak{D}', \Delta') \rangle$
41.   **end-loop**
42. **end** {Monitor}.

IV

ADVANCED TOPICS

Chapter 9

UNCERTAIN OBSERVATIONS

Abstract

In diagnosis of dynamic systems each observation has a logical content, representing what has been observed, a temporal content, specifying when it has been observed, and a source content, stating which is the source component of each piece of observed evidence. The observation of an active system is an untimed DES observation, therefore its logical and temporal contents are the observed symbols and their precedence relationships, respectively, while the source content is the sender component of each observed event, as it can be hypothesized given the observation and the structural model of the system. While the source content of an active system observation is allowed to be ambiguous, the logical and temporal contents are not. This certainty principle, whilst being a useful simplification for a variety of contexts, may become inappropriate for a wide range of real systems, where the communication between the system and the observer, combined with the possibly limited ability of the observer to observe, is either bound to generate spurious observed events, to randomly lose some events, or to lose temporal constraints among them. To cope with these uncertainties, a number of principles affecting both the observations and the modeled behavior of a system are introduced, that are independent of any processing technique. Furthermore, the notion of an uncertain observation for an untimed DES is introduced and accommodated within a graph whose nodes are labeled by logically uncertain observed events, while edges define a partial temporal ordering among them. This way, an uncertain observation implicitly defines a finite set of observations in the traditional sense, i.e. sequences of logically certain observed events. Thus, solving an uncertain diagnostic problem inherent to an active system amounts to solving in one shot several traditional diagnostic problems.

1.   Introduction

The relevance of the temporal dimension within both observations and behavioral models has been constantly acknowledged since the beginning of model-based diagnosis research [Hamscher and Davis, 1984, Guckenbiehl and Schäfer-Richter, 1990, Hamscher, 1991, Friedrich and Lackinger, 1991, Lackinger and Nejdl, 1991]. However, considering the temporal dimension makes model-based diagnosis significantly complex, both from the conceptual and practical points of view [Brusoni et al., 1998].

According to the discussion presented in [Brusoni et al., 1998], in temporal model-based diagnosis, the observation is endowed with both a logical content, expressing *what* pieces of information have been observed, and a temporal content, expressing *when* they have been observed. Dually, the notion of a candidate diagnosis is twofold: it encompasses both the set of faults explaining the logical content of the observation, and the time constraints explaining the temporal location of the observation. The logical and temporal aspects of diagnoses are considered separately in [Brusoni et al., 1998] in order to classify distinct approaches to temporal model-based diagnosis. However, from the computational point of view, they are closely related to each other and cannot be generated separately.

This intuition is substantiated by the notion of an *explanatory diagnosis* [McIlraith, 1997, McIlraith, 1998], according to which a diagnosis is a conjectured sequence of actions or events that accounts for the observed behavior of the system to be diagnosed, rather than the mere identification of a set of faulty components.

In order to cope with the conceptual and computational difficulties of temporal model-based diagnosis, several simplifying assumptions have been made in different approaches.

In a pure discrete-event perspective, the simplest ontology of time adopted for modeling DESs consists just of temporal ordering constraints among state transitions. Thus, as already remarked in Chapter 2, the favorite behavioral models for model-based diagnosis of DESs in the literature are finite untimed nondeterministic automata.

However, while all of the approaches in the literature that exploit untimed automata, including the active system approach as described in previous chapters, deal with observations which are totally temporally ordered sequences of observed events, in this chapter a more structured notion of an observation, called *uncertain observation*, is proposed.

According to it, observed events are related to one another by means of partial precedence relations. This allows one to account for the possible timestamps of observed events, if available, in order to find out their reciprocal order, without, however, considering the exact values of time points, thereby preserving the untimed (and, then, purely discrete-event) nature of the approach.

Moreover, this chapter relaxes another strong assumption of all of the state-of-the-art approaches to model-based diagnosis of DESs, namely, the preciseness of events: observable events may have an imprecise value ranging over a set of labels, namely an *uncertain* value, both in behavioral models and observations.

Besides, another dimension of the observation of a composite system is explored, namely the *source* of an observed event, that is, the component that generated it. This dimension may be uncertain too. The full combination of this uncertain aspect with the other aspects gives rise to a further notion of an observation, called *complex observation*, which will be dealt with in the next chapter.

The revised approach allows for an (intuitive) representation of several real world situations. In fact, DESs may be only partially observable, and affected by masking phenomena. Their observation may be transmitted to the observer/s by means of one or more channels, whose misbehaviors may cause loss and/or noise.

Each observed label can be possibly timestamped, and there may be several (not necessarily synchronized) clocks for generating the timestamps. The observer/s, in turn, may be able to discriminate among observable labels or not.

The extended expressive power of both the modeling primitives and the notion of an observation leads to a larger search space of the diagnostic algorithms. In fact, the solution of a diagnostic problem featuring an uncertain/complex observation is, in principle, the union of the solutions of all of the diagnostic problems inherent to the certain, i.e. (logically, temporally, and source) univocal observations, called *observation instances*, such an observation represents. An observation instance is a totally temporally ordered sequence of precise observed events, each pertaining to a known component of the system, which complies with the logical and temporal constraints of the relevant observation.

The challenge faced in the following sections and in the next chapter is to enable the diagnostic algorithms to cope with any uncertain/complex observation uniformly, without generating or processing the single observation instances.

The chapter is organized as follows: Section 2 presents four meaningful requirements inherent to the modeling primitives of both active system components and observations, so as they can represent real world phenomena. Section 3 introduces the notions of an uncertain observation, as a (partial) answer to the needs listed in the previous section,

and that of an observation index. Finally, Section 4 details both monolithic and modular algorithms for solving diagnostic problems featuring
uncertain observations.

2.   Uncertainty Requirements

This section introduces four uncertainty requirements, namely *loss*,
logical, *source* and *temporal* uncertainty, inherent to the representation
of messages both in the behavioral models and in the observations of
active systems. Although presented by exploiting the active system formalism, such requirements are valid for DESs in general, and they are
independent of the adopted diagnostic technique.

Besides, the above requirements have to be orthogonal to one another,
that is, they have to be allowed to be combined with each other without
any restriction or loss of generality. The fulfillment of this *orthogonality
requirement* leads to two notions of an observation, namely *uncertain
observation* and *complex observation*. They are discussed in detail within
the context of active systems in the present and in the next chapter,
respectively, wherein diagnostic methods based on them are presented.

2.1   Loss uncertainty

Assuming that the logical content of messages cannot be randomly
lost is an over-constrained assumption for a wide range of real systems in
which messages may get lost because of misbehaviors of the transmission
from the system to the observer.

Requirement 9.1. (Loss uncertainty) *In a behavioral model, the set
of messages generated by the observable transitions is divided into two
disjoint subsets: the subset of reliable messages, and the subset of unreliable messages. When a message is unreliable, it may get lost, otherwise
it cannot be lost.*

One may argue that a simple solution imposed by Requirement 9.1 might
be to extend the behavioral models with additional unobservable *shadow*
transitions, one for each transition generating an unreliable message.

For example, if the message generated by transition T_{13} of component
C_1 of system Φ of Figure 4.1 may get lost, we might extend the model
of C_1 by inserting the T'_{13} additional transition:

$$T'_{13} = S_{12} \xrightarrow{(e_1, I)} S_{11}.$$

This solution, however, is somehow cumbersome, as it requires to repeat,
for each observable transition whose message may get lost, a 'shadow'
transition which differs from it only in the lack of the observable event
in the output set.

A cleaner solution is to introduce the notion of a *null label*, denoted by ϵ, and making the message a variable ranging over a domain of two labels, the observable label and the null label. Considering our example, transition T_{13} would be specified as follows:

$$T_{13} = S_{12} \xrightarrow{(e_1,I)|(X\in\{a,\epsilon\},Out)} S_{11}.$$

The semantics is the following: when transition T_{13} is fired, the observer may or may not receive a label. Due to this nondeterminism, the behavior of the system may vary during a reaction in which T_{13} is triggered more than once. For example, if it is triggered twice, the first time the label might be received, while the second time it might not.

If the above solution is adopted, unreliable messages in behavioral models are clearly distinguished from reliable messages. With shadow transitions, instead, such unreliability is not so evident. This is a typical situation when designing formal notations to describe knowledge (in our case, knowledge about the system behavior).

In fact, the same requirement may be specified either by means of a low-level (existing) notation (in our case, shadow transitions), or through a new *ad-hoc* notation especially designed for the purpose (in our example, the message ranging over a domain). In the latter case, the benefit is twofold:

(1) The knowledge designer is provided with a higher-level formalism, where specific requirements are directly supported by the notation, thereby restricting the gap between the knowledge and the way it is specified;

(2) A high-level construct can be treated more efficiently by an *ad-hoc* processing method (a diagnostic method in the case at hand) than its mapping onto low-level, unspecialized constructs.

In other words, the benefit is both conceptually ergonomic (for humans) and practical (for the diagnostic technique). In particular, a diagnostic technique for active systems dealing with the proposed high-level construct (as well as with all of the constructs that will be introduced in order to substantiate the next three uncertainty requirements and their orthogonality) is described in Section 4 of the present chapter and in Section 3 of the next chapter, respectively.

2.2 Logical uncertainty

The uniqueness of the logical content of messages, both in behavioral models and in observations, is another over-simplified assumption for real systems owing to the following reasons.

(1) The value of the message generated by a transition may vary from an occurrence of the transition to another, this depending, for example, on the granularity level of the modeled components.

(2) One or more messages may interfere with a noisy environment that may change their original logical content during the transmission from the system to the observer. In the most general case, owing to noise, given what has been received by an observer, the messages emitted by the system can only be hypothesized with uncertainty.

(3) The observer may be incapable of discriminating the exact value of a received label due, for instance, to sensitivity problems.

Requirement 9.2. (Logical uncertainty) *In behavioral models and/or observations, a message may be ambiguous, that is, its value may be one out of a given set of labels.*

The three causes listed above determine three orthogonal forms of nondeterminism.

(1) The nondeterministic behavior of the system in generating messages can be represented in the behavioral models of components. So, for instance, if an observable transition T_1 sometimes generates a and sometimes c, the message generated by T_1 can be represented as $X \in \{a, c\}$ in the behavioral models.

Or, suppose that transition T_2 sometimes generates a and sometimes nothing (that is, the transition sometimes is observable and sometimes not). The message generated by T_2 can be represented as $X \in \{a, \epsilon\}$. This particular case of logical uncertainty cannot be distinguished from that of loss uncertainty, even if they have different semantics: the former means that label a may be either generated or not by the system, while the latter means that label a is always generated but it may get lost during the transmission.

(2) The nondeterministic behavior of noisy transmission channel(s) can be modeled in the observation. For instance, if, given a received message, the observer draws that the message generated by the system is either a, or b or c, the logical content of the message in the observation can be represented as a variable ranging over these three labels, i.e., $X \in \{a, b, c\}$. We assume that this ambiguous message is given in the observation, that is, this is the starting point for the diagnosis task: we do not deal with any domain-dependent knowledge which can be exploited to hypothesize which is/are the emitted message(s) given something received by the observer. Most importantly,

we put a constraint on the set of hypothesized messages: such a set must always include the message that was actually generated by the system.

When appropriate, the set of messages hypothesized by the observer might also be, for instance

$$X' \in \{a, c, \epsilon\}, \tag{9.1}$$

where the null label denotes the additional possibility that no message at all has been generated by the system, that is, the observer has received just pure noise.

(3) Nondeterminism in message reception can be modeled within the observation. So, for instance, if the observer cannot discriminate whether the received label is b or c, the logical content of the relevant message in the observation is recorded as $X \in \{b, c\}$. Most importantly, we put a constraint on the set of labels within which the observer cannot discriminate: such a set must always include the label that was actually received by the observer.

As seen above, a possible substantiation of the first form of nondeterminism supported by Requirement 9.2 is to extend the notation introduced for Requirement 9.1 in such a way that the message generated by a transition is a variable ranging over a finite set of labels, instead of two labels only. Thus, a variable message X will be defined as follows:

$$X \in \{\ell_1, \ell_2, \ldots, \ell_n\},$$

where each ℓ_i, which may also be the null label, is a possible value for X. For example, transition T_{13} in Figure 4.1 might be defined with an ambiguous message as follows:

$$T_{13} = S_{12} \xrightarrow{(e_1, I) | (X \in \{a, b, c\}, Out)} S_{11}, \tag{9.2}$$

where the actual observable label generated by the transition is either a, b, or c.

Considering Equation (9.2), although in the behavioral model of C_1 message X is ambiguous, each time transition T_{13} is triggered the generated label will be one and only one out of $\{a, b, c\}$. For example, assume that an occurrence of T_{13} produces c. Thus, if there is no problem with transmission or reception, the message received by the observer will be c.

Conversely, a message within an observation may be ambiguous even if the relevant behavioral model does not involve any ambiguous message

at all. In such a case, the ambiguity is generated during the transmission
or the reception.

Finally, it is possible to have a combination of the ambiguities in the
behavioral models and in the observation. For example, suppose again
that an occurrence of T_{13} in Equation (9.2) generated message c and that
there were transmission problems (second form of nondeterminism). At
the reception, the observer hypothesizes, for instance, that the message
generated by the system is either a or c, that is, in the observation the
received message is

$$X' \in \{a, c\}. \tag{9.3}$$

As stated in point 2 above, an important assumption with ambiguous
messages for observations is that the actual generated label will be included in the set relevant to the received message. As a matter of fact,
in Equation (9.3), the generated label c is included within the set of the
received message X'.

Suppose now that there are both transmission and reception problems,
that is the observation is affected by both forms of nondeterminism 2
and 3. Assume, for instance, that the observer, who has received a label,
cannot discriminate whether it is b or c, and that, if b has been received,
then it can be drawn that the message generated by the system is either
a, or b or c, and that, if c has been received, then it can be drawn that
the message generated by the system is c. The solution is that in the
observation the relevant message is $X' \in \{a, b, c\}$.

According to the orthogonality requirement stated in the beginning
of Section 2, Requirements 9.1 and 9.2 have to be orthogonal to each
other. Pragmatically, this means that the set of possible labels of a logically ambiguous message in the behavioral models may include the null
value ϵ for representing the possibility that such a message may get lost.
For example, a different specification of transition T_{13} of Equation (9.2)
might be the following:

$$T_{13} = S_{12} \xrightarrow{(e_1, I)|(X \in \{a,b,c,\epsilon\}, Out)} S_{11},$$

where the inclusion of ϵ within the domain of X means that the triggering
of T_{13} might generate no label at all (logical uncertainty), or that such
a label may get lost during the transmission (loss uncertainty) or both.

2.3   Source uncertainty

When conjecturing what has happened to a composite system, given
the system observation, it is necessary to conjecture which component

generated each message. In the real world, sometimes this conjecture is certain since, for instance, messages emitted by distinct components are transmitted on distinct channels.

Sometimes, instead, the conjecture is uncertain since, for instance, messages coming from distinct components are not received by the observer on distinct channels and messages having the same logical content may be generated by several components.

Requirement 9.3. (Source uncertainty) *The sender component of a message may be ambiguous, that is, it may be one out of a given set of components.*

This kind of uncertainty is already present in linear observations. Consider, for example, once again the linear observation of system Φ taken into account in Example 4.1 of Section 2 of Chapter 4:

$$OBS(\Phi) = \langle b, b, c, a, b, c, a \rangle. \qquad (9.4)$$

Unlike a and c, each of which is generated by a known component, message b may be either generated by C_1 or C_2. This is named a *shared message*.

In the authors' approach to diagnosis of active systems, shared messages are not a problem as far as the reconstruction process is *monolithic*, that is, when it yields the active space in a single step, like in the example of Chapter 4 cited above. However, as shown in Chapter 5, the reconstruction may be performed in a modular way, following several specified steps. Each step focuses on a cluster of the system, and is based on the restriction (see Section 7.3 of Chapter 3) of the system observation on such a cluster.

In our example, we might focus on clusters ξ_1 and ξ_2 of system Φ, incorporating only component C_1 and component C_2, respectively, and, therefore, the system observation of Equation (9.4) should be restricted on such clusters. However, we have two distinct clusters and just one channel transmitting to the observer the messages generated by both clusters. The problem raises when we have to deal with a b message, as we do not know which component actually generated it.

A similar problem raises also in other approaches to model-based diagnosis of DESs, whenever a global observation is given but diagnosis is performed at a local level.

A simple solution might be to express the global observation by enumerating all of the possible combinations, where each combination makes a different assumption on the actual sender component of each of the three b's.

In our example, eight different assumptions should be considered. It is easy to show that the number of possible combinations is exponential

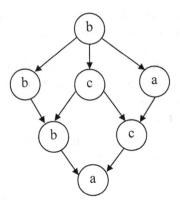

Figure 9.1.   Observation graph.

with the occurrences of the shared message within the global observation. Besides, in the perspective of the authors' approach, the restriction of each combination on the considered clusters has to be obtained, then the reconstruction step for each different observation restriction has to be carried out, and, eventually, the union of the results has to be made.

A more elegant solution is to project each shared message m into a variable message ranging over the domain $\{m, \epsilon\}$. This way, all of the assumptions are implicitly considered without any need for enumeration. Coming back to our example, the restriction of $OBS(\Phi)$ will lead to the following result:

$$OBS_{\langle \xi_1 \rangle}(\Phi) = \langle X, X, a, X, a \rangle$$
$$OBS_{\langle \xi_2 \rangle}(\Phi) = \langle X, X, c, X, c \rangle$$

where $X \in \{b, \epsilon\}$ is the variable message resulting from the restriction of b. Interestingly enough, this shows that the need in the observation for a message represented by means of a variable whose domain includes the null label ϵ comes as well when source uncertainty is concerned, even if we deal with observations without any logical uncertainty.

Substantiating the orthogonality requirement by combining Requirements 9.1 and 9.3 does not need any intervention from the point of view of the modeling primitives. In fact, the constructs inherent to Requirement 9.1 concern behavioral models and those inherent to Requirement 9.3 concern observations, thus, there is no interference between them.

Instead, combining Requirements 9.2 and 9.3 requires considering the representation of observations, since the constructs inherent to Requirement 9.2 concern both behavioral models and observations, and those inherent to Requirement 9.3 concern observations. So, we have to combine logically uncertain messages with source uncertain messages in the observations, in order to obtain messages that are uncertain in both respects.

Pragmatically, this means that the set of possible labels of a logically ambiguous message in the observation may include the null label ϵ as well, this representing the possibility that such a message may have not been generated by the relevant cluster the observation refers to.

For instance, if the logically uncertain message $X \in \{a, c\}$ is also source uncertain, then it has to be represented as $X' \in \{a, c, \epsilon\}$ in the observation restriction inherent to any cluster which may have generated X. In each cluster observation, this case cannot be distinguished from that represented in Equation (9.1).

2.4 Temporal uncertainty

Temporal uncertainty is related to the temporal ordering relationships among messages in observations. Assuming that the total temporal emission order is known in the observation, as done in a linear observation, is unacceptable in many real world situations. Thus, a more flexible ordering structure is needed, that allows a partial ordering relationship among messages.

Requirement 9.4. (Temporal uncertainty) *Messages within an observation are related to one another by means of a partial ordering relationship.*

A natural way for substantiating Requirement 9.4 is to accommodate the messages relevant to an observation within a directed acyclic graph, called an *observation graph*, which corresponds to a *temporally uncertain observation*, where nodes and edges represent the logical content of messages and the partial ordering among them, respectively.

For example, an observation graph is displayed in Figure 9.1. The ordering relationship is not necessarily defined for all of the pairs of nodes, and the acyclicity of an observation graph reflects the acyclicity of the temporal relationship.

Not surprisingly, the notion of a temporally uncertain observation subsumes and generalizes that of a linear observation. At the other hand of the spectrum of temporally uncertain observations there is a *totally* temporally uncertain observation (or *temporally unconstrained observation*), that is, an observation wherein the temporal content of every message is unknown. In between there are several cases of temporally uncertain observations, among which are itemized observations. An itemized observation consists of several totally temporally ordered sequences of messages: this is the case, for instance, when there are several channels

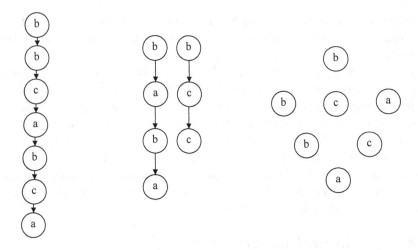

Figure 9.2. Special cases of observation graphs: linear observation (left), itemized observation (center), and temporally unconstrained observation (right).

transmitting messages from the system to the observer, each inherent to a set of components and transmitting one message at a time in the order in which messages are generated.

Shown in Figure 9.2 are three observation graphs, all having the same logical content (which is the same as in Figure 9.1) but featuring distinct temporal contents: from left to right, they represent the linear observation of Equation (9.4), an itemized observation, and a temporally unconstrained observation, respectively. For both the itemized (center) and unconstrained (right) observations, the corresponding graphs are disconnected.

In order to substantiate orthogonality for all of the four uncertainty requirements, we should combine Requirement 9.4 with Requirements 9.1, 9.2, and 9.3.

Combining Requirements 9.1 and 9.4 does not affect the modeling primitives introduced in the previous sections as they concern distinct domains: the former is inherent to behavioral models and the latter to observations.

To combine Requirements 9.2 and 9.4, we have to allow the nodes of an observation graph to be labeled by logically uncertain messages. In other words, a node may contain a variable whose value ranges over a finite set of labels, possibly including the null label ϵ, as displayed in Figure 9.3. Each node has an identifier ω_i, which is unique within the observation graph. Instead, the same message may be duplicated in different nodes. For instance, b is the message associated with both ω_1 and ω_5. The observation graph incorporates three nodes labeled by

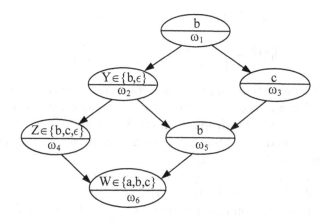

Figure 9.3. Observation graph with logically uncertain messages.

variable messages, specifically, ω_2, ω_4, and ω_6, two of which involve the null label ϵ.

Combining Requirements 9.2, 9.3, and 9.4 needs further attention. In fact, source uncertainty is either independent of temporal uncertainty or not. If the two forms of uncertainty are independent of each other, the temporal relationships between (possibly logically uncertain) messages in the observation are not a function of the sender components of such messages.

Observations like this, which are represented by means of an observation graph as in Figure 9.3, are called *uncertain*: they are the main subject of this chapter. In this case, substantiating the orthogonality of Requirement 9.3 with respect to Requirements 9.2 and 9.4 amounts to providing rules for generating observation restrictions such that they generalize those presented in Section 2.3 for linear observations.

Instead, if source uncertainty and temporal uncertainty are dependent of each other, combining Requirements 9.2, 9.3 and 9.4 leads to the notion of a *complex observation*, which is introduced in next chapter.

3. Uncertain Observation

This section focuses on the formalization of uncertain observations, which is based on a more general notion of a message, namely an *uncertain message*, than that introduced in Section 2 of Chapter 3. In particular, let Λ be a set of non-null *labels* and $\mathcal{V}$ a set of *variables* such that $\forall X \in \mathcal{V}$ the value of X is $\|X\| \subseteq (\Lambda \cup \{\epsilon\})$, where ϵ denotes the *null* label. The set $\mu = \Lambda \cup \mathcal{V}$ is the *domain* of uncertain messages. The *extension* of an uncertain message $m \in \mu$ is the set defined as follows:

$$\|m\| = \begin{cases} \{\ell\} & \text{if } m = \ell, \ell \in \Lambda \\ \|X\| & \text{if } m = X, X \in \mathcal{V}. \end{cases}$$

Intuitively, the extension of an uncertain message is the set of labels associated with the message. Hereafter the adjective 'uncertain' preceding 'message' will be dropped, implicitly assuming that, when not differently specified, the broadest notion of a message, i.e. that of an uncertain message, is adopted.

Given a system/cluster Θ (or a component C of Θ, or a set of components $\mathbf{C}$ of Θ), the set $\Lambda(\Theta)$ (or $\Lambda(C)$ or $\Lambda(\mathbf{C})$) of observable labels inherent to it depends on the considered observer $\Omega(\Theta) = (\mathbb{V}, \mathbb{P})$, as such a set encompasses all of the non-null labels that can be generated by the components in $\mathbb{V}$.

Example 9.1. Considering Figure 9.3, the extension of the message relevant to ω_1 is the singleton $\{b\}$, while the extension relevant to ω_6 is $\{a, b, c\}$. $\diamond$

Let Θ be a system and $\Omega(\Theta) = (\mathbb{V}, \mathbb{P})$ an observer of Θ, where $\mathbb{P} = \{\mathbf{C}_1, \ldots, \mathbf{C}_n\}$. An *observation graph* $obs(\mathbf{C}_i)$ of a set of components $\mathbf{C}_i \in \mathbb{P}$ is a directed (not necessarily connected) acyclic graph

$$obs(\mathbf{C}_i) = (\mathbf{N}, \Upsilon, \mathbf{N}_0, \mathbf{N}_f)$$

such that $\mathbf{N}$ is the set of nodes, where each $\omega \in \mathbf{N}$ is marked with a message $Msg(\omega) \in \mu$, $\Upsilon : \mathbf{N} \mapsto 2^{\mathbf{N}}$ is the set of edges, $\mathbf{N}_0 \subseteq \mathbf{N}$ the set of *roots*, and $\mathbf{N}_f \subseteq \mathbf{N}$ the set of *leaves*.

The '$\prec$' *temporal precedence* relationship among nodes of the graph is defined as follows:

(1) (*Basis*) If $\omega \mapsto \omega' \in \Upsilon$ then $\omega \prec \omega'$;

(2) (*Transitivity*) If $\omega \prec \omega'$ and $\omega' \prec \omega''$ then $\omega \prec \omega''$;

(3) (*Canonicity*) If $\omega \mapsto \omega' \in \Upsilon$ then $\nexists \omega'' \in \mathbf{N} \mid \omega \prec \omega'' \prec \omega'$.

Furthermore, by definition,

(4) (*Generalization*) $\omega \preceq \omega'$ iff $\omega \prec \omega'$ or $\omega = \omega'$;

(5) (*Lower Bound*) $\forall \omega_0 \in \mathbf{N}_0$ ($\nexists \omega \in \mathbf{N} \mid \omega \prec \omega_0$);

(6) (*Upper Bound*) $\forall \omega_f \in \mathbf{N}_f$ ($\nexists \omega \in \mathbf{N} \mid \omega_f \prec \omega$).

Example 9.2. The observation graph shown in Figure 9.3 is such that $\mathbf{N} = \{\omega_1, \ldots, \omega_6\}$, $\mathbf{N}_0 = \{\omega_1\}$, $\mathbf{N}_f = \{\omega_6\}$, and Υ is represented by the edges among nodes. We have $Msg(\omega_1) = b$, while $Msg(\omega_4) = Z$, where $\|Z\| = \{b, c, \epsilon\}$. Since $\omega_1 \mapsto \omega_2 \in \Upsilon$, we have $\omega_1 \prec \omega_2$ and, since $\omega_2 \prec \omega_4$, for transitivity we also have $\omega_1 \prec \omega_4$. $\diamond$

An *uncertain observation* $OBS(\Theta)$ of system Θ is the record of the $n \geq 1$ observation graphs $obs(\mathbf{C}_i)$, $i \in [1 \mathinner{..} n]$, that is,

$$OBS(\Theta) = (obs(\mathbf{C}_1), \ldots, obs(\mathbf{C}_n)).$$

Note that $OBS(\Theta)$, being a record of one or more observation graphs, is an observation graph itself. Moreover, while in itemized observations the order among message coming from the same part in the ordering partition $\mathbb{P}$ is total, in uncertain observations it is partial.

3.1 Uncertain-observation Index

Keeping trace of the messages generated by each reconstructed sequence of transitions is of primary importance in the reconstruction of the system behavior. As detailed in Section 2 of Chapter 4 and Section 4 of Chapter 5, this amounts to including, within each state in the reconstruction process, an appropriate observation index (see Section 7.2 of Chapter 3).

Such an index univocally specifies the set of messages which have already been generated (consumed) if that state is reached. For a linear observation $OBS(\xi) = \langle \ell_1, \ldots, \ell_k \rangle$, such an index has been defined as $\Im(OBS(\xi)) = i$, where i is an integer ranging over $[0 \mathinner{..} k]$ (see Figure 4.2).

In the diagnostic technique based on uncertain observations, $\Im$ is expected to identify the set of messages that have been consumed already in the observation graph inherent to the considered observation. A natural extension from linear observations to uncertain observations of the above approach is to define $\Im$ as the smallest set of nodes such that all of the preceding nodes in the observation graph, along with those in $\Im$, represent the set of consumed messages.

An *index* $\Im$ of an uncertain observation $OBS(\xi) = (\mathbf{N}, \Upsilon, \mathbf{N}_0, \mathbf{N}_f)$ inherent to a system/cluster ξ is therefore a subset of $\mathbf{N}$ such that:

$$\forall \omega \in \Im \ (\nexists \omega' \in \Im \mid \omega' \prec \omega). \tag{9.5}$$

The following functions, *Cons* and *Next*, are defined on $\Im$. $Cons(\Im)$ computes the set of already consumed messages, while $Next(\Im)$ yields the whole set of messages each of which can be consumed at the next reconstruction step.

(1) $Cons(\Im) = \{\omega \mid \omega \in \Omega, \omega' \in \Im, \omega \preceq \omega'\}$;

(2) $Next(\Im) = Front(\Im) \cup Front^{+}(\Im)$, where $Front(\Im)$ includes all the successive nodes of $\Im$. Formally, if $\Im \neq \emptyset$, then

$$Front(\Im) = \{\omega \mid \omega \in \Omega, \omega \notin Cons(\Im), \forall \omega' \mapsto \omega \in \Upsilon \, (\omega' \in Cons(\Im))\}$$

else

$$Front(\Im) = \Omega_0,$$

while $Front^{+}(\Im)$ is the closure of $Front(\Im)$ obtained by recursively including all of the messages reachable from $Front(\Im)$ whose extensions contain the null label. $Front^{+}(\Im)$ is computed by Algorithm A.1, which is reported in Appendix A of this chapter.

$\Im$ is said to be *complete* either when all of the messages have been consumed or all of the remaining messages include the null label, that is, when:

$$(Cons(\Im) = \Omega) \vee (\forall \omega \in (\Omega - Cons(\Im)) \, (\epsilon \in \|Msg(\omega)\|)).$$

Example 9.3. Assuming that the observation graph in Figure 9.3 is the whole uncertain observation $OBS(\Phi)$ of system Φ, a possible index relevant to it is $\Im = \{\omega_2, \omega_3\}$. Condition (9.5) is met, as ω_2 and ω_3 are not involved in any precedence relationship between them. The set of consumed messages identified by such an index is

$$Cons(\Im) = \{\omega_1, \omega_2, \omega_3\}.$$

Moreover, the index also identifies the set of next consistent messages, that is, the set of nodes for which all the preceding nodes have been consumed already, in our case, $Next(\Im) = \{\omega_4, \omega_5\}$.

The $Next$ function is composed of two sets, $Front(\Im)$ and $Front^{+}(\Im)$, where $Front^{+}(\Im)$ represents a sort of ϵ-closure of $Front(\Im)$, that is, the nodes reachable from $Front(\Im)$ under the assumption of null messages. In our case, $Front(\Im) = \{\omega_4, \omega_5\}$ and $Front^{+}(\Im) = \emptyset$.

Instead, if we consider Figure 9.3 with $\Im = \{\omega_1\}$, we will have

$$Front(\Im) = \{\omega_2, \omega_3\},$$
$$Front^{+}(\Im) = \{\omega_4\},$$

under the assumption of the null label for Y. Even if we assume the null label for Z too, we cannot include ω_6 in $Front^{+}(\Im)$, as $\omega_5 \notin Cons(\Im)$. The only complete index in Figure 9.3 is $\{\omega_6\}$.

Finally, if ω_6 included the null label, we would have another complete index, namely $\{\omega_5\}$, as

$$Cons(\{\omega_5\}) = \{\omega_1, \omega_2, \omega_3, \omega_5\},$$
$$\mathbf{N} - Cons(\{\omega_5\}) = \{\omega_4, \omega_6\},$$

where the null label belongs to both the extensions of ω_4 and ω_6.　　　$\diamond$

3.2 Uncertain-observation Restriction

As pointed out in Section 4 of Chapter 5, a modular approach to system behavior reconstruction requires that the system observation be projected (restricted) on the clusters corresponding to a *decomposition* of the system.

A decomposition Ξ of a cluster ξ is a set of disjoint clusters,

$$\Xi = \{\xi_1, \ldots, \xi_m\},$$

such that

$$\bigcup_{i=1}^{m} \mathbb{C}(\xi_i) = \mathbb{C}(\xi).$$

Given a decomposition of a system, it is possible to reconstruct (possibly in parallel) the behavior of the single clusters and eventually to merge these reconstructions into the actual system behavior.

Therefore, the relevant techniques for observation restriction are expected to be extended to observation graphs as well. However, this is not so straightforward as is the case of certain observations.

Given a cluster ξ, an observer $\Omega(\xi)$ of ξ, and a sub-cluster ξ' of ξ, the restriction of an uncertain observation $OBS(\xi)$ on ξ', denoted by $OBS_{\langle \xi' \rangle}(\xi)$, is expected to generate a new uncertain observation, that is, a new observation graph

$$OBS(\xi') = OBS_{\langle \xi' \rangle}(\xi) = (\mathbf{N}', \Upsilon', \mathbf{N}'_0, \mathbf{N}'_f)$$

with, in general, different nodes and edges. Essentially, rules are required for determining the new set of nodes $\mathbf{N}'$ and the new set of edges Υ'.

The new set of roots and leaves, $\mathbf{N}'_0$ and $\mathbf{N}'_f$, respectively, can then be determined from $\mathbf{N}'$ and Υ' as specified in the definition of an uncertain observation (see the beginning of Section 3).

Example 9.4. In order to state rules for the construction of $\mathbf{N}'$, consider our reference example illustrated in Figure 4.1, wherein both the components of system Θ are visible and the ordering partition is a singleton. Remember that labels a and c are generated by components C_1 and C_2, respectively, while label b is shared between them.

Assume to have a node ω in $OBS(\Phi)$ to be restricted on cluster ξ_1, that is, the cluster incorporating the single component C_1. Table 9.1 lists several possible message extensions (first column) along with their restriction on ξ_1 (second column).

For example, the restriction of a does not change the operand, as a is generated by C_1 only. Instead, as already discussed in Section 2.3,

$\|Msg(\omega)\|$	$\|Msg(\omega_{\langle\xi_1\rangle})\|$
a	a
b	b, ϵ
c	ϵ
a, ϵ	a, ϵ
a, b	a, b, ϵ
a, c	a, ϵ
b, c	b, ϵ
b, ϵ	b, ϵ
c, ϵ	ϵ
a, b, c	a, b, ϵ
a, b, ϵ	a, b, ϵ
b, c, ϵ	b, ϵ
a, b, c, ϵ	a, b, ϵ

Table 9.1.   Transformation of message extensions during the restriction on ξ_1.

the restriction of b generates the extension $\{b, \epsilon\}$, as b may have been generated either by C_1 or C_2. In general, when ϵ is in the original extension, it is in the extension of the restriction too.

Another rule is the following: every c is transformed by the restriction into the null label, as c cannot be generated by C_1. For example, the restriction of c produces the null label and $\{a, c\}$ is transformed into $\{a, \epsilon\}$. Finally, the possible duplication of ϵ is removed from the resulting extension. So, the restriction of $\{a, b, c, \epsilon\}$ would produce two extra null labels (one from b and one from c), but these are removed from the result, which is therefore $\{a, b, \epsilon\}$.   ◇

Formally, the restriction $\omega' = \omega_{\langle\xi'\rangle}$, $\omega \in \mathbf{N}$, $\xi' \subset \xi$, is defined by the following rules, where, as known, $\Lambda(\xi)$, $\Lambda(\xi')$, and $\Lambda(\mathbb{C}(\xi) - \mathbb{C}(\xi'))$ depend on the considered observer $\Omega(\xi)$.

(1) If $m \in \|Msg(\omega)\|$ and $m \in \Lambda(\xi')$ then $m \in \|Msg(\omega')\|$;

(2) If $m \in \|Msg(\omega)\|$ and $m \in \Lambda(\xi')$ and $m \in \Lambda(\mathbb{C}(\xi) - \mathbb{C}(\xi'))$ then $\epsilon \in \|Msg(\omega')\|$;

(3) If $m \in \|Msg(\omega)\|$ and $m \notin \Lambda(\xi')$ then $\epsilon \in \|Msg(\omega')\|$;

(4) If $\epsilon \in \|Msg(\omega)\|$ then $\epsilon \in \|Msg(\omega')\|$.

A node ω' resulting from the above rules will belong to $\mathbf{N}'$ if and only if:

$$\|Msg(\omega')\| \neq \{\epsilon\}. \tag{9.6}$$

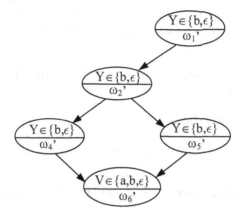

Figure 9.4. Restriction of the observation graph of Figure 9.3 on cluster ξ_1.

The restriction of Υ into Υ' is dictated by the following rule: $\omega_1' \mapsto \omega_2' \in \Upsilon'$, $\omega_1' \in \mathbf{N}'$, $\omega_2' \in \mathbf{N}'$, $\omega_1' = \omega_{1\langle\xi'\rangle}$, $\omega_2' = \omega_{2\langle\xi'\rangle}$ iff

(1) (*Precedence*) $\omega_1 \prec \omega_2$ in $OBS(\xi)$;

(2) (*Canonicity*) $\nexists \omega_3' \in \mathbf{N}'$, $\omega_3' = \omega_{3\langle\xi'\rangle}$, such that $\omega_1 \prec \omega_3 \prec \omega_2$ in $OBS(\xi)$.

Example 9.5. Consider the observation graph outlined in Figure 9.3, namely $OBS(\Phi)$. The restriction $OBS_{\langle\xi_1\rangle}(\Phi)$ is shown in Figure 9.4. The restricted nodes follow the transformations specified in Table 9.1. The restriction of node ω_3 is missing, as it produces the extension $\{\epsilon\}$. According to the rule on the preservation of the canonicity of the graph, the edge $\omega_1' \mapsto \omega_5'$ is missing, as it is derivable by transitivity from $\omega_1' \mapsto \omega_2'$ and $\omega_2' \mapsto \omega_5'$. The set of roots and the set of leaves of the resulting graph are the singletons $\mathbf{N}_0' = \{\omega_1'\}$ and $\mathbf{N}_f' = \{\omega_6'\}$, respectively. $\diamond$

3.3 Uncertain-observation Extension

An uncertain observation implicitly defines a finite set of sequences of logically unambiguous observable labels, called *observation instances*.

Intuitively, an observation instance can be made up from the observation graph of the uncertain observation by choosing a label within the extension of each message, and by putting the selected labels into a sequence that respects the temporal precedence relationship imposed by the graph, and, eventually, by eliminating the null labels.

Example 9.6. Considering Figure 9.3, an instance of $OBS(\Phi)$ is $\langle b, c, b, c, a\rangle$, which corresponds to the following choice:

$$\langle b(\omega_1), \epsilon(\omega_2), c(\omega_3), b(\omega_5), c(\omega_4), a(\omega_6)\rangle,$$

where the ordering relationships within the sequence of labels is consistent with the precedence relationships imposed by the observation graph.
◇

Let $\mathcal{Q}_\omega = \langle \omega_1, \ldots, \omega_p \rangle$ be a sequence of nodes of an uncertain observation $OBS(\xi)$ such that:

(1) $\{\omega_1, \ldots, \omega_p\} = \mathbf{N}(OBS(\xi))$;

(2) $\forall i \in [1 .. p]$, $\forall j \in [1 .. p]$, $i \neq j$ (if $\omega_i \prec \omega_j$ in $OBS(\xi)$ then ω_i precedes ω_j in $\mathcal{Q}_\omega$).

Let $\mathcal{Q}_\ell = \langle \ell_1, \ldots, \ell_p \rangle$, where $\forall i \in [1 .. p]$ ($\ell_i \in \|Msg(\omega_i)\|, \omega_i \in \mathcal{Q}_\omega$). Let

$$\mathcal{Q}'_\ell = \langle \ell'_1, \ldots, \ell'_{p'} \rangle$$

where $p' \leq p$, be the sequence obtained from $\mathcal{Q}_\ell$ by removing the null messages. By definition, $\mathcal{Q}'_\ell \in \|\mathcal{Q}_\omega\|$. $\mathcal{Q}'_\ell$ is called an *instance* of $OBS(\xi)$. The *extension* of $OBS(\xi)$, denoted by $\|OBS(\xi)\|$, is the whole set of the instances of $OBS(\xi)$.

The i-th instance of an observation can be thought of as a special case of an observation, whose graph, denoted by $OBS_{(i)}(\xi)$, is a sequence of nodes corresponding to the messages in the instance. An observation instance $OBS_{(i)}(\xi)$ is not necessarily an observation that can actually be generated by cluster ξ. However, according to the assumptions made in Section 2.2, given the observation graph $OBS(\xi)$, there exists at least one observation instance $OBS_{(i)}(\xi)$ that can be generated by ξ.

4.   Solving Uncertain Diagnostic Problems

In Section 2 of Chapter 4 we presented how an active space can be generated based on a given diagnostic problem $(\Omega(\xi), OBS(\xi), \xi_0)$, where $\Omega(\xi)$ is the observer of cluster ξ, ξ_0 is the state of ξ before the reaction, and $OBS(\xi)$ is a linear observation of ξ. In this section, we present a technique that allows us to cope with an *uncertain diagnostic problem*, where the given observation is uncertain.

4.1   Rough Index Space

In order to reconstruct the system behavior relevant to an uncertain diagnostic problem we need to know all of the possible sets of subsequences of observable labels that are consistent with the observation graph and to use the identifiers of such sub-sequences as the index field within each node of the reconstruction space.

To this end, we define (in three steps) the *index space* of an uncertain observation. The first step is represented by the *rough index space* of an

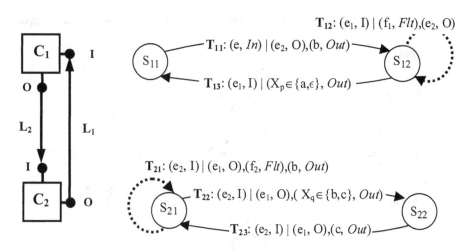

Figure 9.5. System Θ and relevant behavioral models with uncertain messages.

uncertain observation $OBS(\xi) = (\mathbf{N}, \Upsilon, \mathbf{N}_0, \mathbf{N}_f)$, which is a deterministic finite automaton

$$Rspace(OBS(\xi)) = (\mathbb{S}^r, \mathbb{L}^r, \mathbb{T}^r, S_0^r, \mathbb{S}_f^r)$$

where

$\mathbb{S}^r = \{\Im \mid \Im \text{ is an index for } OBS(\xi)\}$ is the set of states;

$\mathbb{L}^r = \mathbf{N}$ is the set of labels;

$\mathbb{T}^r : \mathbb{S}^r \times \mathbb{L}^r \mapsto \mathbb{S}^r$ is the transition function defined as follows: $\Im \xrightarrow{\omega} \Im' \in \mathbb{T}^r$ iff

(1) $\omega \in Front(\Im)$ (see Section 3.1),

(2) $\Im' = (\Im \cup \{\omega\}) - \{\omega' \mid \omega' \in \Im, \omega' \prec \omega\}$;

$S_0^r = \Im_0 = \emptyset$ is the initial state;

$\mathbb{S}_f^r = \{\Im \in \mathbb{S}^r \mid Cons(\Im) = \mathbf{N}\}$ is the set of final states.

Example 9.7. Consider a variation of system Φ, named Θ, shown in Figure 9.5. With respect to Figure 4.1, we have introduced uncertainty in the observable transitions T_{13} and T_{22}. The former generates an unreliable message (see Requirement 9.1), as it involves the null label ϵ, while the latter is now ambiguous, as it may generate either b or c. Consequently, both transitions are labeled by variables, namely X_p and X_q, instead of the original messages a and c, respectively.

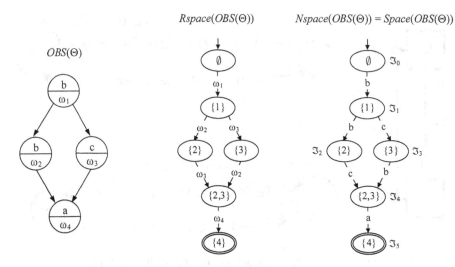

Figure 9.6. Uncertain observation for system Θ (left), relevant rough index space (center) and deterministic index space (right).

An uncertain observation $OBS(\Theta)$ is depicted on the left of Figure 9.6. The rough index space of $OBS(\Theta)$ is shown on the center of the same figure. Each node in $Rspace(OBS(\Theta))$ is labeled by the set $\Im$ of integers identifying the relevant set of nodes in $OBS(\Theta)$. For instance, $\{2,3\}$ is a shorthand for $\{\omega_2, \omega_3\}$. Besides, each edge is marked by one ω_i belonging to $Front(\Im)$, the frontier of $\Im$ in $OBS(\Theta)$. Starting from the initial state $\emptyset$, the only possible observation node is ω_1, as $Front(\emptyset) = \{\omega_1\}$. Thus, the edge is $\emptyset \xrightarrow{\omega_1} \{1\}$. By contrast, since $Front(\{1\}) = \{2,3\}$, two different edges leave node $\{1\}$, which are marked by ω_2 and ω_3, respectively. $\diamond$

An important property of a rough index space is that each path from the root to a final state corresponds to a different sequence in which all of the nodes of the observation graph can be consumed. The whole set of such sequences equals the set of sequences of consumed nodes that are consistent with the precedence relationships within the observation graph. For instance, one of these sequences is $\langle \omega_1, \omega_3, \omega_2, \omega_4 \rangle$, which yields the sequence of messages $\langle b, c, b, a \rangle$.

4.2   Nondeterministic Index Space

The *nondeterministic index space* of an uncertain observation $OBS(\xi)$ is the nondeterministic finite automaton

$$Nspace(OBS(\xi)) = (\mathbb{S}^n, \mathbb{L}^n, \mathbb{T}^n, S_0^n, \mathbb{S}_f^n)$$

obtained from $Rspace(OBS(\xi)) = (\mathbb{S}^r, \mathbb{L}^r, \mathbb{T}^r, S_0^r, \mathbb{S}_f^r)$ as follows:

$\mathbb{S}^n = \mathbb{S}^r$ is the set of states;

$\mathbb{L}^n = \{\ell \mid \ell \in \|Msg(\omega)\|, \omega \in \mathbb{L}^r\}$ is the set of labels;

$\mathbb{T}^n : \mathbb{S}^n \times \mathbb{L}^n \mapsto 2^{\mathbb{S}^r}$ is the nondeterministic transition function defined as follows:

$$S \xrightarrow{\ell} S' \in \mathbb{T}^n \text{ iff } S \xrightarrow{\omega} S' \in \mathbb{T}^r, \ell \in \|Msg(\omega)\|;$$

$S_0^n = S_0^r = \emptyset$ is the initial state;

$\mathbb{S}_f^n = \mathbb{S}_f^r$ is the set of final states.

Example 9.8. Shown on the right of Figure 9.6 is the nondeterministic index space $Nspace(OBS(\Theta))$. Incidentally, this is isomorphic to $Rspace(OBS(\Theta))$, but the transitions are marked by the observable labels of the messages relevant to ω_i.

Since in our example all of the messages are constant labels (either a, b, or c), each ω_i is simply replaced by the corresponding observable label. The resulting automaton is in general nondeterministic, as several transitions marked by the same label may leave the same state. This is not the case for our example, as $Nspace(OBS(\Theta))$ is in fact deterministic. ◇

Example 9.9. Within the nondeterministic index space of the uncertain observation of Figure 9.3, shown on the left of Figure 9.7, several transitions marked by the same label leave the same state. For example, two out of the four transitions leaving state $\{2\}$ are marked by c. ◇

4.3 Deterministic Index Space

The *deterministic index space* or, simply, the *index space* of an uncertain observation $OBS(\xi)$, is the finite automaton

$$Space(OBS(\xi)) = (\mathbb{S}, \mathbb{L}, \mathbb{T}, S_0, \mathbb{S}_f)$$

obtained by transforming

$$Nspace(OBS(\xi)) = (\mathbb{S}^n, \mathbb{L}^n, \mathbb{T}^n, S_0^n, \mathbb{S}_f^n)$$

into a deterministic finite automaton [Aho et al., 1986], where

$\mathbb{S} \subseteq 2^{\mathbb{S}^n}$ is the set of states;

$\mathbb{L} = \mathbb{L}^n$ is the set of labels;

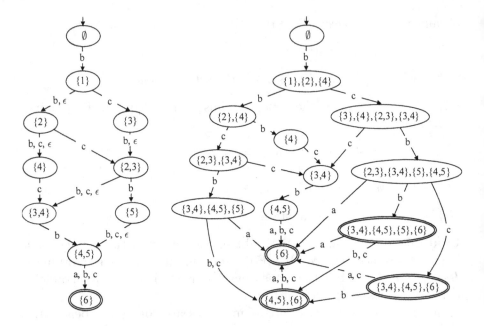

Figure 9.7. Nondeterministic index space (left) and deterministic index space (right) for the uncertain observation in Figure 9.3.

$\mathbb{T} : \mathbb{S} \times \mathbb{L} \mapsto \mathbb{S}$ is the transition function;

S_0 is the initial state;

$\mathbb{S}_f$ is the set of final states.

The *viable labels* of a state S of an index space

$$Space(OBS(\xi)) = (\mathbb{S}, \mathbb{L}, \mathbb{T}, S_0, \mathbb{S}_f)$$

is the set of observable labels which mark the transitions leaving S, that is:

$$Viable(S) = \{\ell \mid S \xrightarrow{\ell} S' \in \mathbb{T}\}.$$

Example 9.10. Considering Figure 9.6, the index space of $OBS(\Theta)$ coincides with the relevant nondeterministic index space. Instead, the index space shown on the right of Figure 9.7 substantially differs from the corresponding nondeterministic index space on the left. In accordance with the *subset construction algorithm* [Aho et al., 1986], which generates the equivalent deterministic automaton, each state is identified by a subset of the states in the nondeterministic automaton.   ◇

The *extension* of $Space(OBS(\xi))$, denoted by $\|Space(OBS(\xi))\|$, is the (finite) set of sequences of observable labels corresponding to the whole set of paths from S_0 to a state $S_f \in \mathbb{S}_f{}^1$. Such an extension is endowed of the remarkable property stated by the following theorem.

Theorem 9.1. *Let Space(OBS(ξ)) be the index space of an uncertain observation OBS(ξ). Then, the extension of the index space equals the extension of the observation, namely:*

$$\| Space(OBS(\xi)) \| \equiv \| OBS(\xi) \|.$$

This theorem opens the way to the use of the index space of an observation, instead of the observation itself, in order to perform a reconstruction compliant with the observation. Operationally, this means that the observation has first to be processed so as to obtain the observation index, then the reconstruction has to be carried out.

4.4 Monolithic Resolution

The central point of the reconstruction of the system behavior based on an index space is that the states of the index space are used as index fields in the search space. Assume an uncertain diagnostic problem $\wp(\xi) = (\Omega(\xi), OBS(\xi), \xi_0)$. Based on the index space

$$Space(OBS(\xi)) = (\mathbb{S}, \mathbb{L}, \mathbb{T}, S_0, \mathbb{S}_f),$$

we can generate the relevant active space by specifying at a given state:

(1) When a transition is triggerable from the point of view of the consistency with the observation;

(2) How the new value $\mathfrak{S}'$ of the index can be computed from the old value $\mathfrak{S}$.

The index of the initial state of the active space is the initial state of the observation index, namely $\mathfrak{S}_0$. Then, the reconstruction algorithm proceeds by looking for all of the transitions that are triggerable from the current state N of the reconstruction space.

Formally, from the point of view of the consistency with the observation, a transition

$$T = S \xrightarrow{\alpha|\beta} S'$$

---

[1]The deterministic automaton $Space(OBS(\xi))$ is equivalent to the nondeterministic one $Nspace(OBS(\xi))$ insofar as the two extensions are equal, that is, using a uniform notation,

$$\| Space(OBS(\xi)) \| = \| Nspace(OBS(\xi)) \|.$$

of component C is triggerable from the current node $N = (\sigma, \Im, D)$ of the search space, where S belongs to σ, only if at least one of the following conditions hold:

(1) T is *not* observable, that is, $(m, Out) \notin \beta$;

(2) C is *not* visible, that is, $C \in Ukn(OBS(\xi))$;

(3) T is both visible and observable but its message may be null, that is, $(m, Out) \in \beta$ and $C \notin Ukn(OBS(\xi))$ and $\epsilon \in \|m\|$;

(4) T is both visible and observable and it can generate a viable label, that is, $(m, Out) \in \beta$ and $C \notin Ukn(OBS(\xi))$ and $(\|m\| \cap Viable(\Im)) \neq \emptyset$.

In particular, the last condition states that an observable and visible transition is triggerable when the extension of the relevant message includes at least an observable label out of those marking the edges leaving the current state of the index space.

The new value $\Im'$ of the index field is generated, starting from the old value $\Im$, based on the following rules, corresponding to the four above conditions, respectively:

(1) and 2) $\Im' = \Im$ and the edge $N \xrightarrow{T} N'$ is created in the reconstruction space[2];

(3) $\Im' = \Im$ and the edge $N \xrightarrow{T(\epsilon)} N'$ is created in the reconstruction space;

(4) $\Im'$, where $\Im \xrightarrow{\ell} \Im' \in \mathbb{T}$, $\ell \in (\|m\| \cap Viable(\Im))$, and the edge $N \xrightarrow{T(\ell)} N'$ is created in the reconstruction space.

Thus, there exists a new value $\Im'$ for each different viable label belonging to the extension of message m.

Example 9.11. With reference to the system Θ depicted in Figure 9.5, consider the uncertain diagnostic problem $\wp(\Theta) = (\Omega(\Theta), OBS(\Theta), \Theta_0)$, where $\Omega(\Theta) = (\mathbb{V}, \mathbb{P})$, $\mathbb{V} = \{C_1, C_2\}$, $\mathbb{P} = \{\{C_1, C_2\}\}$, $OBS(\Theta)$ is displayed on the left of Figure 9.2, and $\Theta_0 = (S_{11}, S_{21})$.

Based on the reconstruction rules stated above, the reconstruction space for $\wp(\Theta)$ is generated as shown in Figure 9.8.

---

[2] A successor node N' of N is created in the reconstruction search space if transition T is triggerable, that is, if T is consistent not only with the observation but also with the situation of the link(s). However, the latter check is omitted here for the emphasis of this chapter is on observation.

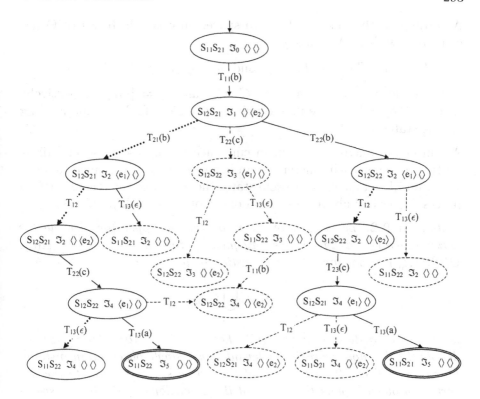

Figure 9.8. Reconstruction space of Example 9.11.

The triggerable transitions from node $N = ((S_{12}, S_{21}), \Im_1, \{\langle\rangle, \langle e_2\rangle\})$ are T_{21} and T_{22}. However, these yield three edges in the search space, one marked by $T_{21}(b)$, and two marked by $T_{22}(c)$ and $T_{22}(b)$, respectively. The last two edges bring to nodes which differ in the index field only, namely $\Im_3$ and $\Im_2$, respectively.

According to the fourth rule above, transition T_{22} is triggerable from N since it is both observable and visible, and

$$((\|X_q\| = \{b, c\}) \cap (Viable(\Im_1) = \{b, c\})) = \{b, c\} \neq \emptyset.$$

Analogously transition T_{21} is triggerable because it is both observable and visible, and

$$(\|b\| \cap (Viable(\Im_1) = \{b, c\})) = \{b\} \neq \emptyset.$$

By contrast, T_{21}, although both observable and visible, is not triggerable from node $((S_{12}, S_{21}), \Im_2, \{\langle\rangle, \langle e_2\rangle\})$ since

$$(\|b\| \cap (Viable(\Im_2) = \{c\})) = \emptyset.$$

According to the generated active space, only two histories of Θ are consistent with $\wp(\Theta)$, namely

$$h_1 = \langle T_{11}, T_{21}, T_{12}, T_{22}, T_{13} \rangle \text{ and } h_2 = \langle T_{11}, T_{22}, T_{12}, T_{23}, T_{13} \rangle,$$

yielding the two diagnoses $\delta_1 = \{C_1, C_2\}$ and $\delta_2 = \{C_1\}$, respectively. Since $\delta_1 \cap \delta_2 = \{C_1\}$, we may conclude that C_1 is faulty, while nothing can be stated about C_2. $\diamond$

An uncertain diagnostic problem cumulatively represents several diagnostic problems, each inherent to an instance of the given uncertain observation. The following theorem states the ability of the reconstruction process described above to solve in one shot all of such problems.

Theorem 9.2. *Let $\Re$ denote the reconstruction operator yielding the active space relevant to a given diagnostic problem. Let $\wp(\xi) = (\Omega(\xi), OBS(\xi), \xi_0)$ be an uncertain diagnostic problem where*

$$\|OBS(\xi)\| = \bigcup_{i=1}^{n} OBS_{(i)}(\xi)$$

is the set of instances of $OBS(\xi)$. Let $\wp_{(i)}(\xi) = (\Omega(\xi), OBS_{(i)}(\xi), \xi_0)$ denote the diagnostic problem relevant to the single observation instance $OBS_{(i)}(\xi)$. Then, the extension of the active space relevant to the uncertain problem equals the union of the extensions of the active spaces relevant to the single problems, that is,

$$\|\Re(\wp(\xi))\| \equiv \bigcup_{i=1}^{n} \|\Re(\wp_{(i)}(\xi))\|.$$

4.5   Modular Resolution

Up to now we have considered *monolithic* reconstructions only, that is, generations of active spaces in a single step. However, as detailed in Chapter 5, the reconstruction of the system reaction can be conveniently performed in a distributed, possibly parallel, way as well.

Intuitively, according to the *modular* approach, the system is first partitioned into a set of subsystems. Then, partial active spaces are generated for subsystems and merged together in a bottom-up fashion until the global active space of the entire system is produced.

This approach presents several advantages, such as distributed and parallel computation, that may increase the efficiency of the diagnostic process.

Even though the focus of this chapter is on the notion of a temporal observation affected by uncertainty, it is nevertheless worthwhile providing some informal hints on the technique of modular reconstruction.

This allows us to state a further remarkable property of index spaces, expressed by Theorem 9.3. To this end, we have first to introduce the notion of an index space restriction.

4.5.1 Index Space Restriction

Let $Space(OBS(\xi)) = (\mathbb{S}, \mathbb{L}, \mathbb{T}, S_0, \mathbb{S}_f)$ be the index space of an observation of a cluster ξ. The *restriction* of $Space(OBS(\xi))$ on a cluster $\xi' \subset \xi$, denoted by $Space_{\langle\xi'\rangle}(OBS(\xi))$, is a deterministic finite automaton $(\mathbb{S}', \mathbb{L}', \mathbb{T}', S_0', \mathbb{S}_f')$ obtained from $Space(OBS(\xi))$ by means of the following two steps:

(1) Replace each observable label ℓ marking transitions in $\mathbb{T}$ with its restriction on ξ', as described in Section 3.2;

(2) Transform the nondeterministic automaton obtained in Step 1 into an equivalent deterministic one.

Example 9.12. Shown on the right of Figure 9.9 are the restrictions of the index space $Space(OBS(\Theta))$, displayed in Figure 9.6, on clusters ξ_1 (top) and ξ_2 (bottom), respectively.

Each index space is associated (on its left) with the corresponding nondeterministic automaton (obtained by accomplishing Step 1). Accordingly, the restriction on ξ_1 follows the rules outlined in Table 3.2: b is transformed into $\{b, \epsilon\}$, c into ϵ, while a is unchanged. States of the (restricted) index spaces for ξ_1 and ξ_2 have been identified by relevant symbols, namely $\Im_{1_i}$ and $\Im_{2_j}$, respectively. ◇

Theorem 9.3. *Let $Space(OBS(\xi))$ be the index space of an uncertain observation of cluster ξ, and $\xi' \subset \xi$ a sub-cluster of ξ. Then, the extension of the index space of the restriction of $OBS(\xi)$ on ξ' equals the extension of the restriction on ξ' of the index space of the observation, that is:*

$$\|Space(OBS_{\langle\xi'\rangle}(\xi))\| \equiv \|Space_{\langle\xi'\rangle}(OBS(\xi))\|.$$

Theorem 9.3 provides two alternative means for generating the index space inherent to a cluster ξ' when the uncertain observation $OBS(\xi)$, $\xi \supseteq \xi'$, is available:

(1) Projection $OBS_{\langle\xi'\rangle}(\xi)$ is computed first, and then the index space of the resulting observation is determined;

(2) The index space of $OBS(\xi)$ is computed and then projected on ξ'.

The latter alternative is more promising than the former from the efficiency point of view.

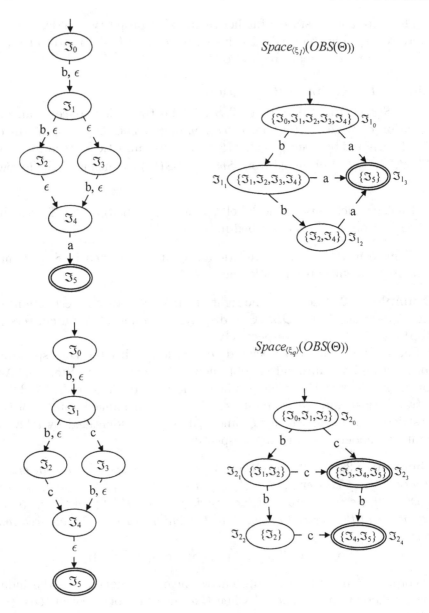

Figure 9.9. Restrictions of *Space*(*OBS*(Θ)) (see Figure 9.6) on ξ₁ (top) and ξ₂ (bottom).

4.5.2   Modular Reconstruction

The technique for the modular generation of active spaces given an uncertain diagnostic problem, which is based on the exploitation of the index space restrictions of the uncertain observation of the problem, is introduced by means of a simple example.

Example 9.13. Consider the uncertain diagnostic problem coped with monolithically in Example 9.11, whose search space is outlined in Figure 9.8. Now, we aim to solve the same problem in a modular way.

Specifically, we divide the original problem

$$\wp(\Theta) = (\Omega(\Theta), OBS(\Theta), \Theta_0)$$

into two sub-problems, one for cluster ξ_1, namely

$$\wp(\xi_1) = (\Omega(\xi_1), OBS(\xi_1), \xi_{1_0}),$$

where $\Omega(\xi_1) = \Omega_{\langle \xi_1 \rangle}(\Theta)$, $OBS(\xi_1) = OBS_{\langle \xi_1 \rangle}(\Theta)$, $\xi_{1_0} = S_{11}$, and one for cluster ξ_2, namely

$$\wp(\xi_2) = (\Omega(\xi_2), OBS(\xi_2), \xi_{2_0}),$$

where $\Omega(\xi_2) = \Omega_{\langle \xi_2 \rangle}(\Theta)$, $OBS(\xi_2) = OBS_{\langle \xi_2 \rangle}(\Theta)$, $\xi_{2_1} = S_{21}$, respectively.

The active space relevant to $\wp(\Theta)$ can then be generated by merging the active spaces relevant to $\wp(\xi_1)$ and $\wp(\xi_2)$. That is, denoting with $\Re$ the reconstruction operator and with $\mathfrak{M}$ the merging operator, the following reconstruction equivalence holds:

$$\Re(\wp(\Theta)) = \mathfrak{M}_{OBS(\Theta)}(\Re(\wp(\xi_1)), \Re(\wp(\xi_2))),$$

where $\mathfrak{M}$ incorporates the observation of Θ as an extra constraining argument. In fact, once generated the (partial) active spaces for ξ_1 and ξ_2, which are based on the (restricted) cluster observation, the merging operator is still constrained by the (uncertain) observation of the global system Θ. This is due to the fact that, generally speaking, several precedence relationships are lost during the restriction, specifically, those relating messages from different clusters[3].

In principle, according to the modular resolution technique, we need to generate:

---

[3]The merging operator $\mathfrak{M}$ behaves like the Join operator of Relational Algebra [Codd, 1970], denoted by $\bowtie$ and defined as follows:

$$r \bowtie_p s = \sigma_p(r \times s)$$

where r and s are relations, σ and $\times$ are the Selection and Cartesian product operators, respectively, while p is a predicate. The role of the predicate is played in $\mathfrak{M}$ by the constraining observation $OBS(\Theta)$.

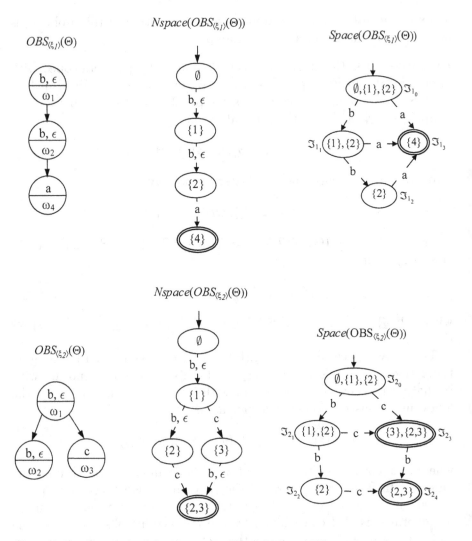

Figure 9.10. Generation of index spaces $Space(OBS_{\langle \xi_1 \rangle}(\Theta))$ and $Space(OBS_{\langle \xi_2 \rangle}(\Theta))$, where $OBS(\Theta)$ is the observation displayed in Figure 9.6.

(1) The observer restrictions $\Omega_{\langle \xi_1 \rangle}(\Theta)$ and $\Omega_{\langle \xi_2 \rangle}(\Theta)$;

(2) The observation restrictions $OBS_{\langle \xi_1 \rangle}(\Theta)$ and $OBS_{\langle \xi_2 \rangle}(\Theta)$, where observation $OBS(\Theta)$ is displayed on the left of Figure 9.6;

(3) The relevant index spaces, $Space(OBS_{\langle \xi_1 \rangle}(\Theta))$ and $Space(OBS_{\langle \xi_2 \rangle}(\Theta))$.

Based on the definition of observer restriction given in Section 6.1 of Chapter 3, $\Omega_{\langle \xi_1 \rangle}(\Theta) = (\mathbb{V}', \mathbb{P}')$, where $\mathbb{V}' = \{C_1\}$ and $\mathbb{P}' = \{\{C_1\}\}$, and $\Omega_{\langle \xi_2 \rangle}(\Theta) = (\mathbb{V}'', \mathbb{P}'')$, where $\mathbb{V}'' = \{C_2\}$ and $\mathbb{P}'' = \{\{C_2\}\}$.

All of the pieces of information for ξ_1 and ξ_2 inherent to points 2 and 3 above are shown on the top and on the bottom of Figure 9.10, respectively. For each cluster, the figure outlines the restricted observation (left), the nondeterministic index space of the restricted observation (center), and the corresponding index space (right). In accordance with Theorem 9.3, such index spaces coincide, in fact, with those determined by restrictions in Figure 9.9.

The reconstruction spaces relevant to $\wp(\xi_1)$ and $\wp(\xi_2)$ are shown on the top of Figure 9.11. Each node of the reconstruction spaces is identified by two fields only, namely σ, the state of the relevant component, and $\Im$, the identifier of a node in the relevant index space. The third field, D, is immaterial, as each cluster incorporates a single component, so that links L_1 and L_2 are outside the clusters. The dangling sets of such links are instead considered in the merging of the local active spaces, as displayed on the bottom of Figure 9.11.

The generation of the final active space requires some explanation. First, the σ field of each node consists of the pairs (ξ_{1_i}, ξ_{2_j}) of node identifiers of the two local active spaces, respectively. Moreover, the index field is represented by a node $\Im_k$ of the index space of the system observation $OBS(\Theta)$, as shown on the right of Figure 9.6. Finally, the dangling set field is included as well.

From the initial state $((\xi_{1_0}, \xi_{2_0}), \Im_0, \{\langle\rangle, \langle\rangle\})$ (which is obtained by concatenating the initial states of the local active spaces, namely ξ_{1_0} and ξ_{2_0}, the initial state $\Im_0$ of the global index space, and the empty dangling set), the only triggerable transition, among those leaving the initial states ξ_{1_0} and ξ_{2_0}, is $T_{11}(b)$. In fact, both T_{21} and T_{22} require the input event e_2, which is not in the dangling set, this being empty. Transition $T_{11}(b)$ is triggerable since the generated message b is in $Viable(\Im_0) = \{b\}$ (see Section 4.3 and Example 9.11).

The new state $((\xi_{1_1}, \xi_{2_0}), \Im_1, \{\langle\rangle, \langle e_2\rangle\})$ is obtained from the initial one by replacing:

(1) ξ_{1_0} with ξ_{1_1}, which is the new state reached by transition $T_{11}(b)$ in $\Re(\wp(\xi_1))$;

(2) $\Im_0$ with $\Im_1$, which is the new state reached by the (only) transition leaving $\Im_0$ in $Space(OBS(\Theta))$ (see Figure 9.6);

(3) The empty dangling set with $\{\langle\rangle, \langle e_2\rangle\}$, as the activation of T_{11} is supposed to generate the e_2 output event on link L_2.

At this point, the candidate transitions are those leaving either ξ_{1_1} or ξ_{2_0}, that is, T_{12}, $T_{13}(\epsilon)$, $T_{13}(a)$, $T_{21}(b)$, $T_{22}(b)$, and $T_{22}(c)$. In determining

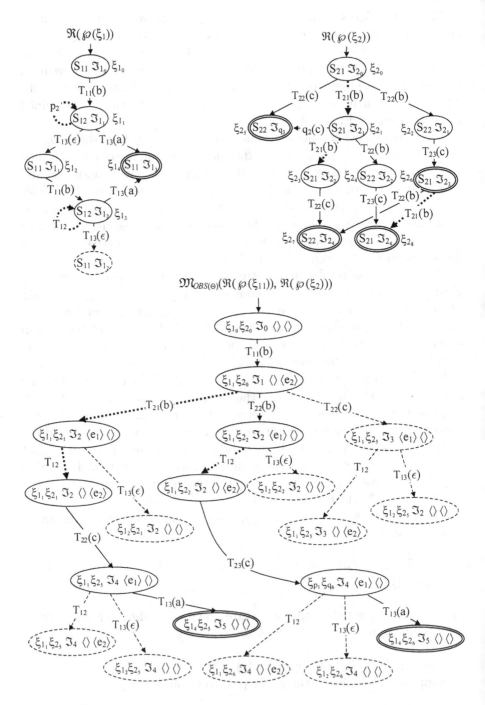

Figure 9.11. Modular reconstruction relevant to Example 9.13.

the triggerable transitions among the candidates, we need to consider two constraints:

(1) The availability of the input event within the dangling set;

(2) The consistency with the current state $\Im_1$ of the index space.

The first constraint restricts the set of candidate transitions to T_{21} and T_{22}. The second one allows for $T_{21}(b)$, $T_{22}(b)$, and $T_{22}(c)$, as $Viable(\Im_1) = \{b, c\}$.

Following this merging pattern, the global active space is eventually completed as highlighted on the bottom of Figure 9.11. As expected, it incorporates the same two histories of the active space generated by the monolithic technique in Example 9.11 and displayed in Figure 9.8. ◇

The equivalence between the monolithic and modular reconstruction is formally stated by Theorem 9.4, where $\mathfrak{M}$ denotes the generalized merging operator (see Example 9.13), that is, the merging of $n \geq 2$ active spaces.

Theorem 9.4. *Let $\wp(\xi) = (\Omega(\xi), OBS(\xi), \xi_0)$ be an uncertain diagnostic problem for a cluster ξ, $\Xi = \{\xi_1, \ldots, \xi_n\}$ a decomposition of ξ, and*

$$\mathbb{R} = \bigcup_{i=1}^{n} \Re(\wp(\xi_i))$$

a set of relevant active spaces, where $\wp(\xi_i) = (\Omega(\xi_i), OBS(\xi_i), \xi_{i_0})$, $\xi_0 = (\xi_{1_0}, \ldots, \xi_{n_0})$, and $\forall i \in [1 .. n]$ $(\Omega(\xi_i) = \Omega_{\langle\xi_i\rangle}(\xi), OBS(\xi_i) = OBS_{\langle\xi_i\rangle}(\xi))$. Then, the extension of the active space yielded by the merging operator applied to $\mathbb{R}$ equals the extension of the active space yielded by the monolithic reconstruction of $\wp(\xi)$, that is,

$$\|\mathfrak{M}_{OBS(\xi)}(\mathbb{R})\| \equiv \|\Re(\wp(\xi))\|.$$

5.   Summary

- *Loss uncertainty requirement*: the expressive power of component behavioral models has to enable the representation of messages that may be lost.

- *Logical uncertainty requirement*: the expressive power of both component behavioral models and observation descriptions has to enable the representation of messages whose values are not univocal.

- *Source uncertainty requirement*: the expressive power of observation descriptions has to enable the representation of messages whose sender components are not univocally known by the observer.

- *Temporal uncertainty requirement*: the expressive power of observation descriptions has to enable the representation of messages whose temporal ordering is not univocally known by the observer.

- *Uncertain observation*: observation featuring logical, source, and temporal uncertainty, where source uncertainty and temporal uncertainty are assumed to be independent one of the other.

- *Observation graph*: pictorial description of an uncertain observation, wherein a node represents a (possibly logically uncertain) message and an arrow a temporal precedence relationship.

- *Uncertain diagnostic problem*: diagnostic problem featuring an uncertain observation.

- *Uncertain-observation index*: index to be used within each state of the reconstruction space, so as to represent the part of the uncertain observation that has already been generated by all the paths leading from the initial state to the current state. Each index value partitions the set of messages of the whole reaction observation into two sets, depending on whether a message has already been generated (consumed) when the current state is reached or not. Given such an index, it is possible to infer which are the next messages to be generated.

- *Uncertain-observation restriction*: projection of an uncertain observation inherent to a cluster on one of its subclusters.

- *Uncertain-observation instance*: totally temporally ordered sequence of observable labels that is compliant with a given uncertain observation.

- *Uncertain-observation extension*: set of all the instances of a given uncertain observation.

- *(Deterministic) index space of an uncertain observation*: deterministic graph produced by processing an uncertain observation. In such a graph each node is an index of the considered uncertain observation and each transition is marked by one or more observable labels such that the reception of any of them causes the same index change.

- *Extension of the index space of an uncertain observation*: set of all the sequences of observable labels compliant with a given index space. It is proven that it equals the extension of the observation itself.

- *Restriction of the index space of an uncertain observation*: projection of the index space of the uncertain observation inherent to a cluster

on one of its subcluster. It is proven that it equals the index space
of the restriction of the observation itself.

Appendix A
Algorithms

Algorithm A.1. (*Closure*) The *Closure* function takes as input an uncertain observation $OBS(\xi)$ and the frontier of the index $\Im$ of $OBS(\xi)$, and yields a set F^+ of nodes in $OBS(\xi)$. The essential part of the algorithm is the loop in Lines 4 to 11, in which F^+ is incrementally updated with those nodes that are either preceded by nodes in $Cons(\Im)$ or by nodes including the null label ϵ.

> **function** *Closure*($OBS(\xi), Front(\Im)$): $F^+ \subseteq \mathbf{N}$
> **input**
> $OBS(\xi) = (\mathbf{N}, \Upsilon, \mathbf{N}_0, \mathbf{N}_f)$: an uncertain observation of cluster ξ,
> $Front(\Im)$: frontier of the index $\Im$;
> 1. **begin**
> 2. $F^+ := \emptyset$;
> 3. $Temp := Front(\Im)$;
> 4.   **repeat**
> 5. $Temp' := \emptyset$;
> 6.   **for each** $\omega \in \mathbf{N}$ **such that**
> $\forall \omega' \mapsto \omega \in \Upsilon$ (($\omega' \in Temp$, $\epsilon \in \|Msg(\omega')\|$) **or** $\omega' \in Cons(\Im)$) **do**
> 7. $Temp' := Temp' \cup \{\omega\}$
> 8.   **end-for**
> 9. $F^+ := F^+ \cup Temp'$;
> 10. $Temp := Temp \cup Temp'$
> 11.   **until** $Temp' = \emptyset$;
> 12.   **return** F^+
> 13. **end.**

Appendix B
Proofs of Theorems

Theorem 9.1 *Let $Space(OBS(\xi))$ be the index space of an uncertain observation $OBS(\xi)$. Then, the extension of the index space equals the extension of the observation, namely:*

$$\|Space(OBS(\xi))\| \equiv \|OBS(\xi)\|.$$

Proof. Since the extension of the deterministic index space equals the extension of the nondeterministic index space, it is enough to prove that

$$\|Nspace(OBS(\xi))\| \equiv \|OBS(\xi)\|.$$

That is, denoting with s a generic sequence of observable labels, we have to prove the following bidirectional entailment:

$$s \in \|Nspace(OBS(\xi))\| \Leftrightarrow s \in \|OBS(\xi)\|.$$

To this end, we extend the notion of an extension to the rough index space as follows: $\|Rspace(OBS(\xi))\|$ is composed of the whole set of sequences of ω_i marking a path from the initial state to a final state of $Rspace(OBS(\xi))$.

Now, assume $s \in \|OBS(\xi)\|$. According to the definition of an uncertain-observation extension (see Section 3.3) , there exists (at least) a relevant sequence of nodes $\mathcal{Q}_\omega = \langle \omega_1, \dots, \omega_p \rangle$ that consume all of the nodes in $\mathbf{N}(OBS(\xi))$ whilst respecting the precedence relationships imposed by $OBS(\xi)$. The string s is obtained by picking out a label $\ell_i \in \|Msg(\omega_i)\|$ and by discarding null messages. The sequence $\mathcal{Q}_\omega$ belongs to $\|Rspace(OBS(\xi))\|$ as, intuitively, the latter represents all of the possible modes in which nodes of $OBS(\xi)$ can be consumed. Thus, since $Nspace(OBS(\xi))$ is obtained by replacing an edge in $Rspace(OBS(\xi))$ marked by ω_i with several edges, each marked by a label in $\|Msg(\omega_i)\|$, s will belong to $\|Nspace(OBS(\xi))\|$ too.

Conversely, assume that $s \in \|Nspace(OBS(\xi))\|$. This means that $s \in \|\mathcal{Q}_\omega\|$, where $\mathcal{Q}_\omega \in \|Rspace(OBS(\xi))\|$. Since a path in $Rspace(\xi)$ represents a mode in which nodes $\omega_i \in \mathbf{N}(OBS(\xi))$ can be consumed, it follows that, from the definition of an uncertain-observation extension (see Section 3.3), $\mathcal{Q}_\omega$ will include s as an instance of $\|OBS(\xi)\|$. This concludes the proof of Theorem 9.1. $\qquad\square$

Theorem 9.2 *Let $\Re$ denote the reconstruction operator yielding the active space relevant to a given diagnostic problem. Let $\wp(\xi) = (\Omega(\xi), OBS(\xi), \xi_0)$ be an uncertain diagnostic problem where*

$$\|OBS(\xi)\| = \bigcup_{i=1}^{n} OBS_{(i)}(\xi)$$

is the set of instances of $OBS(\xi)$. Let $\wp_{(i)}(\xi) = (\Omega(\xi), OBS_{(i)}(\xi), \xi_0)$ denote the diagnostic problem relevant to the single observation instance $OBS_{(i)}(\xi)$. Then, the extension of the active space relevant to the uncertain problem equals the union of the extensions of the active spaces relevant to the single problems, that is,

$$\|\Re(\wp(\xi))\| \equiv \bigcup_{i=1}^{n} \|\Re(\wp_{(i)}(\xi))\|. \tag{B.1}$$

Proof. Proving Equivalence (B.1) amounts to proving the following two entailments, where h denotes a history:

$$h \in \|\Re(\wp(\xi))\| \Rightarrow \exists i \in [1 \mathinner{\ldotp\ldotp} n] \ (h \in \|\Re(\wp_{(i)}(\xi))\|); \tag{B.2}$$

$$h \in \|\Re(\wp_{(i)}(\xi))\|, i \in [1 \mathinner{\ldotp\ldotp} n] \Rightarrow h \in \|\Re(\wp(\xi))\|. \tag{B.3}$$

To prove Formula (B.2), assume $h \in \|\Re(\wp(\xi))\|$, where h is a sequence of transitions T_i associated with a (possibly null) label ℓ_i, that is, $h = \langle T_1(\ell_1), \ldots, T_m(\ell_m) \rangle$. Let $\lambda = \langle \ell'_1, \ldots, \ell'_k \rangle$, $k \leq m$, be the sequence of observable (non-null) labels generated by h. Let $OBS_{(i)}(\xi)$ be the observation instance isomorphic to λ and $\wp_{(i)}(\xi) = (OBS_{(i)}(\xi), \xi_0)$ the relevant diagnostic problem. Note that $Space(OBS_{(i)}(\xi))$ is isomorphic to λ, that is, the former is a linear graph composed of $k + 1$ nodes and k edges, the i-th edge being labeled by ℓ'_i. In other words,

$$\forall i \in [0 \mathinner{\ldotp\ldotp} (k-1)] \ (Viable(\Im_i) = \{\ell'_{i+1}\}),$$

while $Viable(\Im_k) = \emptyset$. By virtue of Theorem 9.1, $\lambda \in \|Space(OBS(\xi))\|$, that is, there exists a path in $Space(OBS(\xi))$ which is marked by the sequence of labels in λ. This implies that there exists in $Space(OBS(\xi))$ a path $P(\lambda)$ isomorphic to $Space(OBS_{(i)}(\xi))$. We prove that $h \in \|\Re(\wp_{(i)}(\xi))\|$ by induction on the transitions of h.

Basis. The initial states of $\|\Re(\wp(\xi))\|$ and $\|\Re(\wp_{(i)}(\xi))\|$ coincide, being $S_0 = (\sigma_0, \Im_0, D_0)$, where $\sigma_0 = \xi_0$, $\Im_0 = \emptyset$, and $D_0 = \emptyset$.

Induction. First, note that, due to the isomorphism of $P(\lambda)$ and $Space(OBS_{(i)}(\xi))$, we may define a state $N_u = (\sigma_u, \Im_u, D_u)$ in $\|\Re(\wp(\xi))\|$ *isomorphic* to a state $N_c = (\sigma_c, \Im_c, D_c)$ in $\|\Re(\wp_{(i)}(\xi))\|$ when $\sigma_u = \sigma_c$, $D_u = D_c$, and $\Im_u$ corresponds to $\Im_c$ in $P(\lambda)$. In such a case, $Viable(\Im_c) \subseteq Viable(\Im_u)$. Let N_u and N_c denote two isomorphic nodes in $\Re(\wp(\xi))$ and $\Re(\wp_{(i)}(\xi))$, respectively. We have to prove that, if $N_u \xrightarrow{T} N'_u$ is in $h \in \|\Re(\wp(\xi))\|$ and N_u is isomorphic to N_c, then $N_c \xrightarrow{T} N'_c$ is in $\Re(\wp_{(i)}(\xi))$, where N'_c is isomorphic to N'_u. This comes from the fact that, since the two states are isomorphic, the set of triggerable transitions is the same. This is clear when T is not observable, since the fact that such a transition is triggerable does not depend on the index, and the other two fields, σ and D, are the same. When T is observable and ℓ is the generated message, if $T(\ell)$ is triggerable from N_u, then it is triggerable from N_c too. In fact, on the one hand, $N'_u = (\sigma'_u, \Im'_u, D'_u)$, where

$\Im'_u$ is the next index in $P(\lambda)$. On the other, $T(\ell)$ is triggerable from N_c since, due to the isomorphism between $Space(OBS_{(i)}(\xi))$ and $P(\lambda)$, $Viable(\Im_c) = \{\ell\}$. Thus, $N'_c = (\sigma'_c, \Im'_c, D'_c)$, where $\sigma'_u = \sigma_c$, $D'_u = D'_c$, and $\Im'_c$ corresponds to $\Im_u$ in $P(\lambda)$, in other words, N'_u is isomorphic to N'_c, which concludes the induction and, therefore, the proof of formula (B.2).

Now we prove Formula (B.3). We assume that h is a history within the active space relevant to a diagnostic problem $\wp_{(i)}(\xi) = (OBS_{(i)}(\xi), \xi_0)$. Let λ be the sequence of observable labels generated by h. From Theorem 9.1, $\lambda \in \|Space(OBS(\xi))\|$. We denote with $P(\lambda)$ the relevant path in $\Re(\wp(\xi))$. $P(\lambda)$ is isomorphic to $Space(OBS_{(i)}(\xi))$. We prove that $h \in \|\Re(\wp(\xi))\|$ by induction on the transitions of h.

Basis. The initial states of $\|\Re(\wp(\xi))\|$ and $\|\Re(\wp_{(i)}(\xi))\|$ coincide.

Induction. We have to prove that, if $N_c \xrightarrow{T} N'_c$ is in $h \in \|\Re(\wp_{(i)}(\xi))\|$ and N_c is isomorphic to N_u, then $N_u \xrightarrow{T} N'_u$ is in $\Re(\wp(\xi))$, where N'_u is isomorphic to N'_c. This comes from the fact that, since the two states N_c and N_u are isomorphic, the set of triggerable transitions is the same. This is clear when T is not observable. When T is observable and ℓ is the generated message, if $T(\ell)$ is triggerable from N_c, then it is triggerable from N_u too. In fact, on the one hand, $N'_c = (\sigma'_c, \Im'_c, D'_c)$, where $\Im'_c$ is the next index in $Space(OBS_{(i)}(\xi))$. On the other, $T(\ell)$ is triggerable from N_u since, due to the isomorphism between $Space(OBS_{(i)}(\xi))$ and $P(\lambda)$, $\ell \in Viable(\Im_u)$. Thus, $N'_u = (\sigma'_u, \Im'_u, D'_u)$, where $\sigma'_c = \sigma_u$, $D'_c = D'_u$, and $\Im'_u$ in $P(\lambda)$ corresponds to $\Im_c$, in other words, N'_c is isomorphic to N'_u, which concludes the induction and, therefore, the proof of formula (B.3), as well as the proof of Theorem 9.2. $\square$

Theorem 9.3 *Let $Space(OBS(\xi))$ be the index space of an uncertain observation of cluster ξ, and $\xi' \subset \xi$ a sub-cluster of ξ. Then, the extension of the index space of the restriction of $OBS(\xi)$ on ξ' equals the extension of the restriction on ξ' of the index space of the observation, that is:*

$$\|Space(OBS_{\langle\xi'\rangle}(\xi))\| \equiv \|Space_{\langle\xi'\rangle}(OBS(\xi))\|. \tag{B.4}$$

Proof. Equivalence (B.4) can be unfolded as follows:

$$\|\mathcal{D}(\mathcal{N}(Rspace(OBS_{\langle\xi'\rangle}(\xi))))\| \equiv \|\mathcal{D}_{\langle\xi'\rangle}(\mathcal{N}(Rspace(OBS(\xi))))\|, \tag{B.5}$$

where $\mathcal{N}$ denotes the operator yielding the nondeterministic index space relevant to the given rough index space, while $\mathcal{D}$ denotes the operator yielding the deterministic index space relevant to the given nondeterministic index space. Equivalence (B.5) can be split into Equivalences (B.6) and (B.7):

$$\|\mathcal{N}(Rspace(OBS_{\langle\xi'\rangle}(\xi)))\| \equiv \|\mathcal{N}_{\langle\xi'\rangle}(Rspace(OBS(\xi)))\|; \tag{B.6}$$

$$\|\mathcal{D}_{\langle\xi'\rangle}(Nspace(OBS(\xi)))\| \equiv \|\mathcal{D}(Nspace_{\langle\xi'\rangle}(OBS(\xi)))\|. \tag{B.7}$$

To prove Equivalence (B.6), we first define the *conservative restriction* of an uncertain observation $OBS(\xi)$ on a cluster $\xi' \subset \xi$, denoted by $OBS_{\ll\xi'\gg}(\xi)$. Such a definition is the same as that of uncertain-observation restriction (see Section 3.2) except that all of the nodes are maintained in the result, rather then discarding those for which Condition (9.6) hold. Thus, roughly, a conservative restriction is a restriction in which all of the restricted nodes are preserved in the resulting observation graph, even the nodes whose message extension is the singleton $\{\epsilon\}$, named *empty nodes*. Based on the notion of a conservative restriction, Equivalence (B.6) can be split on its turn into Equivalences (B.8) and (B.9):

$$\|Nspace_{\langle\xi'\rangle}(OBS(\xi))\| \equiv \|Nspace(OBS_{\ll\xi'\gg}(\xi))\|; \tag{B.8}$$

$$\|Nspace(OBS_{\ll\xi'\gg}(\xi))\| \equiv \|Nspace(OBS_{\langle\xi'\rangle}(\xi))\|. \tag{B.9}$$

Equivalence (B.8) can be understood insofar as the conservative restriction preserves the topology of the observation graph, so that $Rspace(OBS_{\ll\xi'\gg}(\xi))$ will have the same topology and marking labels as $Rspace(OBS(\xi))$. What differs are the extensions of the ω_i as, in contrast with $OBS(\xi)$, in $OBS_{\ll\xi'\gg}(\xi)$ the ω_i are restricted. On the other hand, when generating the nondeterministic index space from $Rspace(OBS_{\ll\xi'\gg}(\xi))$, the marking labels will be exactly those which are obtained by restricting $Nspace(OBS(\xi))$ on ξ'. Therefore, Equivalence (B.8) holds.

Now we prove Equivalence (B.9), which can be rephrased as follows:

$$s \in \|Nspace(OBS_{\ll\xi'\gg}(\xi))\| \Leftrightarrow s \in \|Nspace(OBS_{\langle\xi'\rangle}(\xi))\|. \tag{B.10}$$

To prove Formula (B.10), assume $s \in \|Nspace(OBS_{\ll\xi'\gg}(\xi))\|$. The string s is formed by the observable labels marking a path in the relevant nondeterministic index space. Such labels are picked out from the extensions of the node messages which have been ordered in a sequence Q that respects the precedence relationships of the relevant observation $OBS_{\ll\xi'\gg}(\xi)$. Since $OBS_{\langle\xi'\rangle}(\xi)$ differs from $OBS_{\ll\xi'\gg}(\xi)$ in the missing empty nodes and since these nodes are immaterial to the generation of s, it is possible to find a sequence of nodes in $OBS_{\langle\xi'\rangle}(\xi)$ which is the restriction of Q to the non-empty nodes, and pick out the same observable labels in s. In other terms, $s \in \|Nspace(OBS_{\langle\xi'\rangle}(\xi))\|$. Likewise, assume to have a string $s \in \|Nspace(OBS_{\langle\xi'\rangle}(\xi))\|$, which has been picked out from a sequence Q of nodes in $OBS_{\langle\xi'\rangle}(\xi)$. Consider the sequence Q' of nodes in $OBS_{\ll\xi'\gg}(\xi)$ whose restriction on the non-empty nodes equals Q. Since it is possible to pick out the same observable labels as s, we may claim that $s \in \|Nspace(OBS_{\ll\xi'\gg}(\xi))\|$, which concludes the proof of Equivalence (B.9) as well as the proof of Equivalence (B.6).

Now we prove Equivalence (B.7), which can be rephrased as follows:

$$s \in \|\mathcal{D}_{\langle\xi'\rangle}(Nspace(OBS(\xi)))\| \Leftrightarrow s \in \|\mathcal{D}(Nspace_{\langle\xi'\rangle}(OBS(\xi)))\|.$$

On the one hand, assume $s \in \|\mathcal{D}(Nspace_{\langle\xi'\rangle}(OBS(\xi)))\|$. Due to the equivalence of the deterministic and nondeterministic automata, the string s is generated by a path p_n' in $Nspace_{\langle\xi'\rangle}(OBS(\xi))$, where only observable labels are considered[1]. In its turn, p_n' is obtained as a restriction on ξ' of (at least) a path p_n in $Nspace(OBS(\xi))$. Consider the path p_d in $\mathcal{D}(Nspace(OBS(\xi)))$ generating the same sequence of observable labels as p_n. Among the strings in $\mathcal{D}_{\langle\xi'\rangle}(Nspace(OBS(\xi)))$ obtained by restricting p_d on ξ' is also s, as the same restriction rules are applied (see the definition of an uncertain-observation restriction in Section 3.2) to the labels marking p_n. In other terms, $s \in \|\mathcal{D}_{\langle\xi'\rangle}(Nspace(OBS(\xi)))\|$.

On the other hand, assume $s \in \|\mathcal{D}_{\langle\xi'\rangle}(Nspace(OBS(\xi)))\|$. The string s is generated by a path p_n' in $\mathcal{D}_{\langle\xi'\rangle}(Nspace(OBS(\xi)))$. In its turn, p_n' is obtained as a restriction on ξ' of (at least) a path p_d in $\mathcal{D}(Nspace(OBS(\xi)))$. The path p_d is obtained from

---

[1]Several paths in $Nspace_{\langle\xi'\rangle}(OBS(\xi))$ may give rise to the same s, which only differ in the number and disposition of null labels: p_n' is one of them.

(at least) a path p_n in $Nspace(OBS(\xi))$. The restriction of p_n on ξ' yields, among other paths, a path p'_n in $Nspace_{\langle \xi' \rangle}(OBS(\xi))$ whose associated sequence of observable labels is s. Due to the equivalence between the nondeterministic and deterministic automata, the same path p_d will be included in $\mathcal{D}(Nspace_{\langle \xi' \rangle}(OBS(\xi)))$. In other terms, $s \in \|\mathcal{D}_{\langle \xi' \rangle}(Nspace(OBS(\xi)))\|$, which concludes the proof of Equivalence (B.9) and, consequently, the proof of Equivalence (B.6). The latter brings us to the conclusion of the proof of Equivalence (B.5), that is, the proof of Theorem 9.3. □

Theorem 9.4 *Let* $\wp(\xi) = (\Omega(\xi), OBS(\xi), \xi_0)$ *be an uncertain diagnostic problem for a cluster* ξ, $\Xi = \{\xi_1, \ldots, \xi_n\}$ *a decomposition of* ξ, *and*

$$\mathbb{R} = \bigcup_{i=1}^{n} \Re(\wp(\xi_i))$$

a set of relevant active spaces, where $\wp(\xi_i) = (\Omega(\xi_i), OBS(\xi_i), \xi_{i_0})$, $\xi_0 = (\xi_{1_0}, \ldots, \xi_{n_0})$, *and* $\forall i \in [1 .. n]$ $(\Omega(\xi_i) = \Omega_{\langle \xi_i \rangle}(\xi), OBS(\xi_i) = OBS_{\langle \xi_i \rangle}(\xi))$. *Then, the extension of the active space yielded by the merging operator applied to* $\mathbb{R}$ *equals the extension of the active space yielded by the monolithic reconstruction of* $\wp(\xi)$, *that is,*

$$\|\mathfrak{M}_{OBS(\xi)}(\mathbb{R})\| \equiv \|\Re(\wp(\xi))\|. \tag{B.11}$$

Proof. Theorem 9.4 is the generalization of Theorem 6.1 for uncertain observations. Hereafter, we give some arguments for why it holds. First, we need a more formal definition of an active space, that is, a 5-tuple $(\mathbb{S}, \mathbb{E}, \mathbb{T}, S_0, \mathbb{S}_f)$, where $\mathbb{S}$ is the set of states, $\mathbb{E}$ the set of events, $\mathbb{T}$ the transition function, S_0 the initial state, and $\mathbb{S}_f$ the set of final states. Let $\mathcal{A}_i = (\mathbb{S}_i, \mathbb{E}_i, \mathbb{T}_i, S_{0_i}, \mathbb{S}_{f_i})$, be the active space relevant to $\wp(\xi_i)$, $i \in [1 .. n]$. Let $\mathcal{A} = (\mathbb{S}, \mathbb{E}, \mathbb{T}, S_0, \mathbb{S}_f)$ and $\widehat{\mathcal{A}} = (\widehat{\mathbb{S}}, \widehat{\mathbb{E}}, \widehat{\mathbb{T}}, \widehat{S}_0, \widehat{\mathbb{S}}_f)$ be the active spaces relevant to $\mathfrak{M}_{OBS(\xi)}(\mathbb{R})$ and $\Re(\wp(\xi))$, respectively. The proof of Theorem 9.4 amounts to proving that there exists an isomorphism between $\mathcal{A}$ and $\widehat{\mathcal{A}}$, specifically, the following equivalences:

$$\mathbb{S} \equiv \widehat{\mathbb{S}}, \tag{B.12}$$

$$\mathbb{E} \equiv \widehat{\mathbb{E}}, \tag{B.13}$$

$$\mathbb{T} \equiv \widehat{\mathbb{T}}, \tag{B.14}$$

$$S_0 \equiv \widehat{S}_0, \tag{B.15}$$

$$\mathbb{S}_f \equiv \widehat{\mathbb{S}}_f. \tag{B.16}$$

The proofs of Equivalences (B.13) and (B.15) are quite simple. Then, we should prove Equivalence (B.14), which can be rephrased as follows:

$$\mathbb{T} \subseteq \widehat{\mathbb{T}}, \tag{B.17}$$

$$\widehat{\mathbb{T}} \subseteq \mathbb{T}. \tag{B.18}$$

Each of Equivalences (B.17) and (B.18) can be proved by induction on the nodes of the relevant active spaces, where the basis is in fact Equivalence (B.15).

Consider Equivalence (B.17). The induction is the following:

$$(T \in \mathbb{T}, T = N \xrightarrow{\tau} N', N = \widehat{N}, \widehat{N} \in \widehat{\mathbb{S}}) \Rightarrow (\widehat{T} \in \widehat{\mathbb{T}}, T = \widehat{N} \xrightarrow{\tau} \widehat{N}', \widehat{N}' = N'). \tag{B.19}$$

Assuming the LHS of Formula (B.19), where $N = \widehat{N} = (\sigma, \Im, D)$, means that:

(1) The transition τ is among those marking the edges leaving one of the nodes in $\mathbb{S}_i$, corresponding to the current node $N \in \mathbb{S}$;

(2) If the input E of τ comes from a link L from ξ_i to a different cluster in Ξ, E must be the first event in the queue of events within L;

(3) If τ is a transition of a visible component, the message it generates must be consistent with $OBS(\xi)$;

(4) If τ generates an output E towards a link L from ξ_i to a different cluster in Ξ, E must be consistent with the management policy of L.

Based on these conditions, we have to show that the RHS of Formula (B.19) holds, that is:

(1) The transition τ is among those marking the edges leaving one of the nodes in $\widehat{\mathbb{S}}$;

(2) If the input E of τ comes from a link L within the links of ξ, E must be the first event in the queue of events within L;

(3) If τ is a transition of a visible component, the message it generates must be consistent with $OBS(\xi)$;

(4) If τ generates an output E towards a link L within the links of ξ, E must be consistent with the management policy of L.

Since the proofs of Conditions 1, 2, and 4 do not depend on the observation, we draw our attention to Condition 3, which can be unfolded into the following disjunctive conditions (where $\tau = S \xrightarrow{\alpha \mid \beta} S'$):

(a) $(m, Out) \notin \beta$;

(b) $(m, Out) \in \beta$ and $\epsilon \in \|m\|$;

(c) $(m, Out) \in \beta$ and $(\|m\| \cap Viable(\mathfrak{S})) \neq \emptyset$.

Assuming that τ marks an edge leaving $\widehat{N}$ means that, based on the merging process, τ marks as well an edge leaving one node within the set of nodes $N_i \in \mathbb{S}_i$ corresponding to $\widehat{N}$. When either Condition (a) or (b) is met, τ is triggerable from $\widehat{N}$ too, as no constraints are imposed on the extension of m. On the other hand, when τ is observable and the extension of $\|m\|$ does not include ϵ (Condition (c)), then we pick out a label $\ell \in (\|m\| \cap Viable(\mathfrak{S}))$ and consider transition $\tau(\ell)$, which is assumed to mark an edge leaving N, that is, $N \xrightarrow{\tau(\ell)} N' \in \mathbb{T}$. Since $N = \widehat{N}$, the index field $\mathfrak{S}$ is the same in both N and $\widehat{N}$. Consequently, if $N \xrightarrow{\tau(\ell)} N'$ is consistent in $\mathcal{A}$, then $\widehat{N} \xrightarrow{\tau(\ell)} \widehat{N}'$ will be consistent in $\widehat{\mathcal{A}}$ too. This concludes the proof of Formula (B.19). Consider now Equivalence (B.18). The induction is the following:

$$(\widehat{T} \in \widehat{\mathbb{T}}, \widehat{T} = \widehat{N} \xrightarrow{\tau} \widehat{N}', \widehat{N} = N, N \in \mathbb{S}) \Rightarrow (T \in \mathbb{T}, T = N \xrightarrow{\tau} N', N' = \widehat{N}'). \quad \text{(B.20)}$$

Again, we focus our proof on the consistency of the transition with the observation, as expressed by Conditions (a), (b), and (c) above. Now, the starting point is the assumption that τ is a transition marking an edge leaving a node $\widehat{N} = (\widehat{\sigma}, \widehat{\mathfrak{S}}, \widehat{D})$ within $\widehat{\mathcal{A}}$. Since $\widehat{N} = N$, we have $\widehat{\mathfrak{S}} = \mathfrak{S}$. The proof of Formula (B.20) is supported by the following lemma.

Lemma 9.4.1 *Let* $\widehat{N} = (\widehat{\sigma}, \widehat{\mathfrak{S}}, \widehat{D})$ *be a node in* $\widehat{\mathcal{A}}$ *and* $N_1 = (\sigma_1, \mathfrak{S}_1, D_1), \ldots, N_n = (\sigma_n, \mathfrak{S}_n, D_n)$ *the corresponding nodes in* $\mathcal{A}_1, \ldots, \mathcal{A}_n$, *respectively, from which* $\widehat{N}$ *is* made up by means of the merging operation. Let each $Space(OBS(\xi_i))$ be obtained, by virtue of Theorem 9.3, as the restriction $Space_{\langle\xi_i\rangle}(OBS(\xi))$. Let $\|\Im_i\|$ denote the subset of indexes that, according to the subset construction algorithm, identify $\Im_i$ in the restricted index space. Then,

$$\forall i \in [1 .. n] \ (\widehat{\Im} \in \|\Im_i\|). \tag{B.21}$$

Proof. Let $\mathcal{P} = \widehat{S}_0 \rightsquigarrow \widehat{N}$ be a path in $\widehat{\mathcal{A}}$, that is, a prefix of a history in $\|\widehat{\mathcal{A}}\|$ involving node $\widehat{N}$. Consider an active space $\mathcal{A}_i$ and the sequence λ_i obtained from $\mathcal{P}$ as follows. For each edge in $\mathcal{P}$, let $\tau(\ell)$ be the relevant marking transition. If τ is relevant to a component in ξ_i, then insert ℓ into λ_i, otherwise insert ϵ into λ_i. Two cases are possible:

(1) λ_i includes ϵ symbols only. This means that the nondeterministic automaton $\mathcal{N}$ obtained in Step 1 of the definition of an index space restriction (see Section 4.5.1) will have a null path, that is, a path composed of ϵ labels only, from the root $\widehat{\Im}_0$ to $\widehat{\Im}$. Consequently, $\Im_i$ will coincide with the root of $Space_{\langle\xi_i\rangle}(OBS(\xi))$ and $\widehat{\Im}$ will belong to $\|\Im_i\|$, since $\|\Im_i\|$ will include all of the indexes on the null path, from $\widehat{\Im}_0$ to $\widehat{\Im}$.

(2) $\lambda_i = \gamma a \eta$, where γ is a sequence of either observable or ϵ labels, a an observable label, and η a (possibly empty) sequence of ϵ labels. Let $\widehat{\Im}_a$ be the state reached in $\mathcal{N}$ with γa. $\mathcal{N}$ will have a null path from $\widehat{\Im}_a$ to $\widehat{\Im}$. Besides, due to the subset construction algorithm, $\widehat{\Im}_a$ will be included in $\|\Im_i\|$. Due to the null path η, $\widehat{\Im}$ will be included in $\|\Im_i\|$ as well.

Thus, we have proved Formula (B.21), that is, Lemma 9.4.1.

Based on Lemma 9.4.1, we may conclude the proof of Formula (B.20) as follows. Consider the active space $\mathcal{A}_i$ such that ξ_i incorporates the component relevant to transition τ. Consider as well the node $N_i = (\sigma_i, \Im_i, \mathcal{D}_i) \in \widehat{S}_i$ corresponding to $\widehat{N}$. By virtue of Lemma 9.4.1, $\widehat{\Im} \in \Im_i$. Thus, if ℓ is the observable label associated with τ, ℓ will be a marking label for an edge leaving $\Im_i$ as well, in other words, τ will be a consistent transition marking an edge leaving N_i. Therefore, τ will be a consistent transition marking an edge leaving $\mathcal{N}$ too. This concludes the proof of Formula (B.20) as well as the proof of Formula (B.18), which brings to the conclusion of the proof of Equivalence (B.14).

The proofs of Equivalences (B.12) and (B.16) follow directly from the way in which Equivalence (B.14) has been proved, as detailed in the proof of Theorem 6.1. This brings us to the conclusion of the proof of Theorem 9.4. $\qquad\square$

Chapter 10

COMPLEX OBSERVATIONS

Abstract

An uncertain observation of a DES can be ambiguous in the logical content of observed events, in their temporal relationships, and in their source components. However the precedence relationships involving each observed event of an uncertain observation are assumed to be independent of its sender component. When this assumption does not hold, the observation cannot be defined as uncertain any more. Thus the notion of an uncertain observation is further generalized to that of a complex observation and then exploited by a diagnostic approach pertinent to active systems, which is similar to that proposed when the diagnostic problem features an uncertain observation. In fact, the diagnostic approach processes the given complex observation in order to express by means of a concise deterministic graph, called an observation index space, all the sequences of (both logically and source certain) observable events that are consistent with the given observation, where each of them is called an observation instance. Then, the index space of the observation instead of the observation itself is exploited during history reconstruction.

1. Introduction

In the previous chapters, we have implicitly assumed that, given a system Θ and an observer $\Omega(\Theta) = (\mathbb{V}, \mathbb{P})$ of such a system, any observation $OBS(\Theta)$ be compliant with the following two constraints:

- Visible components belonging to distinct parts of $\mathbb{P}$ do not share any message (i.e. each visible component belonging to a part does not emit any homonymous message with respect to visible components belonging to all of the other parts);

- Given an observation, the temporal characterization of the messages coming from each single part of $\mathbb{P}$ is independent of the specific sender component of each message (that is, only one set of temporal relations among messages coming from the same part is given and not several, depending on the sender components of specific messages).

In particular, the second constraint implies that the sender component of each distinct occurrence of a shared message (where such occurrences could so far be emitted only by components belonging to the same part of $\mathbb{P}$) is not specified within the observation.

These constraints are relaxed by the notion of a *complex observation* that will be given in the next section. Therefore, such a notion is actually a generalization of the notions of observations considered up to now in this book.

Based on the informal discussion of Section 2.4 of Chapter 9, complex observations result from the combination of all of the three uncertainty requirements affecting observations, provided that source and temporal uncertainties depend on each other. When this condition holds, the precedence relationships involving source uncertain messages vary depending on the sender components of such messages.

Assume that message m_1 is source uncertain, while message m_2 is generated by component C_2. It may occur that m_1 precedes m_2 if we assume that the sender component of m_1 is C_2, while the reciprocal temporal order of the two messages is unknown if we assume that the sender of m_1 is C_1. This is the case, for example, when the messages emitted by C_1 and C_2 are stamped by distinct clocks and the temporal distance between the timestamps of m_1 and m_2 is less than the synchronization error of the two clocks.

Or assume, for instance, that

- The transmission means from the system to the observer be such that the reciprocal emission order of two occurrences of a shared message equals the reception order in case they come from the same component while it is unknown in case the two sender components are distinct;

- The observer does not know exactly which is the sender component of a received shared message.

Then, from the point of view of the observer, the emission order of the two occurrences varies depending on the hypothesized sender components.

Thus, in theory, a complex observation is represented by several distinct observation graphs, that is, several distinct uncertain observations, each pertaining to a distinct assumption on the sender components of source uncertain messages. In practice, a concise representation of all of such observation graphs has been envisaged.

The chapter is organized as follows: Section 2 introduces the notion of a complex observation. Section 3 defines the concept of an index of a complex observation, which, analogously to the index of an uncertain observation, is a deterministic automaton obtained by processing the observation prior to history reconstruction. Then, such an index is exploited for both monolithic and modular resolutions of diagnostic problems featuring complex observations.

2. Complex Observation

Like an uncertain observation, a *complex observation* is represented by a single (not necessarily connected) DAG, called *complex-observation graph*. Each node of the DAG represents a received message and is in general marked by a set of contextual nodes, each denoting a distinct assumption on the sender component of such a message. A contextual node is, on its turn, marked by a (either constant or variable) message associated with a set of components. Such components represent the possible source of the message.

Each node in the DAG, since representing a message received by the observer, has to comply with the constraint expressed in Section 2.2 of Chapter 9, according to which the set of labels hypothesized by the observer on arrival of a message must include the label that was actually generated by the system. In the context of complex observations, this means that, for each node in the DAG, one (and only one) of the labels within the union of all of the message extensions inherent to its (inner) contextual nodes is assumed to have been generated by the system. No assumption is made on the intersection of the extensions of messages within the same node. Consequently, the same label may be associated with different sets of components.

Precedence relationships are defined on contextual nodes. A node of an uncertain observation is a particular case of a node of a complex observation, where the set of contextual nodes is a singleton, whose associated source is in fact the whole set of visible system components.

A *complex diagnostic problem* relevant to a cluster ξ is a triple $(\Omega(\xi),$ $OBS(\xi), \xi_0)$, where $\Omega(\xi)$ is the observer of ξ, ξ_0 is the initial state of ξ, while $OBS(\xi)$ is a complex observation. To solve a complex diagnostic problem we have to generate the index space of the relevant complex observation. The generation of such an index space requires some kind of additional automated reasoning. However, once such an index space

has been generated, the reconstruction technique is essentially the same
as that for uncertain observations.

Formally, let Θ be a system and $\Omega(\Theta) = (\mathbb{V}, \mathbb{P})$ an observer of Θ,
where $\mathbb{P} = \{\mathbf{C}_1, \ldots, \mathbf{C}_n\}$. Let μ be the domain of the messages of a
part $\mathbf{C}_i \in \mathbb{P}$ (see the definition of a message given at the beginning of
Section 3 of Chapter 9). Let $\mathbb{M} \subseteq \mu \times (2^{\mathbf{C}_i} - \emptyset)$ be the (non-empty)
domain of *contextual messages* of $\mathbf{C}_i$.

A *complex observation graph* $obs(\mathbf{C}_i)$ of a set of components $\mathbf{C}_i \in \mathbb{P}$
is a 5-tuple

$$obs(\mathbf{C}_i) = (\mathbf{N}, \mathbb{N}, \Upsilon, \mathbb{N}_0, \mathbb{N}_f),$$

where

$\mathbf{N}$ is the set of *noncontextual nodes*;

$\mathbb{N} = \{(\omega, M) \mid \omega \in \mathbf{N}, M \in \mathbb{M}\}$ is the set of *contextual nodes* such that,
denoting with $\|\omega\|$ the set of contextual nodes within $\omega \in \mathbf{N}$, namely

$$\|\omega\| = \{N \mid N \in \mathbb{N}, N = (\omega, M)\},$$

the following condition hold:

$$\forall \omega \in \mathbf{N}, \forall N = (\omega, M), \forall N' = (\omega, M'), N \neq N' \ (M \neq M');$$

$\Upsilon : \mathbb{N} \mapsto 2^{\mathbb{N}}$ is a mapping function among contextual nodes;

$\mathbb{N}_0 \subseteq \mathbb{N}$ is the set of *roots*;

$\mathbb{N}_f \subseteq \mathbb{N}$ is the set of *leaves*.

The '$\prec$' *temporal precedence* relationship among contextual nodes is defined as follows:

(1) (*Basis*) If $N \mapsto N' \in \Upsilon$ then $N \prec N'$;

(2) (*Transitivity*) If $N \prec N'$ and $N' \prec N''$ then $N \prec N''$;

(3) (*Canonicity*) If $N \mapsto N' \in \Upsilon$ then $\nexists N'' \in \mathbb{N} \ (N \prec N'' \prec N')$.

Furthermore, by definition,

(4) (*Generalization*) $N \preceq N'$ iff $N \prec N'$ or $N = N'$;

(5) (*Lower Bound*) $\forall N_0 \in \mathbb{N}_0 \ (\nexists N \in \mathbb{N} \mid N \prec N_0)$;

(6) (*Upper Bound*) $\forall N_f \in \mathbb{N}_f \ (\nexists N \in \mathbb{N} \mid N_f \prec N)$;

(7) (*Separation*) $\forall (\omega, M) \mapsto (\omega', M') \in \Upsilon \ (\omega \neq \omega')$;

(8) *(Acyclicity)* $\nexists N \in \mathbb{N} \, (N \prec N)$.

A *complex observation* $OBS(\Theta)$ of system Θ is a record of n complex observation graphs $obs(\mathbf{C}_i)$, $i \in [1 .. n]$, $n \geq 1$, that is:

$$OBS(\Theta) = (obs(\mathbf{C}_1), \ldots, obs(\mathbf{C}_n)).$$

Note that $OBS(\Theta)$, being a record of one or more complex observation graphs, is a complex observation graph itself. Moreover, the order among messages coming from the same part $\mathbf{C}_i$ in the ordering partition of the observer is partial and it may vary depending on the sender components of specific messages. Finally, there may be shared messages coming from distinct parts of $\mathbb{P}$.

A complex observation $OBS(\Theta)$ is a *degenerate complex observation* when there exists an isomorphism between $\mathbf{N}$ and $\mathbb{N}$, that is, when:

$$\forall \omega \in \mathbf{N} \, (\|\omega\| = \{N\}).$$

A degenerate complex observation $OBS(\Theta)$ is a *strictly degenerate complex observation* when (*i*) the complex observation graphs inherent to distinct parts in $\mathbb{P}$ do not share any observable label, and (*ii*) no restricting assumptions are made on the set of components associated with the contextual messages, that is, when

$$\forall N \in \mathbb{N} \, (N = (\omega, M), M = (m, \mathbb{V})).$$

Owing to the isomorphism between $\mathbf{N}$ and $\mathbb{N}$, and to the lack of constraints on the set of sender components in the contextual message, a strictly degenerate complex observation

$$OBS(\Theta) = (\mathbf{N}, \mathbb{N}, \Upsilon, \mathbb{N}_0, \mathbb{N}_f)$$

is in fact an uncertain observation, represented by the observation graph $(\mathbb{N}, \Upsilon, \mathbb{N}_0, \mathbb{N}_f)$. Thus, denoting with $\mathbb{OBS}$, $\mathbb{OBS}^d$, and $\mathbb{OBS}^u$ the classes of complex observations, degenerate complex observations, and uncertain observations, respectively, the following relationships hold:

$$\mathbb{OBS}^u \subset \mathbb{OBS}^d \subset \mathbb{OBS}.$$

Complex observations in $(\mathbb{OBS} - \mathbb{OBS}^d)$ are called *strictly complex observations*.

Example 10.1. Shown in Figure 10.1 is a complex observation for system Φ depicted in Figure 4.1, where

$$\Omega(\Phi) = (\mathbb{V}, \mathbb{P}), \mathbb{V} = \{C_1, C_2\}, \mathbb{P} = \{\{C_1, C_2\}\}.$$

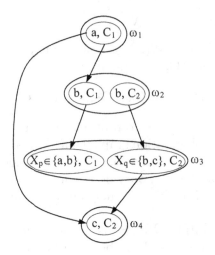

Figure 10.1.   Complex-observation graph relevant to Example 10.1.

The graph is such that:

$$\mathbf{N} = \{\omega_1, \ldots, \omega_4\},$$

$$\mathbb{N} = \{(\omega_1, (a, \{C_1\})), (\omega_2, (b, \{C_1\})), (\omega_2, (b, \{C_2\})), (\omega_3, (X_p, \{C_1\})),$$
$$(\omega_3, (X_q, \{C_2\})), (\omega_4, (c, \{C_2\}))\},$$

$$\mathbb{N}_0 = \{(\omega_1, (a, \{C_1\})), (\omega_2, (b, \{C_2\}))\},$$

$$\mathbb{N}_f = \{(\omega_3, (X_p, \{C_1\})), (\omega_4, (c, \{C_2\}))\}.$$

According to the above definition of a complex observation, there cannot be two identical contextual messages within the same noncontextual node.

Conditions 1 to 6 of the definition of a complex observation mirror those in the definition of an uncertain observation (see the beginning of Section 3 of Chapter 9), except that the precedence relationship here refers to contextual nodes. Condition 7 prevents precedence relationships between contextual nodes within the same noncontextual node ω. The reason for it is that each contextual node within the same noncontextual node corresponds to a different assumption as to the same uncertain message, which is unique. Finally, Condition 8 establishes the acyclicity of the (directed) graph of contextual nodes. $\diamond$

2.1   Complex-observation Index

The peculiarity of complex observations is reflected by the notion of a complex-observation index $\mathfrak{S}$ as well. According to the definition given in Section 3.1 of Chapter 9, the index of an uncertain observation is a set of nodes identifying the set of consumed messages within the observation graph. When complex observations are considered, $\Im$ is expected to include contextual nodes and, above all, to maintain information about the assumptions made in the course of message consummation.

Example 10.2. Consider the complex observation displayed in Figure 10.2. We adopt the following notational shorthand: if ω_i includes a single contextual node, the latter is identified by N_i, otherwise the contextual node $(\omega_i, (\ell, C))$ is identified by N_i^C, where $\ell \in \{a, b, c\}$ and $C \in \{C_1, C_2\}$.

If the set of consumed (contextual) nodes is $\{N_1, N_2^{C_2}, N_3\}$, then $\Im = \{N_3\}$ is in principle enough to identify such nodes. Conversely, if the consumed nodes are $\{N_1, N_2^{C_1}, N_3\}$, then $\Im = \{N_2^{C_1}, N_3\}$ might as well identify such nodes, with the implicit assumption that node $N_2^{C_2}$ is not included in $Cons(\Im)$, even if $N_2^{C_2} \prec N_3$. As a matter of fact, for the observation graph in Figure 10.2, the definition of the index might be inherited from that of the index of an uncertain observation (see Section 3.1 of Chapter 9), with the additional rule of mutual exclusion of contextual nodes relevant to the same ω.

However, this method does not work in general. To show this, consider the complex observation in Figure 10.1 and the index $\Im = \{N_3^{C_1}, N_4\}$. Without any additional information, such an index is ambiguous: even though, owing to the inclusion of $N_3^{C_1}$, we can exclude $N_3^{C_2}$ from $Cons(\Im)$, there is still the ambiguity on which contextual node in ω_2 is part of $Cons(\Im)$. Consequently, we are forced to include either $N_2^{C_1}$ or $N_2^{C_2}$ in $Cons(\Im)$. The ambiguity stems from the fact that both solutions are consistent with $\Im$, namely

$$Cons(\Im) = \{N_1, N_2^{C_1}, N_3^{C_1}, N_4\},$$
$$Cons(\Im) = \{N_1, N_2^{C_2}, N_3^{C_1}, N_4\}.$$

Such an ambiguity can be eliminated by defining the index of a complex observation as a pair

$$\Im = (\mathcal{P}, \mathfrak{A}),$$

where, roughly, $\mathcal{P}$ is a set of contextual nodes governed by the rules established in the definition of the index of an uncertain observation (see Section 3.1 of Chapter 9), while $\mathfrak{A}$ is the set of assumptions made throughout message consummation. For example, the ambiguity discussed above can be overcome by specifying either

$$\Im' = (\{N_3^{C_1}, N_4\}, \{\neg N_2^{C_2}, \neg N_3^{C_2}\})$$

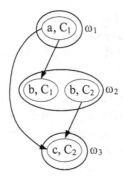

Figure 10.2.   Complex-observation graph of system Φ depicted in Figure 4.1.

or

$$\Im'' = (\{N_3^{C_1}, N_4\}, \{\neg N_2^{C_1}, \neg N_3^{C_2}\}),$$

where $\neg N$ means that the contextual node N is assumed *not* to be in $Cons(\Im)$[1]. For the sake of conciseness, in the following, the symbol $\neg$ will be omitted from the elements of the assumption set.   ◇

An *index* of a complex observation $OBS(\xi) = (\mathbf{N}, \mathbb{N}, \Upsilon, \mathbb{N}_0, \mathbb{N}_f)$ is a pair of disjoint sets

$$\Im = (\mathcal{P}, \mathfrak{A})$$

where:

$\mathcal{P} \subseteq \mathbb{N}$ is the *positive set*, such that $\forall N \in \mathcal{P}$ ($\nexists N_i \in \mathcal{P} \mid N_i \prec N$);

$\mathfrak{A} \subseteq \mathbb{N}$ is the *assumption set* such that, $\forall \omega \in \mathbf{N}$, the following conditions hold:

(1) $\exists N \in (\|\omega\| - \mathfrak{A})$;

(2) if $N \in (\|\omega\| - \mathfrak{A}), N \preceq N', N' \in \mathcal{P}$ then $(\|\omega\| - \{N\}) \subseteq \mathfrak{A}$.

The $Cons(\Im)$ function is defined as follows:

$$Cons(\Im) = \{N \mid N \in (\mathbb{N} - \mathfrak{A}), N \preceq N', N' \in \mathcal{P}).$$

---

[1] In principle, the assumption $\neg N_3^{C_2}$ might be derived implicitly from $\Im$. However, we prefer expressing it explicitly for uniformity reasons and, above all, as shown later, because it supports the systematic generation of index spaces.

The *Front*($\Im$) function is defined as follows:

$$Front(\Im) = \{(N, \mathcal{A}) \mid \omega \in \mathbf{N}, N \in (\|\omega\| - Cons(\Im)),$$
$$(N \in (\mathbf{N}_0 - \mathfrak{A}) \vee \forall N' \in \|\omega'\|, \omega' \in \mathbf{N}, N' \prec N \ (N' \in (Cons(\Im) \cup \mathfrak{A}) \vee$$
$$(N' \notin Cons(\Im),$$
$$\|\omega'\| \not\subseteq (\mathfrak{A} \cup \{N'' \mid N'' \in \|\omega'\|, N'' \notin Cons(\Im), N'' \prec N\}))),$$
$$\mathcal{A} \in 2^{\mathbf{N}}, \mathcal{A} = (\|\omega\| - \{N\}) \cup \{N' \mid N' \notin (\mathfrak{A} \cup Cons(\Im)), N' \prec N\}\}.$$
$$\text{(10.1)}$$

$\Im$ is said to be *complete* when:

$$\{\omega \mid (\omega, M) \in Cons(\Im)\} = \mathbf{N}. \tag{10.2}$$

Intuitively, in the definition of $\mathfrak{A}$ given above, Conditions 1 and 2 have the following meaning, respectively:

(1) Not all of the contextual nodes within the same noncontextual node can be included in the assumption set, as one contextual message is supposed to have been generated by the system;

(2) If a contextual node N does not belong to the assumption set and it either belongs to the positive set or precedes another contextual node N' belonging to the positive set, then all of the other contextual nodes in the same noncontextual node of N must belong to the assumption set.

Function $Cons(\Im)$ yields the set of consumed messages. These are identified by all of the contextual nodes, not in the assumption set, which precede or coincide with a contextual node within the positive set.

The frontier function, *Front*($\Im$), yields the set of contextual nodes (along with relevant assumptions) each of which can be consumed at the next step without violating the precedence relationships. According to Formula (10.1), a pair $(N, \mathcal{A})$ belongs to the frontier when:

1) N is not consumed and either it is a root that is not included in the assumption set $\mathfrak{A}$ of $\Im$, or, for each contextual node N' preceding N, either:

   (a) N' has been consumed already, or

   (b) N' belongs to the assumption set, or

   (c) N', along with all of the other contextual nodes N'' relevant to the same noncontextual node ω' and preceding N, can be assumed not to have been generated without violating the constraint that at least a contextual node in $\|\omega'\|$ will never belong to the assumption set;

2) $\mathcal{A}$ is the subset of $\mathbb{N}$ including:

   (a) All of the contextual nodes in ω but N, and

   (b) All of the contextual nodes N' that are not consumed, and do not belong to the assumption set, and precede N.

Example 10.3. Consider the above definition of a complex-observation index in the context of the complex observation outlined in Figure 10.1. A possible index is

$$\Im = (\{N_2^{C_1}, N_4\}, \{N_2^{C_2}, N_3^{C_2}\}),$$

for which $Cons(\Im) = \{N_1, N_2^{C_1}, N_4\}$.

Consider $\Im = \Im_0 = (\emptyset, \emptyset)$, that is, the initial (empty) index. Determining $Front(\Im_0)$ amounts to finding the first contextual nodes that can be consumed. All of the nodes in the set of roots $\mathbb{N}_0$ meet the condition for N in Formula (10.1), namely N_1 and $N_2^{C_2}$ (see Example 10.1).

Another contextual node which can be consumed first is $N_3^{C_2}$, provided that we assume $N_2^{C_2}$ not being generated. Such an assumption does not violate Point 1(c) above, as

$$(\|\omega'\| = \{N_2^{C_1}, N_2^{C_2}\}) \not\subseteq (\emptyset \cup \{N_2^{C_2}\} \cup \emptyset).$$

If we had an extra precedence relationship $N_2^{C_1} \prec N_3^{C_2}$ in Figure 10.1, Point 1(c) would be violated. Intuitively, this is due to the fact that, on the one hand, by assuming $N_2^{C_2}$ not being generated, we are forced to assume that $N_2^{C_1}$ is the actual generated message. On the other, since $N_2^{C_1}$ precedes $N_3^{C_2}$, the former is expected to be consumed already, which is not the case.

Assuming $N = N_3^{C_2}$, according to Formula (10.1), we can compute the relevant set of assumptions as

$$\mathcal{A} = (\{N_3^{C_1}, N_3^{C_2}\} - \{N_3^{C_2}\}) \cup \{N_2^{C_2}\} = \{N_2^{C_2}, N_3^{C_1}\}.$$

Eventually, the frontier of the initial index $\Im_0$ will be

$$Front(\Im_0) = \{(N_1, \emptyset), (N_2^{C_2}, \{N_2^{C_1}\}), (N_3^{C_2}, \{N_2^{C_2}, N_3^{C_1}\})\}.$$

As another example, we can compute the frontier of

$$\Im_1 = (\{N_2^{C_2}\}, \{N_2^{C_1}\})$$

and find the following result:

$$Front(\Im_1) = \{(N_1, \emptyset), (N_3^{C_2}, \{N_3^{C_1}\})\}.$$

According to Equation (10.2), an index $\Im$ is complete when all of the noncontextual nodes are involved in the set of consumed messages. In our example, the set of complete indexes includes

$$(\{N_2^{C_1}, N_4\}, \{N_2^{C_2}, N_3^{C_1}\}),$$
$$(\{N_3^{C_1}, N_4\}, \{N_2^{C_2}, N_3^{C_2}\}),$$
$$(\{N_3^{C_1}, N_4\}, \{N_2^{C_1}, N_3^{C_2}\}),$$
$$(\{N_4\}, \{N_2^{C_1}, N_3^{C_1}\}).$$

$\diamond$

2.2 Complex-observation Restriction

The advantages of a modular reconstruction can be exploited for solving complex diagnostic problems only if the information as to the observation of the considered system can be projected on subsystems. In fact, if a modular reconstruction of the given complex diagnostic problem is adopted, such a problem is hierarchically decomposed into subproblems, where each subproblem is inherent to a subsystem.

Each subproblem is solved by generating the relevant active space by means of an (adapted) monolithic reconstruction procedure.

Each reconstruction search space state inherent to a subproblem contains a reference to the observation of the relevant subsystem. In this perspective, analogously to Section 3.2 of Chapter 9, this section introduces the notion of a complex-observation restriction.

The projection of a complex observation $OBS(\xi)$ of a cluster

$$\xi = (\mathbb{C}, \mathbb{L}, \mathbb{D})$$

on a subcluster

$$\xi' = (\mathbb{C}', \mathbb{L}', \mathbb{D}'), \mathbb{C}' \subset \mathbb{C},$$

is obtained by performing the following steps:

(1) All of the non-null labels inherent to components that do not belong to $\mathbb{C}'$ are removed from the contextual nodes of $OBS(\xi)$;

(2) All of the components that do not belong to $\mathbb{C}'$ are removed from the contextual nodes of $OBS(\xi)$;

(3) A null label ϵ is added to each contextual node of $OBS(\xi)$ that has been deprived of one or more labels and/or components in the previous two steps;

(4) A null label ϵ is added to each contextual node of $OBS(\xi)$ that includes one or more observable labels shared between $\mathbb{C}'$ and $(\mathbb{C} - \mathbb{C}')$;

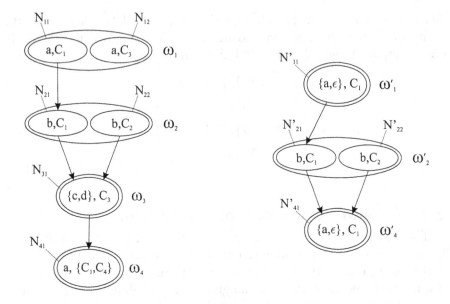

Figure 10.3. Complex observation and a relevant projection.

(5) All empty contextual nodes are removed;

(6) All empty noncontextual nodes are removed;

(7) Precedence relations between the surviving contextual nodes are updated, that is, the precedence relations of a canceled contextual node
are inherited by its predecessor contextual nodes.

Step 3 guarantees that the subcluster observation encompasses also the
case when no label coming from the subcluster has been observed by
the observer in correspondence to the considered contextual node, since
the observer observed something coming from components that do not
belong to $\mathbb{C}'$, they belong instead to $(\mathbb{C} - \mathbb{C}')$.

   According to Step 4, in the restriction of the observation, each contex-
tual node including one or more labels shared between $\mathbb{C}'$ and $(\mathbb{C} - \mathbb{C}')$
contains ϵ as well, thus guaranteeing that the subcluster observation
involves also the case when every shared label does not come from ξ'.

Example 10.4. Let ξ be a cluster whose set of components is

$$\mathbb{C}(\xi) = \{C_1, C_2, C_3, C_4\},$$

and ξ' be the subcluster of ξ whose set of components is

$$\mathbb{C}(\xi') = \{C_1, C_2\}.$$

Let

$$\Omega(\xi) = (\mathbb{V} = \{C_1, C_2, C_3, C_4\}, \mathbb{P} = \{\{C_1, C_2, C_3, C_4\}\})$$

be the observer of ξ, and

$$\Lambda(\xi) = \{a, b, c, d\}, \Lambda(\xi') = \{a, b\},$$

the sets of observable labels of ξ and ξ', respectively. The complex observation $OBS(\xi)$ gathered by $\Omega(\xi)$ is shown on the left of Figure 10.3. The projection of $OBS(\xi)$ on ξ' is depicted on the right of the same figure.

Contextual nodes N_{12} and N_{31} have no counterpart in $OBS_{\langle\xi'\rangle}(\xi)$ since they are inherent to components that do not belong to $\mathbb{C}(\xi')$. N_{11}' has been obtained from N_{11} by adding the null label since N_{12}, which in $OBS(\xi)$ belongs to the same noncontextual node ω_1 as N_{11}, has been deleted in the restriction.

N_{41}' has been created from N_{41} by removing component C_4 since it does not belong to $\mathbb{C}(\xi')$, and by adding ϵ so as to encompass the case when the a observed in correspondence to the noncontextual node ω_4 comes from C_4 and, therefore, nothing coming from ξ' is observed.

Node N_{31}' has disappeared since empty: its precedence relation (it has to precede node N_{41}') has been inherited by all of its predecessors, that is, N_{21}' and N_{22}', which, in fact, both precede N_{41}'.

Noncontextual node ω_3' does not exist as the projection of its contextual node is empty. $\diamond$

Formally, let $\xi = (\mathbb{C}, \mathbb{L}, \mathbb{D})$ be a cluster, $\xi' = (\mathbb{C}', \mathbb{L}', \mathbb{D}')$ a subcluster, $\mathbb{C}' \subset \mathbb{C}$, and $\Omega(\xi)$ an observer of ξ. Given a contextual node N of complex observation

$$OBS(\xi) = (\mathbf{N}, \mathbb{N}, \Upsilon, \mathbb{N}_0, \mathbb{N}_f),$$

where

$$N \in \mathbb{N}, N = (\omega, M), M = (m, \mathbb{C}_m),$$

the restriction of N on $\xi' \subset \xi$,

$$N' = N_{\langle\xi'\rangle} = (\omega, M'), M' = (m', \mathbb{C}_m'),$$

is derived as follows:

(1) If $\ell \in \|m\|$ and $\ell \in \Lambda(\xi')$ and $\mathbb{C}_m \cap \mathbb{C}' \neq \emptyset$ then $\ell \in \|m'\|$;

(2) If $\ell \in \|m\|$ and $\ell \in \Lambda(\xi')$ and $\ell \in \Lambda(\mathbb{C} - \mathbb{C}')$ and $\mathbb{C}_m \cap (\mathbb{C} - \mathbb{C}') \neq \emptyset$ then $\epsilon \in \|m'\|$;

(3) If $\ell \in \|m\|$ and $\ell \notin \Lambda(\xi')$ then $\epsilon \in \|m'\|$;

(4) If $\epsilon \in \|m\|$ then $\epsilon \in \|m'\|$;

(5) $\mathbb{C}'_m = \mathbb{C}_m \cap \mathbb{C}'$.

Let

$$OBS_{\langle \xi' \rangle}(\xi) = (\mathbf{N}', \mathbb{N}', \Upsilon', \mathbb{N}'_0, \mathbb{N}'_f)$$

be the restriction of $OBS(\xi)$ on ξ'. A node N' resulting from the above rules will belong to $\mathbf{N}'$ if and only if the following conditions hold:

$$\|m'\| \neq \{\epsilon\},$$
$$\mathbb{C}'_m \neq \emptyset.$$

A noncontextual node $\omega \in \mathbf{N}$ will belong to $\mathbf{N}'$ if and only if $\|\omega\| \neq \emptyset$. The restriction of Υ into Υ' is dictated by the following rule:

$$N'_1 \mapsto N'_2 \in \Upsilon', N'_1 \in \mathbf{N}', N'_2 \in \mathbf{N}', N'_1 = N_{1\langle \xi' \rangle}, N'_2 = N_{2\langle \xi' \rangle},$$

if and only if the following conditions hold:

(1) (*Precedence*) $N_1 \prec N_2$ in $OBS(\xi)$;

(2) (*Canonicity*) $\nexists N'_3 \in \mathbf{N}'$, $N'_3 = N_{3\langle \xi' \rangle}$, such that $N_1 \prec N_3 \prec N_2$ in $OBS(\xi)$.

2.3   Complex-observation Extension

A *contextual label* is a pair (ℓ, C), where ℓ is an observable label of component C. A contextual label (ℓ, C) is null if $\ell = \epsilon$.

Let $N = (\omega, M)$, $M = (m, \mathbb{C})$, be a contextual node of a complex observation. The *extension* of N, denoted by $\|N\|$, is the set of contextual labels complying with M, that is,

$$\|N\| = \{(\ell, C) \mid \ell \in \|m\|, C \in \mathbb{C}\}.$$

A complex observation implicitly defines a finite set of sequences of both logically and source unambiguous contextual labels. Each of such sequences is called *observation instance*. Intuitively, an observation instance can be made up by first picking up a contextual node within each noncontextual node of the given observation and creating a sequence of these contextual nodes that respects the temporal precedence relationship imposed by the observation graph. From this sequence a new sequence is drawn by replacing each contextual node with a contextual label belonging to the extension of a contextual node itself. Finally, the null contextual labels are removed.

Let

$$\mathcal{Q}_N = \langle N_1, \ldots, N_p \rangle$$

be a sequence of contextual nodes of a complex observation $OBS(\xi)$ such that:

(1) $\{\omega \mid N = (\omega, M), N \in \mathcal{Q}_N\} = \mathbf{N}(OBS(\xi));$

(2) $\forall i \in [1..p], \forall j \in [1..p], i \neq j,$

$$\text{if } N_i \prec N_j \text{ in } OBS(\xi) \text{ then } N_i \text{ precedes } N_j \text{ in } \mathcal{Q}_N.$$

Let

$$\mathcal{Q}_\ell = \langle (\ell_1, C_1), \ldots, (\ell_p, C_p) \rangle,$$

where

$$\forall i \in [1..p] \, ((\ell_i, C_i) \in \|N_i\|, N_i \in \mathcal{Q}_N).$$

Let

$$\mathcal{Q}'_\ell = \langle (\ell'_1, C'_1), \ldots, (\ell'_{p'}, C'_{p'}) \rangle,$$

where $p' \leq p$, be the sequence obtained from $\mathcal{Q}_\ell$ by removing all of the null contextual labels. By definition,

$$\mathcal{Q}'_\ell \in \|\mathcal{Q}_N\|.$$

$\mathcal{Q}'_\ell$ is called an *instance* of $OBS(\xi)$. The *extension* of $OBS(\xi)$, denoted by $\|OBS(\xi)\|$, is the whole set of the instances of $OBS(\xi)$.

The i-th instance of an observation can be thought of as a special case of an observation, whose graph, denoted by $OBS_{(i)}(\xi)$, is a sequence of p' noncontextual nodes, each containing just one contextual node, where such a contextual node corresponds exactly to a contextual label in the instance.

An observation instance $OBS_{(i)}(\xi)$ is not necessarily an observation that can actually be generated by cluster ξ. However, the assumptions is made that, given the observation graph $OBS(\xi)$, there exists at least one observation instance $OBS_{(i)}(\xi)$ that can be generated by ξ, and among the observation instances in $\|OBS(\xi)\|$ there exists the one actually generated by the system.

3. Solving Complex Diagnostic Problems

Given a complex diagnostic problem, the index space of the complex observation gathered during the system reaction is generated beforehand so as to improve the efficiency of the behavior reconstruction process. As already remarked when dealing with uncertain observations, if the reconstruction of the system behavior is based on an index space, then each search space state contains the identifier of a state of the index space.

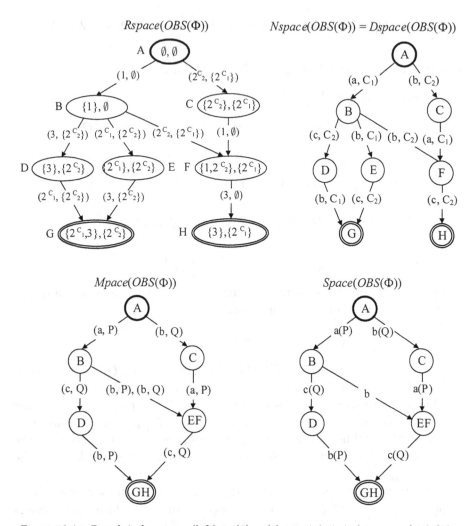

Figure 10.4. Rough index space (left) and (non)deterministic index space (right) for the complex observation displayed in Figure 10.2.

3.1   Rough Index Space

The *rough index space* of a complex observation

$$OBS(\xi) = (\mathbf{N}, \mathbb{N}, \Upsilon, \mathbb{N}_0, \mathbb{N}_f)$$

is a deterministic finite automaton

$$Rspace(OBS(\xi)) = (\mathbb{S}^r, \mathbb{L}^r, \mathbb{T}^r, S_0^r, \mathbb{S}_f^r)$$

where

$\mathbb{S}^r = \{\mathfrak{I} \mid \mathfrak{I} \text{ is an index for } OBS(\xi)\}$ is the set of states;

$\mathbb{L}^r \subseteq \mathbb{N} \times 2^{\mathbb{N}}$ is the set of labels;

$\mathbb{T}^r : \mathbb{S}^r \times \mathbb{L}^r \mapsto \mathbb{S}^r$ is the transition function defined as follows:

$$\mathfrak{I} \xrightarrow{(N,\mathcal{A})} \mathfrak{I}' \in \mathbb{T}^r,$$

where

$$\mathfrak{I} = (\mathcal{P}, \mathfrak{A}), \mathfrak{I}' = (\mathcal{P}', \mathfrak{A}'),$$

if and only if the following conditions hold:

(1) $(N, \mathcal{A}) \in \mathit{Front}(\mathfrak{I})$,
(2) $\mathcal{P}' = (\mathcal{P} \cup \{N\}) - \{N' \mid N' \in \mathcal{P}, N' \prec N\}$,
(3) $\mathfrak{A}' = \mathfrak{A} \cup \mathcal{A}$;

$\mathbb{S}_0^r = \mathfrak{I}_0 = (\mathcal{P}_0, \mathbb{N}_0) = (\emptyset, \emptyset)$ is the initial state;

$\mathbb{S}_f^r = \{\mathfrak{I} \mid \mathfrak{I} \in \mathbb{S}^r, \mathfrak{I} \text{ is complete }\}$ is the set of final states.

Example 10.5. The rough index space of the complex observation of Figure 10.2 is depicted on the left of Figure 10.4. Contextual node identifiers are denoted by i and i^C, which are a shorthand for N_i and N_i^C, respectively, where $i \in [1..3]$ and $C \in \{C_1, C_2\}$.

According to the above definition of a rough index space of a complex observation, each state of the rough index space is an index for the observation. Specifically, the initial state is the initial index $(\emptyset, \emptyset)$.

Each edge leaving a state $\mathfrak{I}$ is marked by an element $(N, \mathcal{A})$ of the frontier of $\mathfrak{I}$. Thus, there are two edges leaving the initial state, that are marked by $(N_1, \emptyset)$ and $(N_2^{C_2}, \{N_2^{C_1}\})$, respectively.

According to the definition of the transition function $\mathbb{T}^r$, the states reached by these two edges are computed, on the one hand, by inserting into $\mathcal{P}$ the contextual node N marking the edge and by removing all of the redundant contextual nodes N' in $\mathcal{P}$, on the other, by including $\mathcal{A}$ into the assumption set $\mathfrak{A}$. For example, the following transition (from F to H) holds:

$$(\{N_1, N_2^{C_2}\}, \{N_2^{C_1}\}) \xrightarrow{(N_3, \emptyset)} (\{N_3\}, \{N_2^{C_1}\})$$

where both N_1 and $N_2^{C_2}$ have been eliminated from the final positive set, as $N_1 \prec N_3$ and $N_2^{C_2} \prec N_3$. As usual, final states are denoted by double circles. ◇

Example 10.6. The rough index space relevant to the complex observation of Figure 10.1 is displayed in Figure 10.5. ◇

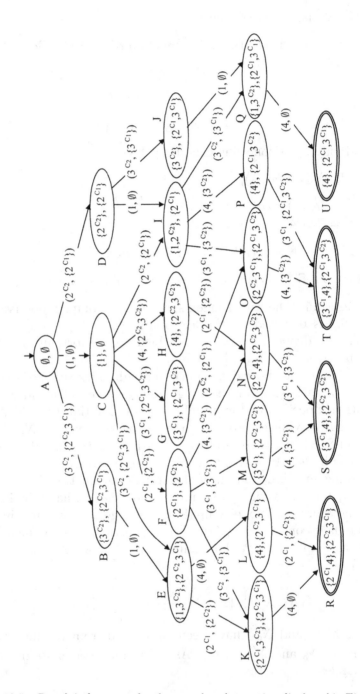

Figure 10.5.   Rough index space for the complex observation displayed in Figure 10.1.

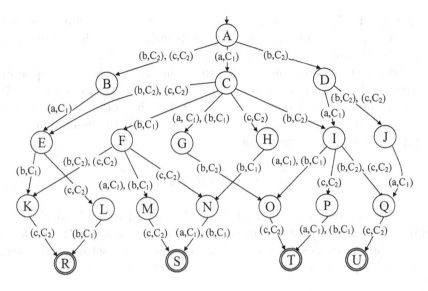

Figure 10.6. Nondeterministic index space for the complex observation displayed in Figure 10.1.

3.2 Nondeterministic Index Space

The *nondeterministic index space* of a complex observation $OBS(\xi)$ is the nondeterministic finite automaton

$$Nspace(OBS(\xi)) = (\mathbb{S}^n, \mathbb{L}^n, \mathbb{T}^n, S_0^n, \mathbb{S}_f^n)$$

obtained from $Rspace(OBS(\xi)) = (\mathbb{S}^r, \mathbb{L}^r, \mathbb{T}^r, S_0^r, \mathbb{S}_f^r)$ as follows:

$\mathbb{S}^n = \mathbb{S}^r$ is the set of states;

$\mathbb{L}^n = \{(\ell, \mathbb{C}) \mid (N, \mathcal{A}) \in \mathbb{L}^r, N = (\omega, M), M = (m, \mathbb{C}), \ell \in \|m\|\}$ is the set of labels;

$\mathbb{T}^n : \mathbb{S}^n \times \mathbb{L}^n \mapsto 2^{\mathbb{S}^n}$ is the nondeterministic transition function defined as follows:

$$S \xrightarrow{(\ell, \mathbb{C})} S' \in \mathbb{T}^n$$

if and only if

$$S \xrightarrow{(N, \mathcal{A})} S' \in \mathbb{T}^r, N = (\omega, M), M = (m, \mathbb{C}), \ell \in \|m\|;$$

$S_0^n = S_0^r$ is the initial state;

$\mathbb{S}_f^n = \mathbb{S}_f^r$ is the set of final states.

Example 10.7. The transformation from a rough to a nondeterministic index space essentially replaces the transition function $\mathbb{T}^r$ with $\mathbb{T}^n$, maintaining, however, the same set of nodes ($\mathbb{S}^n = \mathbb{S}^r$).

Figure 10.6 outlines the nondeterministic index space corresponding to the rough index space displayed in Figure 10.5. Nodes in Figure 10.6 are identified by the labels $A, \ldots, U$ marking the nodes in Figure 10.5.

According to the definition of a nondeterministic index space of a complex observation, each edge

$$S \xrightarrow{(N,\mathcal{A})} S' \in \mathbb{T}^r$$

is replaced by one or several edges in $\mathbb{T}^n$, each of which is marked by a pair $(\ell, \mathbb{C})$, where ℓ belongs to the extension of the message m relevant to the contextual node N, while $\mathbb{C}$ is the associated set of components.

For example, transition

$$A \xrightarrow{(N_3^{C_2}, \{N_2^{C_2}, N_3^{C_1}\})} B$$

in Figure 10.5 is replaced in Figure 10.6 by two transitions, namely

$$A \xrightarrow{(b,\{C_2\})} B, A \xrightarrow{(c,\{C_2\})} B,$$

as

$$N = N_3^{C_2} = (\omega_3, M), M = (X_q, \{C_2\}), \|X_q\| = \{b, c\}.$$

The automaton in Figure 10.6 is nondeterministic because it includes nodes that are left by several transitions marked by the same label. One of these nodes is A, from which two exiting different transitions, directed to nodes B and D, respectively, are marked by the same label $(b, \{C_2\})$.

There is a subtle peculiarity in the way nondeterminism is governed in index spaces relevant to complex observations. Due to the contextualization of messages, two transitions marked by labels $(\ell, \mathbb{C}_i)$ and $(\ell, \mathbb{C}_j)$, respectively, where $i \neq j$, does not lead to nondeterminism as they differ in the elements $\mathbb{C}_i$ and $\mathbb{C}_j$.

For example, in Figure 10.6 there are two transitions leaving F which are marked by $(b, \{C_2\})$ and $(b, \{C_1\})$, respectively. Intuitively, in the reconstruction process, candidate observable transitions are expected to conform not only to the observable label in the index space but to the contextual part as well. That is, the generator of the candidate transition is required to be in the relevant set of components.   ◇

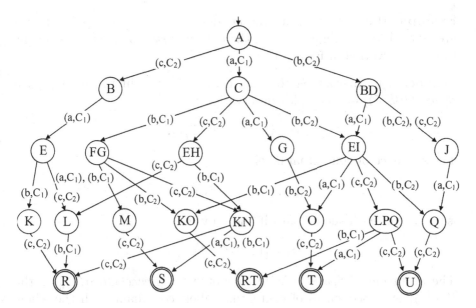

Figure 10.7. Deterministic index space for the complex observation displayed in Figure 10.1.

3.3 Deterministic Index Space

The *deterministic index space* or, simply, the *index space* of a complex observation $OBS(\xi)$,

$$Space(OBS(\xi)) = (\mathbb{S}, \mathbb{L}, \mathbb{T}, S_0, \mathbb{S}_f),$$

is the finite automaton obtained by transforming the nondeterministic index space

$$Nspace(OBS(\xi)) = (\mathbb{S}^n, \mathbb{L}^n, \mathbb{T}^n, S_0^n, \mathbb{S}_f^n)$$

into a deterministic finite automaton. This transformation has the same formal definition as that from the nondeterministic to the deterministic index space of an uncertain observation (see Section 4.3 of Chapter 9), although the content of $\mathbb{L}^n$ is different in complex observations with respect to uncertain observations.

Example 10.8. Displayed on the right of Figure 10.4 is the deterministic index space relevant to the observation in Figure 10.2. This is in fact what we obtain by transforming the rough index space into the nondeterministic index space, that is, in the particular example of Figure 10.4,

$$Nspace(OBS(\Phi)) = Space(OBS(\Phi)).$$

◇

Example 10.9. Drawn from the nondeterministic index space of Figure 10.6, shown in Figure 10.7 is the deterministic index space relevant to the observation of Figure 10.1. $\diamond$

Consider a path from S_0 to a state $S_f \in \mathbb{S}_f$ within the index space $Space(OBS(\xi))$ of a complex observation :

$$\langle (\ell_1, \mathbb{C}_1), \ldots, (\ell_p, \mathbb{C}_p) \rangle.$$

The sequence of contextual labels

$$\langle (\ell_1, C_1), \ldots, (\ell_p, C_p) \rangle.$$

is compliant with such a path if and only if

$$\forall i \in [1 .. p](C_i \in \mathbb{C}_i).$$

The *extension* of $Space(OBS(\xi))$, denoted by $\|Space(OBS(\xi))\|$, is the (finite) set of sequences of contextual labels compliant with the whole set of paths from S_0 to a state $S_f \in \mathbb{S}_f$.

Given this definition of the extension of the index space, specific for a complex observation, we claim that the property expressed by the following proposition holds.

Proposition 10.1. *Theorem 9.1 (for uncertain observations) is valid for complex observations too.*

As for uncertain observations, this property allows for the exploitation of a processed form of the complex observation, i.e. the index space, instead of the observation itself during the reconstruction of histories in order to solve a complex diagnostic problem.

3.4　　Monolithic Resolution

Assume a complex diagnostic problem $\wp(\xi) = (\Omega(\xi), OBS(\xi), \xi_0)$ and the relevant index space

$$Space(OBS(\xi)) = (\mathbb{S}, \mathbb{L}, \mathbb{T}, S_0, \mathbb{S}_f).$$

The index of the initial state of the active space inherent to $\wp(\xi)$ is the initial state of the complex-observation index space, namely $\mathfrak{S}_0$. Then, the reconstruction algorithm proceeds by looking for all of the transitions that are triggerable from the current state N of the reconstruction space.

From the point of view of the consistency with the complex observation, a transition

$$T = S \xrightarrow{\alpha|\beta} S'$$

of component C is triggerable from the current node

$$N = (\sigma, \mathfrak{S}, D)$$

of the search space, where S belongs to σ, only if at least one of the following conditions hold:

(1) T is *not* observable, that is, $(m, Out) \notin \beta$;

(2) C is *not* visible, that is, $C \in Ukn(OBS(\xi))$;

(3) T is both visible and observable but its message may be null, that is, $(m, Out) \in \beta$ and $C \notin Ukn(OBS(\xi))$ and $\epsilon \in \|m\|$;

(4) T is both visible and observable and it can generate a viable contextual label, that is,

$$(m, Out) \in \beta, C \notin Ukn(OBS(\xi)), (ContExt(m, C) \cap Viable(\mathfrak{S})) \neq \emptyset,$$

where $ContExt(m, C)$ is the *contextual extension* of message m emitted by component C, defined as follows

$$ContExt(m, C) = \{(\ell, C) \mid \ell \in \|m\|, \ell \neq \epsilon\},$$

and the definition of $Viable(\mathfrak{S})$ has been adapted from that given in Section 4.3 of Chapter 9 for an uncertain-observation index, that is, for a complex observation

$$Viable(\mathfrak{S}) = \{(\ell_v, C_v) \mid \mathfrak{S} \xrightarrow{(\ell_v, \mathbb{C})} \mathfrak{S}' \in \mathbb{T}, C_v \in \mathbb{C}\},$$

the only difference being that, while for uncertain observations the set $Viable(\mathfrak{S})$ includes observable labels, for complex observations $Viable(\mathfrak{S})$ is a set of contextual labels.

In particular, the last condition states that an observable and visible transition of component C is triggerable when the contextual extension of the message generated by the transition includes at least one out of the contextual labels whose sender component is C that are compatible with the edges leaving the current state of the index space.

If T is triggerable[2], a new node $N' = (\sigma', \mathfrak{S}', D')$ is added to the reconstruction search space, where σ' and D' are determined as usual,

---

[2] T is triggerable if it is consistent not only with the given complex observation but also with the dangling sets of the links. However, as in the previous chapter, only the first check is considered, as the focus is on observations, while consistency with respect to dangling sets has already been dealt with in Section 2 of Chapter 4 for monolithic reconstruction, and in Section 4 of Chapter 5 for modular reconstruction.

while the new value $\Im'$ of the index field is generated, starting from the old value $\Im$, based on the following rules, corresponding to the four above conditions, respectively:

(1) and (2) $\Im' = \Im$ and the edge $N \xrightarrow{T} N'$ is created in the reconstruction space[3];

(3) $\Im' = \Im$ and the edge $N \xrightarrow{T(\epsilon)} N'$ is created in the reconstruction space;

(4) $\Im'$, where

$$\Im \xrightarrow{(\ell,C)} \Im' \in \mathbb{T}, C \in \mathbb{C}, (\ell,C) \in (ContExt(m,C) \cap Viable(\Im)),$$

and the edge $N \xrightarrow{T(\ell)} N'$ is created in the reconstruction space.

According to rule 4, there exists a new value $\Im'$ for each different viable contextual label such that the sender component is that performing the relevant transition T and the observable label belongs to the extension of the message emitted by the transition itself. Note that the above conditions 1 to 3 and the corresponding rules are the same as for uncertain diagnostic problems.

Given the definition of a complex-observation instance provided in Section 2.3, the above conditions for checking whether a transition is triggerable, and the above rules for generating a successor node during the reconstruction process, we claim what follows.

Proposition 10.2. *Theorem 9.2 (for uncertain diagnostic problems) is valid for complex diagnostic problems too.*

Proposition 10.2 supports the monolithic resolution of complex diagnostic problems.

3.5   Modular Resolution

Given a complex diagnostic problem inherent to a system/cluster ξ, such a problem can be recursively split into subproblems according to a reconstruction graph. Each node of the reconstruction graph is a subproblem inherent to a subsystem of ξ: such a subproblem is the projection of the given problem on the relevant subsystem.

---

[3] Transition identifiers are used as labels for the edges within behavior reconstruction search spaces. Throughout the book it is always implicitly assumed that, during the behavior reconstruction process, each of such labels univocally identifies both a transition within a behavioral model and the specific component performing that transition.

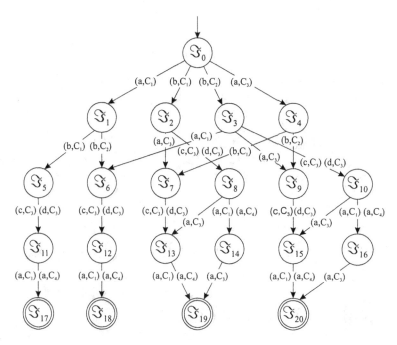

Figure 10.8. Index space of the complex observation depicted on the left of Figure 10.3.

This projection is possible also for complex diagnostic problems since in Section 2.2 it has been explained how to determine the restriction of a complex observation inherent to ξ on any subsystem of ξ.

In order to solve each complex subproblem, a process of (adapted) monolithic reconstruction (see Section 3.4) is applied. Such a process requires that a complex-observation index space, relevant to the considered subproblem observation, is available. This is why we introduce the notion of an index space restriction in the following section.

3.5.1   Index Space Restriction

Let $Space(OBS(\xi)) = (\mathbb{S}, \mathbb{L}, \mathbb{T}, S_0, \mathbb{S}_f)$ be the index space of a complex observation of a cluster $\xi = (\mathbb{C}, \mathbb{L}, \mathbb{D})$, where $OBS(\xi)$ has been gathered by observer $\Omega(\xi)$.

The *restriction* of $Space(OBS(\xi))$ on a subcluster $\xi' = (\mathbb{C}', \mathbb{L}', \mathbb{D}')$, $\mathbb{C}' \subset \mathbb{C}$, is a deterministic finite automaton

$$Space_{(\xi')}(OBS(\xi)) = (\mathbb{S}', \mathbb{L}', \mathbb{T}', S_0', \mathbb{S}_f')$$

obtained from $Space(OBS(\xi))$ by means of the following two steps:

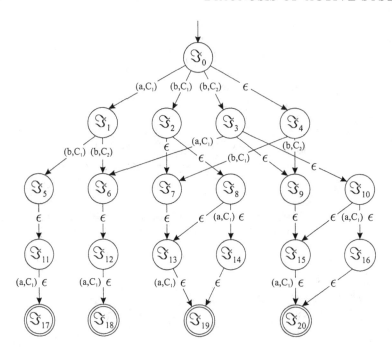

Figure 10.9. Intermediate step while restricting the index space of Figure 10.8.

(1) Replace each transition T in $\mathbb{T}$, marked by the pair $(\ell_v, \mathbb{C}_v)$, as follows, where $\mathbb{C}_{int} = \mathbb{C}_v \cap \mathbb{C}'$:

   (a) if $\mathbb{C}_{int} = \emptyset$, replace $T(\ell_v, \mathbb{C}_v)$ with $T(\epsilon)$;

   (b) if $\mathbb{C}_{int} \neq \emptyset$ and $\nexists C \in (\mathbb{C}_v - \mathbb{C}_{int}) \mid (\ell_v \in \Lambda(C))$, replace $T(\ell_v, \mathbb{C}_v)$ with $T(\ell_v, \mathbb{C}_{int})$;

   (c) if $\mathbb{C}_{int} \neq \emptyset$ and $\exists C \in (\mathbb{C}_v - \mathbb{C}_{int}) \mid (\ell_v \in \Lambda(C))$, replace $T(\ell_v, \mathbb{C}_v)$ with two transitions: $T(\epsilon)$ and $T(\ell_v, \mathbb{C}_{int})$;

(2) Transform the nondeterministic automaton obtained in Step 1 into an equivalent deterministic one.

Example 10.10. Figure 10.8 shows the index space $Space(OBS(\xi))$ inherent to the complex observation $OBS(\xi)$ of Example 10.4, displayed on the left of Figure 10.3. In order to determine the restriction of the same index space on the subcluster ξ' of Example 10.4, Step 1 is applied, as depicted in Figure 10.9. Finally, Step 2 is applied, thus obtaining $Space_{(\xi')}(OBS(\xi))$, which is displayed in Figure 10.10.   ◇

Given the notion of the index space of a complex observation and that of its restriction, we claim that the following proposition holds.

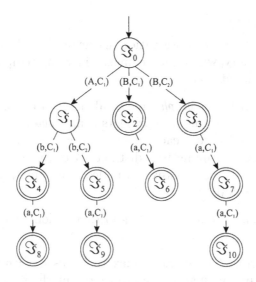

Figure 10.10. Restriction of the index space of Figure 10.8.

Proposition 10.3. *Theorem 9.3 (for uncertain observations) is valid for complex observations too.*

Based on Proposition 10.3, the index space to be exploited for generating the active space inherent to a subproblem $\wp_{\langle \xi' \rangle}(\xi)$ can be generated either as $Space(OBS_{\langle \xi' \rangle}(\xi))$ or as $Space_{\langle \xi' \rangle}(OBS(\xi))$.

Embracing the first choice means first computing $OBS_{\langle \xi' \rangle}(\xi)$, by following the method described in Section 2.2, and then determining the index space of $OBS_{\langle \xi' \rangle}(\xi)$, by following the three-step process described in Sections 3.1, 3.2, and 3.3. All these activities have to be performed for each subproblem of the reconstruction graph.

Making the second choice means determining once for all the index space of $OBS(\xi)$ and, for each subproblem of the reconstruction graph, projecting such a space on the relevant cluster.

3.5.2 Modular Reconstruction

The technique for a modular reconstruction in order to solve complex diagnostic problems is analogous to that already introduced for uncertain diagnostic problems. Moreover, also the equivalence between monolithic and modular reconstruction holds, as claimed by the following proposition.

Proposition 10.4. *Theorem 9.4 (for uncertain diagnostic problems) is valid for complex diagnostic problems too.*

4.   Summary

- *Complex observation*: observation featuring logical, source, and temporal uncertainty, where source uncertainty and temporal uncertainty can be dependent on each other.

- *Complex-observation graph*: pictorial description of a complex observation, wherein a node represents a message and contains one or more contextual nodes, each inherent to a (possibly logically uncertain) message pertaining to a distinct set of sender components. An arrow drawn between contextual nodes represents a temporal precedence relationship.

- *Complex diagnostic problem*: diagnostic problem featuring a complex observation.

- *Complex-observation index*: index to be used within each state of the reconstruction space, so as to represent the contextual nodes of the complex observation that have already been generated by all the paths leading from the initial state to the current state. Each index value partitions the set of contextual nodes of the whole reaction observation into two sets, depending on whether the message relevant to the contextual node has already been generated (consumed) when the current state is reached or not. Given such an index, it is possible to infer which are the next contextual nodes to be generated.

- *Complex-observation restriction*: projection of a complex observation inherent to a cluster on one of its subclusters.

- *Contextual label*: a pair consisting of an observable label and its sender component.

- *Complex-observation instance*: totally temporally ordered sequence of contextual labels that is compliant with a given complex observation.

- *Complex-observation extension*: set of all the instances of a given complex observation.

- *(Deterministic) index space of a complex observation*: deterministic graph produced by processing a complex observation. In such a graph each node is an index of the considered complex observation and each transition is marked by one or more contextual labels such that the reception of any of them causes the same index change.

- *Extension of the index space of a complex observation*: set of all the sequences of contextual labels compliant with the index space of a complex observation. It is claimed that it equals the extension of the observation itself.

- *Restriction of the index space of a complex observation*: projection of the index space of the complex observation inherent to a cluster on one of its subclusters. It is claimed that it equals the index space of the restriction of the observation itself.

Chapter 11

UNCERTAIN EVENTS

This chapter has been contributed by Roberto Garatti
garatrob@tin.it

Abstract

The expressive power of the behavioral model of an active system component is enhanced by introducing the general notion of an uncertain event, so as each event in a model can be an ambiguous label ranging over a set of symbols. This allows one to represent real world situations wherein a component of a distributed system exhibits a non-deterministic behavior in the generation not only of observable events but also of the events exchanged with other components, as it may occur owing to faults. Updating the modeling primitives of the active system approach requires in turn the diagnostic approach to be updated. The intervention is not straightforward since the notion of the dangling set of a link interconnecting two components has changed from that of a buffer containing unambiguous labels to that a buffer containing ambiguous, possibly null, labels. Three alternative way for tackling these new dangling sets during history reconstruction are discussed.

1. Introduction

In Chapter 9 four distinct causes of uncertainty inherent to message generation and/or reception were discussed. In order to represent the forms of uncertainty affecting message generation, namely loss uncertainty and logical uncertainty, within the modeling primitives of the active system approach, the behavioral models of components were extended so as to support the notion of an *uncertain message*. As a message is a kind of event, it is quite natural to generalize this extension to all of the events within behavioral models, thus producing the notion of an *uncertain event*, according to which an event is a discrete variable whose domain possibly includes the *null* value.

While Chapter 9 took into account the sources of uncertainty inherent to messages only, this chapter takes into account the sources of uncertainty affecting events other than messages, that is, input events and (unobservable) output events. Both modeling primitives and the reconstruction process are updated, the former in order to allow for the representation of uncertain events, the latter to manage this new form of nondeterminism as efficiently as possible.

The chapter is organized as follows: Section 2 updates two of the uncertainty requirements presented in Chapter 9, thus increasing the nondeterminism of the behavioral models of components. Section 3 formally defines the concept of an uncertain event and discusses the pros and cons of three alternative methods for performing history reconstruction when behavioral models encompass uncertain events exchanged between components.

2.   Uncertainty Requirements

This section introduces two uncertainty requirements inherent to the representation of events in the behavioral models of active systems. They are the extensions (i.e. they have to be added) to the two homonymous uncertainty requirements inherent to messages presented in Section 2 of Chapter 9. In both cases such requirements are valid for DESs in general, are independent of the adopted diagnostic technique, and have to be orthogonal to one another.

2.1   Loss uncertainty

Assuming that the logical content of an event exchanged between components cannot be randomly lost is an over-constrained assumption for a wide range of real systems in which events may get lost because of misbehaviors of the transmission from a component to another.

Requirement 11.1. (Loss uncertainty) *In a behavioral model, the set of output events generated by transitions is divided into two disjoint subsets: the subset of* reliable *events, and the subset of* unreliable *events. When an event is unreliable, it may get lost, otherwise it cannot be lost.*

One may argue that loss uncertainty affecting messages was bound to be modeled within the behavioral models of components while loss uncertainty affecting (unobservable) output events is not. In fact, in the active system approach, there is no primitive for representing the transmission channels from components to the observer, while there is a modeling primitive, namely the link, for representing the communication means between components. Therefore, the possibility of losing events exchanged on links could be represented in link models: a link model should be characterized by a toggle, say *reliability*, such that, if the link is reliable, no event transmitted on that link is lost, whereas, if the link is unreliable, each of the events transmitted may be lost.

The above requirement, instead, points to a different modeling choice, which keeps link models unchanged: each single output event of a component is either reliable or unreliable, where its reliability is specified in the component model. This brings two advantages over the alternative of specifying reliable and unreliable links:

- Use of modeling primitives for uncertain events analogous to that for uncertain messages;

- Greater expressive power since each event can be modeled as either certain or uncertain, independently of the link on which it is conveyed and on the other events conveyed on the same link. If, instead, the alternative solution were adopted, every event conveyed on an unreliable link would be uncertain. This flexibility allows us to model some real-world circumstances that cannot be described by means of the other solution, for instance the case when the same physical channel may lose some types of events while it never loses others (e.g. for they are transmitted by means of fault correction codes).

In order to fulfill Requirement 11.1, an uncertain event of a component is represented within the component model as a variable whose value is ranging over two labels, one of which is the null label. For instance, a transition T, which is triggered by event e on terminal I and generates event e_1 on output terminal O, where such an event can be lost, is modeled as follows:

$$T = S_1 \xrightarrow{(e,I)|(\{e_1,\epsilon\},O)} S_2.$$

2.2 Logical uncertainty

The uniqueness of the logical content of input and output events in behavioral models is another over-simplified assumption for real systems owing to the following reasons.

(1) The value of an event generated by a transition may vary from an occurrence of the transition to another. This depends, for example, on the granularity level of the model of the relevant component, or on some internal phenomena of the component, difficult to understand and/or to model in a deterministic way, that may randomly change

the expected value of an event produced as output or even prevent its production.

(2) The value of an output event, if transmitted on a noisy link, may be corrupted and, therefore, the value received by the target component may be different with respect to that generated by the source component.

(3) A state transition may be triggered by several distinct input events, that is, the input event may vary from an occurrence of the transition to another, this depending, for example, on the granularity level of the model of the relevant component, or on the actual tolerance of the physical device as regards its input.

Requirement 11.2. (Logical uncertainty) *In behavioral models, each input and output event may be uncertain, that is, its value may be one out of a given set of values.*

The three causes listed above determine three orthogonal forms of non-determinism, which are all represented within the behavioral models of components.

(1) The representation of the nondeterministic behavior of the system in generating output events is achieved by assigning a set of values, possibly including the null value, to each uncertain output event. For instance, suppose that each occurrence of transition T, which is triggered by event e on terminal I, generates either the output event e_1, or e_2, or e_3, or nothing. Such a transition is modeled as follows:

$$T = S_1 \xrightarrow{(e,I)|(\{e_1,e_2,e_3,\epsilon\},O)} S_2. \tag{11.1}$$

It is worth noting the different semantics of the null value ϵ in case of loss uncertainty with respect to logical uncertainty: in the former case, every occurrence of the relevant transition produces an event that, sometimes, may be lost during the transmission; in the latter, some occurrences of the transition do not generate any event.

(2) The nondeterministic change of value of an event that is caused by the noisy link on which the event is transmitted is modeled in the behavioral model of the source component. For instance, suppose that each occurrence of transition T, which is triggered by event e on terminal I, generates output event e_1. Suppose also that such an event is transmitted on a link that may transform it into either e_2, or e_3, or e_4. Transition T is modeled as follows:

$$T = S_1 \xrightarrow{(e,I)|(\{e_1,e_2,e_3,e_4\},O)} S_2.$$

(3) The nondeterminism of the input event of a transition is modeled by using an input event that ranges over several values. For instance, if each occurrence of transition T, which does not generate any output event, is triggered by either event e_1 or e_2, T is modeled as follows:

$$T = S_1 \xrightarrow{(\{e_1,e_2\},I)} S_2.$$

A modeling constraint is that the set of the possible values of an uncertain input event whose terminal is not the standard input cannot include any ϵ, this meaning that a component cannot take a null event from a link.

It could be objected that a transition, fed by a link and having an uncertain input event including the null event, could be interpreted as a transition that sometimes is spontaneous. We do not allow this interpretation of behavioral models since, in our view, spontaneous transitions can be triggered only by events coming from the outside world. Thus, if a state change sometimes is triggered by an internal input event and sometimes is spontaneous, such a state change has to be modeled by means of two parallel transitions, one triggered by the internal event and the other by an event on the standard input.

According to the syntactic rules of behavioral models that have been enforced in the active system approach, the set of possible values of an uncertain input event can include the null event only if the terminal relevant to the event is the standard input.

Indeed, from a theoretical point of view, this difference between the set of values of uncertain events depending on the terminal is not strictly necessary. If it were assumed that no uncertain input event can include the null event, independently of its terminal, no restriction of the expressive power of the modeling primitives would be caused.

In fact, all of the reconstruction methods of the active system approach assume that a transition that is triggered by an event on the standard input, i.e. an event coming directly from the outside world, can be fired at any moment. Thus all input transitions are spontaneous in the same way, independently of the name, either ϵ or not, of the triggering event.

Substantiating the orthogonality of Requirements 11.1 and 11.2 does not need any intervention. In fact, point 3 of Requirement 11.2, which is inherent to input events, is orthogonal to Requirement 11.1 since the latter does not affect input events; points 1 and 2 of Requirement 11.2, which are inherent to output events, are orthogonal to Requirement 11.1 since they need the same modeling primitives and both points can be combined freely with each other and with loss uncertainty. For instance, given transition T of Equation 11.1, it is impossible to distinguish between the following interpretations:

(a) T generates either output event e_1, or e_2, or e_3, or nothing;

(b) T generates either output event e_1, or e_2, or e_3, and the generated event may be lost during the transmission on the link;

(c) T generates output event e_1 $(e_2)(e_3)$, which, during the transmission on the relevant link, may be lost or transformed into either e_2 or e_3 $(e_1$ or $e_3)(e_1$ or $e_2)$;

(d) T generates either output event e_1 or e_2 $(e_1$ or $e_3)(e_2$ or $e_3)$, and the generated event, during the transmission on the relevant link, may be lost or transformed into e_3 $(e_2)(e_1)$.

Interpretation (a) assumes that the transition features only the kind of logical nondeterminism described at point 1. Interpretation (b) assumes that T is affected by both the kind of logical nondeterminism described at point 1 and loss nondeterminism. According to interpretation (c), T is affected both by the kind of logical nondeterminism described at point 2 and loss nondeterminism. Finally, interpretation (d) assumes that T is affected by both kinds of logical nondeterminism involving output events and loss nondeterminism as well.

Example 11.1. Figure 11.1 shows the structural model of system Θ (top) and the behavioral models of its two components (bottom), which are affected by loss and logical uncertainty. In particular, transitions T_{13} generates an uncertain event on terminal O_1, transition T_{19} generates an uncertain event on terminal O_1 and another on terminal O_2, faulty transition T_{23} generates an uncertain event on terminal O, and transition T_{22} is triggered by an uncertain event on terminal I.

In Figure 11.2 the two behavioral models of the same components are represented by means of modeling primitives devoid of any specific notation for uncertain events, that is, by means of the modeling primitives used in the previous chapters. It is interesting noting that the number of transitions almost doubles that needed when the new primitives are available.

Incorporating in one transition several transitions that have the same source and target component states and that generate output events on the same terminals, as allowed by the uncertainty modeling primitives,

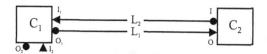

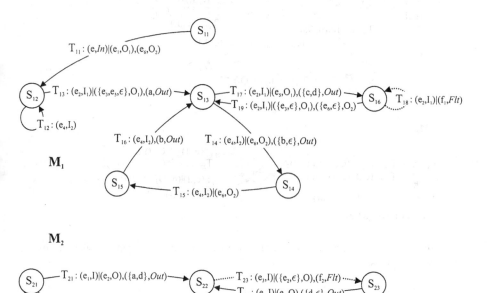

Figure 11.1. System with uncertain events.

produces a beneficial effect on the reconstruction of the histories of the system since it is checked only once whether such a 'unifying' transition is triggerable instead of several times. ◇

3. Reconstruction

The formal definition of an uncertain event is the same as that of an uncertain message given in Section 3 of Chapter 9. In fact, an uncertain message is indeed an instantiation of the concept of an uncertain event. In particular, let Λ be a set of non-null *labels* and $\mathcal{V}$ a set of *variables* such that $\forall X \in \mathcal{V}$ the value of X is

$$\|X\| \subseteq (\Lambda \cup \{\epsilon\}),$$

where ϵ denotes the *null* label, as usual. The set

$$\gamma = \Lambda \cup \mathcal{V}$$

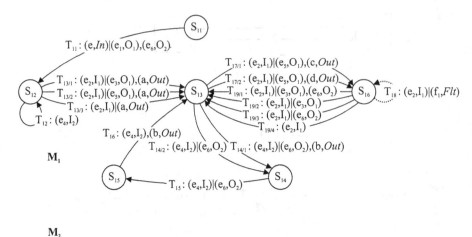

Figure 11.2. Behavioral models with uncertain events represented without uncertainty primitives.

is the *domain* of uncertain events. The *extension* of an uncertain event $E \in \gamma$ is the set defined as follows:

$$\|E\| = \begin{cases} \{e\} & \text{if } E = e, e \in \Lambda \\ \|X\| & \text{if } E = X, X \in \mathcal{V}. \end{cases}$$

Intuitively, the extension of an uncertain event is the set of its associated labels.

The reconstruction method for the generation of the active space inherent to a (sub)problem has to be updated in order to deal with (sub)systems having uncertain events. Since what has been altered is the nature of the events exchanged on links, what needs to be updated is the management of the dangling sets of links within the search space.

In the following sections three alternative kinds of management are presented and their advantages and disadvantages are discussed. Each solution is illustrated by means of an example, wherein the reconstruction of the universal space of the system Θ of Example 11.1 is carried out.

As already remarked in Section 2.2 of Chapter 4, since the reconstruction of the universal space is not constrained by any observation, each search space state does not contain any observation index.

This is beneficial to our purpose. In fact, our aim in this section is to underline how uncertain events exchanged between components are managed during the reconstruction, while the management of uncertain messages, which is orthogonal, has already been discussed in Chapter 9.

3.1 Enumeration of Dangling Sets

Given the current node of the reconstruction search space state, a transition is triggerable if its input event is either

- On the standard input terminal, or

- On a dangling input terminal, or

- On a terminal connected with a link such that the consumable event of this link belongs to the extension of the input event.

From the point of view of its output events, instead, a transition is triggerable if, given the current situation of the dangling sets and the saturation policies of the links, each link connected with the terminals of the output events of the transition can host a new event.

Given a triggerable transition generating an uncertain output event on a terminal connected with a link L, such a transition produces as many distinct successor states in the search space as the cardinality of the extension of the uncertain event. Each of such successor differs from the others in the dangling set of link L, which has been obtained by properly adding a distinct value of the uncertain output event to the content of the link.

Given a triggerable transition whose input terminal is connected with link L, in every successor state the candidate event of $\|L\|$ has been consumed.

Example 11.2. Figure 11.3 displays the search performed for reconstructing by enumeration the universal space of the system Θ of Example 11.1. Here and in the following it is assumed that links L_1 and L_2 are characterized by capacity $\chi = 1$ and saturation policy *WAIT*.

Each application of transitions T_{13}, T_{19}, and T_{23} produces 3, 2, and 2 successor states, respectively, since they generate an uncertain event on the relevant link whose extension has cardinality 3, 2, and 2, respectively. Each edge in the figure, besides being marked by the name of the relevant transition, in case the transition generates an uncertain output event is labeled also by the value of the hypothesized output event in brackets.

In the reconstruction search space, transition T_{13} is triggerable only from states wherein the dangling set of link L_2 contains event e_2. Starting from any of such search space states, T_{13} leads to three different

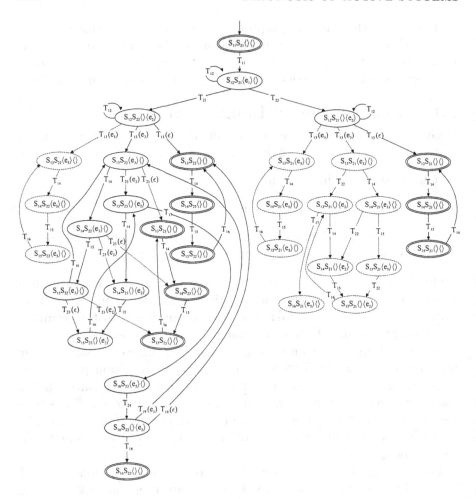

Figure 11.3.   Reconstruction by enumeration of dangling sets.

states depending on which event the transition is assumed to have generated, either e_3, or e_5 or ϵ.

In case T_{13} is assumed to have generated ϵ, the successor state in Figure 11.3 is always a final one since the transition consumes the only event in link L_2 and does not generate any event on link L_1.

Transition T_{22}, instead, can be triggered by either event belonging to the extension of its uncertain input event on link L_1. Figure 11.3 contains just two occurrences of T_{22}, one triggered by event e_1, the other by e_3.
$\diamond$

If a triggerable transition generates several uncertain output events E_1, ..., E_r on r distinct links, then the maximum number of successors it produces in the reconstruction search space equals the product

$$|E_1| * \cdots * |E_r|.$$

Each successor differs from the others in the content of a dangling set at least.

3.2 Uncertain Dangling Sets

In order to reduce the size of the reconstruction search space with respect to that of the reconstruction by enumeration of dangling sets, it is necessary to prevent the reconstruction process from creating several distinct successor states when a transition produces an uncertain output event. To this end, we now introduce the concept of an *uncertain dangling set*, which is a sequence of uncertain events representing several distinct certain dangling sets, called *dangling set instances*. Let

$$D = \langle E_1, \ldots, E_n \rangle$$

be the uncertain dangling set of a link L, that is, a sequence of n uncertain events, where n is the length of D, denoted by $|D|$. The i-th element of D is denoted by $D[i]$. From D it is drawn

$$D' = \langle e_1, \ldots, e_n \rangle,$$

where $\forall i \in [1 \mathinner{..} n](e_i \in \|E_i\|)$. Then

$$D'' = \langle e'_1, \ldots, e'_q \rangle,$$

where $q \leq n$, is obtained from D' by removing all of the null events, that is,

$$\forall h \in [1 \mathinner{..} q](e_h \neq \epsilon, \nexists k \in [1 \mathinner{..} n] \mid (e_k \neq \epsilon, e_k \in D', e_k \notin D'')).$$

By definition, D'' is an instance of D. The whole set of instances of the uncertain dangling set D inherent to link L is called the *extension* of L, which is denoted by $\|L\|$.

Adopting uncertain dangling sets means using them within each state of the reconstruction search space state. This way, each triggerable transition having an uncertain output event (and no other output events) produces just one successor in the reconstruction search space state, the same as a transition that produces a certain event.

Given an uncertain dangling set, its candidate events, that is, the events that can be consumed, are all of the first events of the instances of the considered dangling set. A transition is triggerable if its input event is either

- On the standard input terminal, or

- On a dangling input terminal, or

- On a terminal connected with a link and the set of consumable events of such a link is not disjoint with respect to the extension of the input event.

A reconstruction search space state is final if each of its uncertain dangling sets is either empty or its extension includes an instance at least that is empty.

Example 11.3. Figure 11.4 displays the search performed for the reconstruction of the universal space of the system Θ of Example 11.1 when uncertain dangling sets are adopted. It is straightforward noting how smaller the size of this search space is with respect to that for solving the same problem with reconstruction by enumeration of dangling sets, shown in Figure 11.3.

Indeed, each state of the search space with uncertain dangling sets represents several states of the search space with the complete enumeration of certain dangling sets, where each of the certain dangling sets of the latter is an instance of the uncertain dangling set of the former. For example, state

$$((S_{13}, S_{22}), ((\{e_3, e_5, \epsilon\}), \langle\rangle))$$

of Figure 11.4 represents three states of Figure 11.3, namely

$$((S_{13}, S_{22}), ((\langle e_3 \rangle), \langle\rangle)),$$
$$((S_{13}, S_{22}), ((\langle e_5 \rangle), \langle\rangle)),$$
$$((S_{13}, S_{22}), ((\langle\rangle), \langle\rangle)).$$

Each of the three transitions that in Example 11.2 produce several successors, i.e. T_{13}, T_{19}, and T_{23}, produces just one successor here. For instance, each edge marked by transition T_{13} produces a successor wherein the (uncertain) dangling set of link L_1 is $\langle\{e_3, e_5, \epsilon\}\rangle$, which means that L_1 either contains e_3 only, or e_5 only, or is empty. Since in each of such successor states the dangling set of link L_2 is empty and the dangling set of link L_1 may be empty, all of such states are final. $\diamond$

Although each search space state of the reconstruction process adopting uncertain dangling sets represents several search space states of the reconstruction process by enumeration of certain dangling sets, it does not produce the minimum-size active space, as proven by the following example.

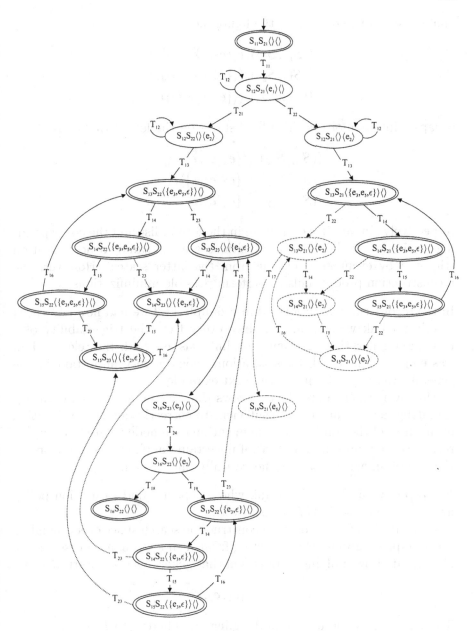

Figure 11.4. Reconstruction by adopting uncertain dangling sets.

Example 11.4. The size of the universal space displayed in Figure 11.4
is not minimal. In fact, if a minimization is carried out, each of the three

final states on the bottom of the figure, i.e.

$$((S_{13}, S_{22}), ((\{e_3, \epsilon\}), \langle\rangle)),$$
$$((S_{14}, S_{22}), ((\{e_3, \epsilon\}), \langle\rangle)),$$
$$((S_{15}, S_{22}), ((\{e_3, \epsilon\}), \langle\rangle)),$$

collapses into one of the three final states on the left of the figure, i.e.

$$((S_{13}, S_{22}), ((\{e_3, e_5, \epsilon\}), \langle\rangle)),$$
$$((S_{14}, S_{22}), ((\{e_3, e_5, \epsilon\}), \langle\rangle)),$$
$$((S_{15}, S_{22}), ((\{e_3, e_5, \epsilon\}), \langle\rangle)),$$

respectively. In fact, for each state in the former list and the corresponding in the latter the exiting transitions are the same as no transition triggered by event e_5 is exiting from the latter states. However, the reconstruction process fails to perform possible minimizations.   ◇

In the reconstruction performed in Example 11.3 we had never to check whether a link were saturated since every time the triggerability of a transition generating an event on a link was assessed, the relevant link was empty. This check for saturation, which is very important in the general case, is dealt with in the next example.

Moreover, the example below shows that, in the general case, uncertain dangling sets alone are insufficient to correctly define the situation of the links of the considered system within each node created during the reconstruction process, and that, if uncertain dangling sets are adopted, the size of an active space is theoretically unbounded.

Example 11.5. Let L be a link whose capacity and saturation policy are $\chi = 2$ and $\pi = WAIT$, respectively.

In the initial state S_0 of the reconstruction search space every dangling set is empty. Suppose that, starting from such a state, a successor state S_1 is created by applying a transition that generates the uncertain event

$$E_1 = \{e_1, e_2, \epsilon\}.$$

Therefore, in S_1 the uncertain dangling set inherent to L is

$$D = \langle\{e_1, e_2, \epsilon\}\rangle.$$

Suppose that a successor state of S_1, namely S_2, is obtained by applying a transition that generates the uncertain event

$$E_2 = \{e_3, e_4, \epsilon\}.$$

Therefore, in S_2 the uncertain dangling set inherent to L becomes

$$D = \langle \{e_1, e_2, \epsilon\}, \{e_3, e_4, \epsilon\} \rangle.$$

Such a dangling set cannot be reckoned as saturated since its extension includes the empty instance and four instances containing just one certain event as well. Thus, if a transition that produces the uncertain event

$$E_3 = \{e_5, e_6\}$$

is applied to S_2, a state S_3 is obtained wherein

$$D = \langle \{e_1, e_2, \epsilon\}, \{e_3, e_4, \epsilon\}, \{e_5, e_6\} \rangle.$$

Since the reconstruction process can access the models of links, it knows that the capacity of link L is 2 and, therefore, it knows that, although the uncertain dangling set contains 3 uncertain events, it represents only instances whose length is up to 2.

Based on this case, we could be tempted to state this rule: an uncertain dangling set D of a generic link L cumulatively represents all the instances belonging to $\|L\|$ such that their length is less than or equal to $\chi(L)$. In fact, all of the instances whose length is greater than the capacity of the link are physically impossible. This rule, however, is incorrect, as illustrated here below.

Suppose that a transition, which consumes the certain event e_1, is applied to S_3, thus generating a successor state S_4. Which is the dangling set of L within S_4? The most natural answer is

$$\langle \{e_3, e_4, \epsilon\}, \{e_5, e_6\} \rangle.$$

However, if the above rule were valid, such an uncertain dangling set would represent a lot of instances whose length is 2: this is apparently incorrect since a link of capacity equal to 2 cannot contain 2 values after one has been extracted, instead it can contain 1 value at most. This means that the above rule is wrong. We try and fix it as follows: an uncertain dangling set D of a link L cumulatively represents all of the instances belonging to $\|L\|$ such that their length is less than or equal to the maximum number of certain events currently buffered in the link, denoted by *max*.

This means that a further piece of information, i.e. *max*, has to be saved together with each uncertain dangling set within the states of the reconstruction search space. In the initial state of the reconstruction search space *max* equals 0 for all of the dangling sets since links are supposed to be initially empty. Every time a transition generating an event E on link L is applied to the current search space state, in the successor state E is added to D and max is incremented by 1 only if $max < \chi(L)$, otherwise the value of max keeps on being $\chi(L)$.

However, in order to characterize the situation of a link max is not enough. In fact, max is the maximum number of certain events contained in the link. But which is the minimum number of such events? For, if the link actually includes a number of certain events equal to its capacity, it is saturated and, therefore, no further (either certain or uncertain) event can be sent through the link. Thus, for each dangling set, let us save within each search space state another piece of information, namely the minimum number of certain events contained in the link, denoted by min.

In the initial state of the reconstruction search space min is set to zero. Every time an event E is added to the relevant link, min is incremented by 1 only if $\epsilon \notin E$. Note that, in every search space state, necessarily $max \geq min$. Link L is saturated if $min = \chi(L)$.

So far in this example we have considered how to add events to links. Now we will discuss how events are extracted from links. Suppose that T is a transition triggered by event E on a terminal connected with link L.

In order to check whether such a transition is triggerable from the point of view of the input event, it is necessary to check the extension of such an event against the first event stuffed in the link. If

$$E_{\mathrm{in}} = \|D[1]\| \cap \|E\| \neq \emptyset,$$

then T is triggerable and, in order to produce the relevant successor states, event $D[1]$ has to be extracted. If $\epsilon \notin \|D[1]\|$, no further check has to be performed. If, instead, $\epsilon \in \|D[1]\|$ and $|D| > 1$, then the first event could be null and the event triggering T could be extracted from the second position of the uncertain dangling set.

Then, E has to be checked against $D[2]$, and so on, until the k-th position is reached such that either an event $D[k]$ is found whose extension does not include ϵ or all of the events in the uncertain dangling set have been checked, that is, $k = |D|$. Therefore, from the point of view of the input event of a transition, as many successor states can be produced as the number of E_{in} computed by the iterative process described here.

So there will be k distinct successor values of D, depending on whether the extracted event is either $D[1]$, or ..., or $D[k]$. Suppose that event $D[i]$ is extracted.

In the successor state of the current search space state, obtained by applying T and hypothesizing that the extracted event is the i-th,

- min is decremented by 1 only if $\epsilon \notin \|D[i]\|$;

Table 11.1. Situation of the reconstruction search space states (see Example 11.5).

State	$D(L)$	$max(L)$	$min(L)$
S_0	$\langle\rangle$	0	0
S_1	$\langle\{e_1, e_2, \epsilon\}\rangle$	1	0
S_2	$\langle\{e_1, e_2, \epsilon\}, \{e_3, e_4, \epsilon\}\rangle$	2	0
S_3	$\langle\{e_1, e_2, \epsilon\}, \{e_3, e_4, \epsilon\}, \{e_5, e_6\}\rangle$	2	1
S_4	$\langle\{e_3, e_4, \epsilon\}, \{e_5, e_6\}\rangle$	1	1

- All of the events in D up to $D[i]$ included are removed from D;

- *max* is decremented by 1, as, in fact, all of the events preceding the one that is really consumed are bound to be null.

Based on the revised rule, the situation of the reconstruction search space states considered so far is outlined in Table 11.1.

However, the description of the reconstruction process given above is still incorrect as regards the management of extracted events. In fact, suppose that, starting from S_3, a transition is applied, thus obtaining the successor state S_4', where such a transition is triggered by event e_5. According to the reconstruction process described above, the situation in S_4' would be

$$(D(L) = \langle\rangle, max(L) = 1, min(L) = 0),$$

which is apparently wrong since the maximum number of certain events stuffed in the link cannot be 1 if the link is empty. Therefore, we have to change once again the reconstruction algorithm in order to fix this incorrect management of *max*.

We do it as follows: every time a transition extracting event $D[i]$ from link L is applied to the current search space state, in the resulting successor state the new value of *max* is the minimum between $(max - 1)$ and $|D(L)|$. This way, the situation in S_4' becomes

$$(D(L) = \langle\rangle, max(L) = 0, min(L) = 0),$$

which is correct.

All the same, the approach still does not work. In fact, consider, for instance, a transition that is triggered by an event on either the standard input or a dangling input terminal, where such a transition produces the uncertain event $E = \{e_1, \epsilon\}$ on link L. This transition can be applied to

Table 11.2. Creation of an infinite set of successor states (see Example 11.5).

$D(L)$	$max(L)$	$min(L)$
$\langle\{e_1, \epsilon\}\rangle$	1	0
$\langle\{e_1, \epsilon\}, \{e_1, \epsilon\}\rangle$	2	0
$\langle\{e_1, \epsilon\}, \{e_1, \epsilon\}, \{e_1, \epsilon\}\rangle$	2	0
$\langle\{e_1, \epsilon\}, \{e_1, \epsilon\}, \{e_1, \epsilon\} \cdots\rangle$	2	0

every state in the reconstruction search space. Suppose to apply it to state S_0, and afterwards repeatedly to the generated successors states. The situation in such states would be as outlined in Table 11.2, that is, infinite successor states would theoretically be created, since the link is never reckoned as saturated.   ◇

The size of the active space produced by the reconstruction process informally described in Example 11.5 is unbounded. Since the means that could be devised for preventing this explosion are expensive to manage, the reconstruction process adopting uncertain dangling sets is discarded hereafter.

3.3   Hybrid Approach

An approach to reconstruction that is located midway between the ones discussed in the previous two sections is obtained if the null output event ϵ is dealt with in a special way and not in the same way as the other events, as instead it is done in the previous two approaches.

The domain of an uncertain event that includes the null event in its extension is now taken as the union of two domains, called *unambiguous domains*: the domain including non-null values only and the domain containing the null event only. Note that the domain of an uncertain event that does not include the null event is unambiguous itself.

During the reconstruction process, from the point of view of the events sent through links, a transition produces as many distinct dangling sets of successor states in the search space as the cardinality of the Cartesian product of the unambiguous domains of the output events. For instance, assuming that terminals O_1 and O_2 of a component C are both connected to empty links, transition

$$T = S_1 \xrightarrow{(e_1, I)|(\{e_2, e_3, \epsilon\}, O_1)(\{e_4, \epsilon\}, O_2)} S_2$$

of C always produces four successor states, characterized by the following dangling sets:

$$\langle \{e_2, e_3\}\rangle \langle e_4\rangle,$$
$$\langle \{e_2, e_3\}\rangle \langle \rangle,$$
$$\langle \rangle \langle e_4\rangle,$$
$$\langle \rangle \langle \rangle.$$

Such dangling sets are the result of the Cartesian product

$$\{\{e_2, e_3\}, \{\epsilon\}\} \times \{\{e_4\}, \{\epsilon\}\},$$

where the two operands are the sets of unambiguous domains of the uncertain events ($\{e_2, e_3, \epsilon\}$ and $\{e_4, \epsilon\}$, respectively. This way, within search space states, each dangling set contains only uncertain events whose extension does not include any ϵ, which makes the detection of saturation quite easy: as in the previous chapters, a link is saturated if it contains a number of events equal to the capacity of the link.

Also the extraction from a link of the input event E, be it certain or not, of a transition T is quite easy since only the first event in the dangling set of the link has to be considered: if

$$\|E\| \cap \|D[1]\| \neq \emptyset,$$

then, from the point of view of the input event, T is triggerable and the first event has to be dequeued. In case the input event of T is coming from the standard input or from a dangling input terminal, from the point of view of the input event T is triggerable as well and the dangling set D is unchanged in successor states.

Example 11.6. Figure 11.5 displays the search performed by the hybrid reconstruction process of the universal space of the system Θ of Example 11.1. Given a transition T of the components of Θ that generates one or more uncertain output events on links, let $X(T)$ be the Cartesian product of the unambiguous domains of such events. Each edge of the reconstruction search space that represents an occurrence of T is marked by the identifier of T and by the relevant element of $X(T)$ between brackets. Each of such transitions produces several successor states.

For instance, since transition T_{13} of Θ generates just an uncertain event $\{e_3, e_5, \epsilon\}$, which is sent to link L_1, the Cartesian product $X(T_{13})$ equals the set of unambiguous domains of event $\{e_3, e_5, \epsilon\}$, that is,

$$X(T_{13}) = \{\{e_3, e_5\}, \{\epsilon\}\}.$$

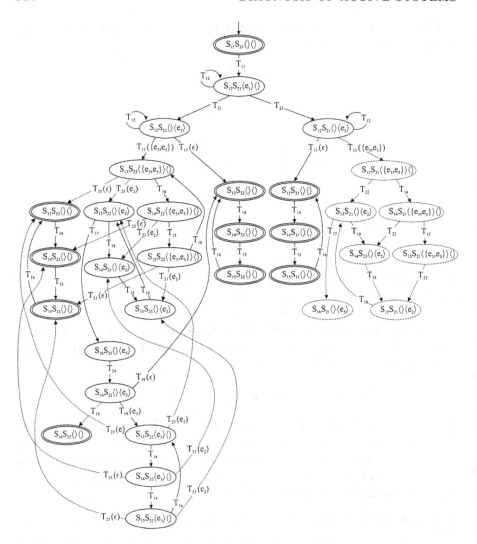

Figure 11.5.   Reconstruction by hybrid approach.

Thus, there are always two occurrences of transition T_{13} exiting from the same search space state, one marked by $T_{13}(\{e_3, e_5\})$ and the other by $T_{13}(\epsilon)$.   ◇

The hybrid approach, while solving the problem of saturation detection, which is critical in the approach adopting uncertain dangling sets, does not minimize the size of the search space, the same as the uncertain dangling sets approach.

3.4 Discussion

In the previous sections we have introduced three distinct methods for managing the contents of links while performing a (either monolithic or modular) reconstruction when the behavioral models of the components of the considered system include uncertain events. However, dangling sets are just one aspect of the nondeterminism affecting the search process, the other being uncertain/complex observations. As dealt with in Chapters 9 and 10, the latter form of nondeterminism is managed by means of observation indexes.

Suppose that we are performing a reconstruction inherent to the complex problem

$$\wp(\xi) = (\Omega(\xi), OBS(\xi), \xi_0).$$

Consider a transition T, triggerable from the current search space state and generating r output events $E_1, \ldots, E_r$ on r (distinct) relevant links (as known, a link is relevant if it is an internal link of ξ in case a monolithic reconstruction is performed, or if it is an interface link of ξ, in case a modular reconstruction is carried out). Suppose that T generates an uncertain message that complies with m distinct new observation indexes.

In case a reconstruction by complete enumeration of dangling sets is adopted, transition T produces $n * m$ successor states, where the maximum value of n is

$$n_{\max} = |E_1| * |E_2| \cdots * |E_n|.$$

If, instead, the hybrid approach is adopted, the number of successor stated produced by T is $n' * m$, where n' is the cardinality of the Cartesian product of the unambiguous domains inherent to the unobservable output events of T sent on the relevant links. Since

$$n' = n_1 * n_2 * \cdots * n_r,$$

where

$$n_i = \begin{cases} 2 & \text{if } \epsilon \in \|E_i\| \text{ (in which case, } |E_i| \geq 2) \\ 1 & \text{otherwise} \end{cases}$$

it follows that $n' \leq n_{\max}$, where the larger is the extension of uncertain events, the smaller is n'. Thus, in the hybrid approach the number of successor states is usually smaller than the worst case of the reconstruction by enumeration of dangling sets.

4. Summary

- *Loss uncertainty requirement*: the expressive power of component behavioral models has to enable the representation of output events that may be lost.

- *Logical uncertainty requirement*: the expressive power of component behavioral models has to enable the representation of input and output events whose values are not univocal.

- *Uncertain event*: event whose value ranges over a set of labels, called extension of the uncertain event, possibly including the null label.

- *Reconstruction by enumeration of dangling sets*: history reconstruction process wherein, in front of an uncertain event sent on a link, a distinct next (certain) dangling set situation is produced for each non-null label in the extension of the event.

- *Uncertain dangling set*: dangling set representing a sequence of uncertain events.

- *Reconstruction by adopting uncertain dangling sets*: history reconstruction process such that the situation of links within each search space state is described by uncertain dangling sets.

- *Hybrid reconstruction*: history reconstruction process such that the situation of links within each search space state is described by uncertain dangling sets wherein uncertain events do not include any null label in their extensions.

Chapter 12

DISTRIBUTED OBSERVATIONS

Abstract

A critical aspect of model-based diagnosis of dynamic systems, in general, and of DESs, in particular, is whether observations are accessible in a centralized way or not. In a real context the type of observation is mainly dictated by the physical layout of the system components as well as other organizational constraints of the system. This motivates the need for decentralized observations, which is addressed by introducing multiple observers, each of which gathers a complex observation, called a view. The set of all views is a distributed observation. In addition, the notion of an observer of an active system is updated so as to make the granularity of visibility finer-grained. Finally, the concept of an observation index is extended to distributed observations and exploited both for monolithic and modular history reconstruction.

1.   Introduction

Observations are an integral part of any diagnostic problem and the input of any diagnostic process: the active system approach does not make any exception in this respect. In particular, in Chapter 3 a first notion of an observation, namely an *itemized observation*, and a specialization of it, namely a *linear observation*, were introduced.

Itemized observations, wherein every message is logically certain and temporally related to other messages by one precedence relationship at most, have been considered throughout Part II of the book. In the first chapter of Part IV (Chapter 9) such a notion was extended by that of an *uncertain observation*, according to which each message is allowed to be both logically uncertain and temporally related to other messages by means of multiple precedence relationships.

In Chapter 10, uncertain observations were generalized into *complex observations*, thus relaxing the two constraints implicit in all the previous notions (see the introduction of Chapter 10). As such, the notion of a complex observation is the more general introduced thus far and subsumes all the others.

Any definition of an observation depends on the notion of the observer that gathers the observation itself. All of the notions of an observation presented in previous chapters are based on the definition of an observer given in Chapter 3.

According to this definition, the observer can see *all* of the messages emitted by the visible subset of the components of the considered system, and the messages that make up the observation are received in distinct 'parcels' such that:

(1) It is known to the observer the set of components each parcel comes from;

(2) The temporal relations among messages within different parcels is completely unknown.

In this chapter, the notion of an observer will be generalized by allowing an observer to see only a subset of the messages emitted by each component. This means that, while so far the granularity of visibility was defined on a component by component basis, from now on it is defined on the basis of each single message of each component.

Moreover, so far it has always been assumed that the observation is gathered by a single observer. In this chapter also this assumption will be dropped, that is, in the general case the considered system will be allowed to be simultaneously observed by multiple observers.

The chapter is organized as follows: Section 2 formally defines the new notion of an observer, and it allows for multiple observers to be considered, each catching a complex view. Section 3 introduces the notion of a distributed observation and that of its relevant index. Based on the new notions of observers and observations, Section 4 straightforwardly defines what is a distributed diagnostic problem, and, finally, Section 5 discusses how to solve it, by presenting three alternative exploitations of observation indexes.

2.   Observer

This section provides a notion of an observer that is different with respect to that given in Section 6 of Chapter 3. Let C be a component of a system/cluster ξ, and $\Lambda_{out}(C)$ the set of non-null labels it can send to the standard output. Let $\mathbb{M}$ be the domain of contextual labels inherent to all of the components in ξ, defined as follows:

$$\mathbb{M}(\xi) = \bigcup_{C \in \mathbb{C}(\xi)} \{(\ell, C) \mid \ell \in \Lambda_{\text{out}}(C)\}.$$

An *observer* $\Omega(\xi)$ of ξ is a set of contextual labels included in $\mathbb{M}(\xi)$, $\Omega(\xi) \subseteq \mathbb{M}(\xi)$. If $(\ell, C) \in \Omega(\xi)$, this means that observer $\Omega(\xi)$ can see the ℓ's emitted by component C. There is no restriction in the composition of contextual labels in the definition of an observer. For instance, it may occur that the same observer can see the a's emitted by component C_1 and not the a's emitted by C_2, while seeing the b's emitted by C_2.

An observer is independent of any ordering partition or, in other words, the implicit ordering partition is always a singleton containing all of the system components. Such a characterization aims to define an observer based on what such an observer can observe. In particular, it catches two orthogonal aspects that together determine a real observer's ability to observe: the physical ability of an observer to detect some events, and the communication between the system to be observed and the observer. In fact, even if an observer is potentially able to detect label e, an e generated by component C can be detected only if it is received, that is, if there are some communication means that convey this category of labels from C to the observer.

The set

$$\mathbb{C}_{\text{obs}}(\Omega(\xi)) = \{C \mid (\ell, C) \in \Omega(\xi)\}$$

encompasses all of the components of ξ that are *observable* by observer $\Omega(\xi)$. $\mathbb{C}_{\text{obs}}$ has not to be confused with the visible set of the old notion of an observer: according to the new notion, a component C is observable by $\Omega(\xi)$, that is, $C \in \mathbb{C}_{\text{obs}}(\Omega(\xi))$, if $\Omega(\xi)$ can see a category of labels at least out of those emitted by C, while, according to the old notion, if a component C is visible, that is, $C \in \mathbb{V}(\Omega(\xi))$, then the observer can see *all* of the categories of labels emitted by C.

2.1 Observer Restriction

Let $\Omega(\xi)$ be an observer of cluster $\xi = (\mathbb{C}, \mathbb{L}, \mathbb{D})$. The *restriction* of this observer on a subcluster $\xi' = (\mathbb{C}', \mathbb{L}', \mathbb{D}')$, $\mathbb{C}' \subset \mathbb{C}$, is the subset of all and only the contextual labels of $\Omega(\xi)$ that are inherent to components of ξ', that is,

$$\Omega_{\langle \xi' \rangle}(\xi) = \{(\ell, C) \mid (\ell, C) \in \Omega(\xi), C \in \mathbb{C}'\}.$$

2.2 Multiple Observers

Given a cluster, distinct observers may have different capabilities to observe each cluster component, that is, each observer may be able to detect a distinct subset of all of the categories of labels generated by the

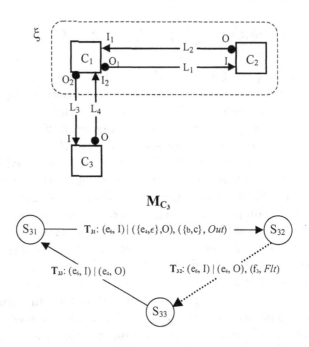

Figure 12.1.   System relevant to Example 12.1.

component. The subset of labels of a specific component that can be observed by a specific observer may possibly be null or overlap the domain of another observer. If a category of output events of a component cannot be observed by any observer, then it is not observable at all.

Example 12.1. Let Θ be the system displayed on top of Figure 12.1, wherein component C_3 has the behavioral model displayed on the bottom of the same figure, cluster ξ is an instance of the system displayed in Figure 2.2 of Chapter 11, and links L_3 and L_4 have the same model as links L_1 and L_2 of cluster ξ, that is, $\chi = 1$ and $\pi = WAIT$.

Three possible observers of Θ are defined here below:

$$\Omega_1(\Theta) = \{(a, C_1), (b, C_1), (b, C_3), (c, C_3)\},$$
$$\Omega_2(\Theta) = \{(a, C_1), (b, C_1), (c, C_1), (d, C_1)\},$$
$$\Omega_3(\Theta) = \{(a, C_2), (d, C_2)\}.$$

Observers Ω_1 and Ω_2 share labels a and b generated by component C_1. An observer may be able to detect only some labels of an (uncertain) message produced by a transition. For instance, observer Ω_1 can see label c of transition T_{17} and not label d; Ω_2, instead, can see all of the labels generated by all of the transitions of C_1.   ◇

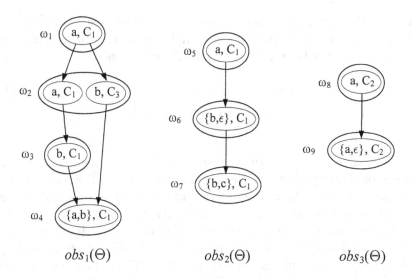

$$obs_1(\Theta) \qquad obs_2(\Theta) \qquad obs_3(\Theta)$$

Figure 12.2. Complex-observation graphs of three views of the system displayed in Figure 12.1.

2.3 View

Since the most general notion of an observation is a complex observation, out of generality, an observer gathers a complex observation. Let $obs(\xi)$ be the complex-observation graph inherent to a cluster ξ, representing the observation gathered by observer $\Omega(\xi)$. The pair formed by a (complex) observation (graph) of ξ and its observer is a *view* $W(\xi)$, formally

$$W(\xi) = (\Omega(\xi), obs(\xi)).$$

There is a constraint affecting each view: among all of the observation instances belonging to $\|obs(\xi)\|$, there is the 'real' one, that is the instance containing all and only the contextual labels actually generated by the system and observable by the observer.

Example 12.2. Suppose that system Θ of Example 12.1, displayed in Figure 12.1, has performed a reaction whose history is $H(\Theta)$, where

$$H(\Theta) = \langle T_{11}, T_{21}, T_{13}, T_{23}, T_{31}, T_{14}, T_{32}, T_{15}, T_{33}, T_{16} \rangle,$$

and the sequence of messages generated by history $H(\Theta)$ during this reaction is

$$\langle a(T_{21}), a(T_{13}), b(T_{31}), b(T_{14}), b(T_{16}) \rangle,$$

which corresponds to the following signature (sequence of contextual labels):

$$\langle (a, C_2), (a, C_1), (b, C_3), (b, C_1), (b, C_1) \rangle.$$

This signature give rise to the following observation instances, containing all and only the contextual labels actually generated by Θ and observable by observers $\Omega_1(\Theta)$, $\Omega_2(\Theta)$, and $\Omega_3(\Theta)$ defined in Example 12.1, respectively:

$$\langle (a, C_1), (b, C_3), (b, C_1), (b, C_1) \rangle, \tag{12.1}$$

$$\langle (a, C_1), (b, C_1), (b, C_1) \rangle, \tag{12.2}$$

$$\langle (a, C_2) \rangle. \tag{12.3}$$

Unfortunately, what the three observers observed is not directly a sequence of contextual labels, instead, owing to the various causes of uncertainty affecting the transmission and reception of observable events, each of them received a complex observation. Three possible complex-observation graphs of system Θ, denoted (from left to right) by $obs_1(\Theta)$, $obs_2(\Theta)$, and $obs_3(\Theta)$, inherent to observers $\Omega_1(\Theta)$, $\Omega_2(\Theta)$, and $\Omega_3(\Theta)$, respectively, are displayed in Figure 12.2. Thus three views are obtained: $W_1(\Theta) = (\Omega_1(\Theta), obs_1(\Theta))$, $W_2(\Theta) = (\Omega_2(\Theta), obs_2(\Theta))$, and $W_3(\Theta) = (\Omega_3(\Theta), obs_3(\Theta))$.

$\Omega_1(\Theta)$ cannot see label a coming from C_2 since $(a, C_2) \notin \Omega_1(\Theta)$. Suppose that $\Omega_1(\Theta)$ sometimes cannot distinguish an a coming from C_1 from a b coming from C_3: this is the case with noncontextual node ω_2 of $obs_1(\Theta)$. According to noncontextual node ω_4 of $obs_1(\Theta)$, $\Omega_1(\Theta)$ could find out the sender component of the fourth received label but it could not recognize whether such a label were either a or b.

$\Omega_2(\Theta)$ can receive only the labels coming from component C_1. Noncontextual node ω_6 of $obs_2(\Theta)$ highlights that the observer could not find out whether the received label were b or just pure noise. Noncontextual node ω_7 of $obs_2(\Theta)$, instead, is characterized by a situation of logical uncertainty analogous to that of ω_4 of $obs_1(\Theta)$.

$\Omega_3(\Theta)$ has received only the label a coming from component C_2. According to noncontextual node ω_8 of $obs_3(\Theta)$, such an observer has detected this a, then, according to ω_9 of $obs_3(\Theta)$, something else was received, but the observer could not find out whether it were an a (necessarily coming from component C_2) or just pure noise (as it actually is).

The extension of the observation of each of the three views includes, as required, the relevant 'real' observation instance, that is $\| obs_1(\Theta) \|$ includes instance (12.1), $\| obs_2(\Theta) \|$ includes instance (12.2), and $\| obs_3(\Theta) \|$ includes instance (12.3). $\diamond$

2.4   View Restriction

Let $W(\xi) = (\Omega(\xi), obs(\xi))$ be a view of cluster $\xi = (\mathbb{C}, \mathbb{L}, \mathbb{D})$. The *restriction* of this view on a subcluster $\xi' = (\mathbb{C}', \mathbb{L}', \mathbb{D}')$, $\mathbb{C}' \subset \mathbb{C}$, is defined as follows:

$$W_{\langle \xi' \rangle}(\xi) = (\Omega_{\langle \xi' \rangle}(\xi), obs_{\langle \xi' \rangle}(\xi)),$$

where $\Omega_{\langle \xi' \rangle}(\xi)$ is the restriction of the relevant observer on the subcluster, as defined in Section 2.1, and $obs_{\langle \xi' \rangle}(\xi)$ is the restriction of the relevant observation on the subcluster. Although $obs(\xi)$ is a complex observation, its restriction cannot be computed in the way described in Section 2.2 of Chapter 10 since the concept of an observer has changed.

The projection of the complex observation $obs(\xi) = (\mathbf{N}, \mathbb{N}, \Upsilon, \mathbb{N}_0, \mathbb{N}_f)$ on subcluster ξ' is obtained by performing the following steps:

(1) All of the non-null labels that cannot be generated by any component belonging to $\mathbb{C}_{obs}(\Omega_{\langle \xi' \rangle}(\xi))$ are removed from the contextual nodes of $obs(\xi)$;

(2) All of the components that do not belong to $\mathbb{C}_{obs}(\Omega_{\langle \xi' \rangle}(\xi))$ are removed from the contextual nodes of $obs(\xi)$;

3) A null label ϵ is added to each contextual node of $obs(\xi)$ that has been deprived of one or more labels and/or components in the previous two steps;

(4) A null label ϵ is added to each contextual node of $obs(\xi)$ that includes one or more observable labels shared between $\mathbb{C}'$ and $(\mathbb{C} - \mathbb{C}')$ and visible by $\Omega_{\langle \xi' \rangle}(\xi)$ and $\Omega(\xi)$, respectively;

(5) All empty contextual nodes are removed;

(6) All empty noncontextual nodes are removed;

(7) Precedence relations between the surviving contextual nodes are updated, that is, the precedence relations of a canceled contextual node are inherited by its predecessor contextual nodes.

Step 3 guarantees that the subcluster observation encompasses also the case when no visible label coming from an observable components of the subcluster has been observed by the observer in correspondence to the considered contextual node, since the observer observed something coming from components that do not belong to $\mathbb{C}'$, they belong instead to $(\mathbb{C} - \mathbb{C}')$.

According to Step 4, in the restriction of the view observation, each contextual node including one or more labels shared between $\mathbb{C}'$ and $(\mathbb{C} - \mathbb{C}')$ contains ϵ as well, thus guaranteeing that the subcluster observation involves also the case when every shared label does not come from ξ'.

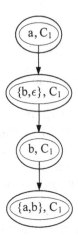

Figure 12.3.   Restriction of the complex-observation graph $obs_1(\Theta)$ displayed in Figure 12.2.

Example 12.3. Let

$$W_1(\Theta) = (\Omega_1(\Theta), obs_1(\Theta)),$$
$$W_2(\Theta) = (\Omega_2(\Theta), obs_2(\Theta)),$$
$$W_3(\Theta) = (\Omega_3(\Theta), obs_3(\Theta)),$$

be the three views defined in Example 12.2, and ξ be the cluster of system Θ defined in Example 12.1.

The restriction of $W_1(\Theta)$ on ξ is

$$W_{1_{(\xi)}}(\Theta) = (\Omega_{1_{(\xi)}}(\Theta), obs_{1_{(\xi)}}(\Theta)),$$

where

$$\Omega_{1_{(\xi)}} = \{(a, C_1), (b, C_1)\},$$

and $obs_{1_{(\xi)}}(\Theta))$ is displayed in Figure 12.3.

The restrictions $W_{2_{(\xi)}}(\Theta)$ and $W_{3_{(\xi)}}(\Theta)$ equal $W_2(\Theta)$ and $W_3(\Theta)$, respectively, since in both cases the set of components observable by the relevant observer is a subset of the components of ξ, that is,

$$\mathbb{C}_{\text{obs}}(\Omega_2(\Theta)) \subset \mathbb{C}(\xi), \mathbb{C}_{\text{obs}}(\Omega_3(\Theta)) \subset \mathbb{C}(\xi).$$

$\diamond$

3.   Distributed Observation

A *distributed observation* of a cluster ξ is the composition of v distinct views of ξ, $v \geq 1$, that is

$$OBS(\xi) = \{W_1(\xi), .., W_v(\xi)\}, \qquad (12.4)$$

where $W_i(\xi) = (\Omega_i(\xi), obs_i(\xi))$. Two distinct views $W_h(\xi)$ and $W_k(\xi)$, $h \neq k$, belonging to the same distributed observation, $W_h(\xi) \in OBS(\xi)$, $W_k(\xi) \in OBS(\xi)$, are *overlapping* in case their observers can see one or more common contextual labels, that is,

$$\Omega_h(\xi) \cap \Omega_k(\xi) \neq \emptyset.$$

3.1 Unknown Set

Given the concept of a distributed observation, which, differently from all previous observations, features several observers, where each single observer is defined in the new way provided in Section 2, the notion of the *unknown set* of a system observation has to be updated with respect to that given in Section 7.1 of Chapter 3.

Let $OBS(\xi) = \{W_1(\xi), .., W_v(\xi)\}$, where $W_i(\xi) = (\Omega_i(\xi), obs_i(\xi))$, be a distributed observation of a cluster $\xi = (\mathbb{C}, \mathbb{L}, \mathbb{D})$. The *unknown set* of $OBS(\xi)$ is the set of components of ξ that are not observable by any observer of ξ, that is,

$$Ukn(OBS(\xi)) = \mathbb{C} - \bigcup_{i=1}^{v} \mathbb{C}_{\text{obs}}(\Omega_i(\xi)).$$

3.2 Distributed-observation Index

The *index* of a distributed observation is expected to play a role similar to that of the indexes of uncertain and complex observations. This means that a distributed-observation index has to univocally identify the set of messages that have already been consumed in the complex observation graph of all of the views of the considered distributed observation. Thus, the index of a distributed observation is the composition of the indexes of all of the complex observation graphs of the observation views, where each of such indexes is a complex-observation index, as defined in Section 2.1 of Chapter 10.

This definition of a distributed-observation index, although theoretically correct, is not that actually adopted by the (monolithic or modular) reconstruction techniques for solving distributed diagnostic problems. In these processes, for efficiency reasons, an equivalent notion of an observation index is exploited, as it will be discusses in Section 5.

3.3 Distributed-observation Restriction

Let $OBS(\xi) = \{W_1(\xi), .., W_v(\xi)\}$ be a distributed observation of a cluster ξ. The *restriction* of $OBS(\xi)$ on a subcluster $\xi' \subset \xi$, denoted by

$$OBS_{\langle \xi' \rangle}(\xi),$$

is a distributed observation

$$OBS'(\xi) = \{W_1'(\xi), .., W_v'(\xi)\}$$

such that

$$W_j'(\xi) = W_{j\langle\xi'\rangle}(\xi).$$

3.4   Distributed-observation Extension

A distributed observation, the same as a complex observation, implicitly defines a finite set of sequences of both logically and source unambiguous contextual labels. Each of such sequences is called *observation instance*. Intuitively, an instance of a distributed observation can be made up by selecting a complex observation instance for each view of the given observation and then by creating a sequence of all of the contextual labels of the v instances. The obtained sequence has to comply with all of the temporal precedence relationships imposed by each single view and it has to be such that each contextual label shared by several views is instantiated just once.

Formally, let $OBS(\xi) = \{W_1(\xi), .., W_v(\xi)\}$, $W_i = (\Omega_i(\xi), obs_i(\xi))$, be a distributed observation of a cluster ξ. Let $\Omega_{ij}(\xi) = \Omega_i(\xi) \cap \Omega_j(\xi)$, $i \neq j$, be the set of contextual labels shared by two distinct observers, and $\Omega_{i_{exc}}(\xi)$ the (possibly null) set of contextual labels that are observable exclusively by observer $\Omega_i(\xi)$, that is, $\Omega_{i_{exc}}(\xi) = \Omega_i(\xi) - \bigcup_{j=1}^{v} \Omega_{ij}(\xi)$, where, by definition, $\Omega_{ii}(\xi) = \emptyset$.

Let $\mathcal{Q}_i = \langle c_{i1}, \ldots, c_{ip} \rangle$, $i \in [1 .. v]$, be a generic instance belonging to the extension of the complex observation $obs_i(\xi)$, that is, $\mathcal{Q}_i \in \|obs_i(\xi)\|$.

An instance $\mathcal{Q}$ of $OBS(\xi)$, $\mathcal{Q} = \langle c_1, \ldots, c_p' \rangle$, is a single sequence of contextual labels such that the following conditions hold:

- Each contextual label in $\mathcal{Q}_i$, $i \in [1 .. v]$, corresponds to one contextual label in $\mathcal{Q}$ and there is no contextual label in $\mathcal{Q}$ that does not correspond to any contextual label in a $\mathcal{Q}_i$ at least;

- $\forall c_k \in \mathcal{Q}$, the following conditions hold:

  (1) If $\exists \Omega_{h_{exc}}(\xi) \mid (c_k \in \Omega_{h_{exc}}(\xi))$, then contextual label c_k corresponds exactly to one contextual label $c_{hn} \in \mathcal{Q}_h$, where $c_k = c_{hn}$, and $\forall c_{hm} \mid (c_{hm}$ precedes c_{hn} in $\mathcal{Q}_h)$, c_k precedes $c_{hm'}$ in $\mathcal{Q}$, where $c_{hm'} \in \mathcal{Q}$ is the (only) contextual label corresponding to $c_{hm} \in \mathcal{Q}_h$;

  (2) If $\nexists \Omega_{h_{exc}}(\xi) \mid (c_k \in \Omega_{h_{exc}}(\xi))$, let $S\Omega(c_k) = \{\Omega_i(\xi) \mid c_k \in \Omega_i(\xi)\}$ be the set of all the observers of ξ that can see contextual label c_k. Then, $\forall \Omega_j(\xi) \in S\Omega(c_k)$, contextual label c_k corresponds to

a contextual label $c_{jn} \in Q_j$, where $c_k = c_{jn}$, and $\forall c_{jm} \mid (c_{jm}$ precedes c_{jn} in Q_j), c_k precedes $c_{jm'}$ in Q, where $c_{jm'} \in Q$ is the contextual label corresponding to $c_{jm} \in Q_j$.

4.   Distributed Diagnostic Problem

Most intuitively, a *distributed diagnostic problem* $\wp(\xi)$ inherent to a cluster ξ is a pair $\wp(\xi) = (OBS(\xi), \xi_0)$, where $OBS(\xi)$ is a distributed observation, and ξ_0 is the state of ξ before the reaction started.

The previous notion of a diagnostic problem, defined in Section 8 of Chapter 3 and taken so far, was a triple since it included the considered observer, while now the identifier of one or more observers is contained in the expression of the distributed observation $OBS(\xi)$.

4.1   Diagnostic Problem Restriction

The *restriction* of $\wp(\xi) = (OBS(\xi), \xi_0)$ on a subsystem $\xi' \subseteq \xi$, denoted by

$$\wp_{\langle \xi' \rangle}(\xi),$$

is a diagnostic problem $\wp(\xi') = (OBS(\xi'), \xi_0')$ such that:

$OBS(\xi') = OBS_{\langle \xi' \rangle}(\xi),$

ξ_0' is the projection of ξ_0 on the components in ξ'.

5.   Solving Distributed Diagnostic Problems

In order to solve a distributed diagnostic problem, it is first necessary to define the *index space(s)* for distributed observations.

Given a diagnostic problem whose distributed observation consists of v distinct, possibly overlapping, views, how many index spaces should be used? In the next sections three possible answers are discussed: either one index space for all of the views, or v index spaces, one for each view, or i index spaces, $1 \leq i \leq v$, each of which is inherent to one or more views.

5.1   Index Space per View

Given a distributed observation $OBS(\xi)$, as defined in Equation (12.4), if this option is selected, for each of the v complex observation graphs $obs_i(\xi)$ a distinct index space $Space(obs_i(\xi))$ is created, by following the steps described in Section 2.1 of Chapter 10. During the (monolithic) reconstruction of the behavior of the system based on $OBS(\xi)$, each active space state will contain v indexes, one for each index space.

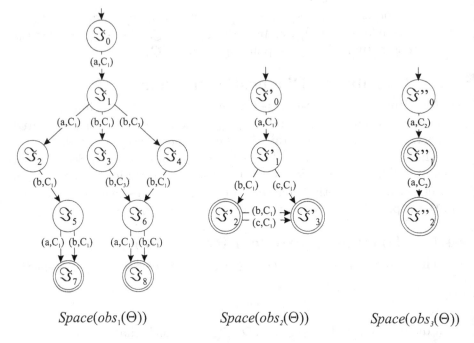

$$Space(obs_1(\Theta)) \qquad Space(obs_2(\Theta)) \qquad Space(obs_3(\Theta))$$

Figure 12.4.   Index spaces of the observation graphs displayed in Figure 12.2.

Example 12.4. Let $OBS(\Theta)$ be a distributed observation inherent to the system Θ shown in Figure 12.1. Suppose that such an observation consists of the three views defined in Example 12.2, that is,

$$OBS(\Theta) = \{W_1(\Theta), W_2(\Theta), W_3(\Theta)\}.$$

Figure 12.4 shows (from left to right) the index spaces $Space(obs_1(\Theta))$, $Space(obs_2(\Theta))$, and $Space(obs_3(\Theta))$ of the three complex-observation graphs displayed in Figure 12.2.   ◇

In other words, embracing this choice means implementing in the reconstruction algorithms the theoretical definition of an index for a distributed observation given in Section 3.2.

5.2   Single Index Space

If this option is selected, during the reconstruction of the behavior of the system based on a distributed observation $OBS(\xi)$, as defined in Equation (12.4), each search space state will contain just one index, this being the identifier of a state within a global index space, which is inherent to the whole observation. The global index space is generated by composing the v index spaces inherent to the v views in $OBS(\xi)$. Then, as in the previous case, first of all, for each observation graph $obs_i(\xi)$, $i \in [1 .. v]$, it is necessary to create a distinct index space $Space(obs_i(\xi))$, by following the steps described in Section 2.1 of Chapter 10.

If there are no overlapping views, then the index space inherent to the whole observation is the Cartesian product of such v index spaces.

If there are two or more overlapping views, the initial state of the global index space is the composition of the initial states of the v view index spaces, and each edge exiting from a global index space state is the copy of an edge exiting from a view index state. If a local edge is marked by a contextual label that is shared by (the observers of) several views, then it becomes a global edge only if the same edge is exiting from all of the current view index space states of the sharing views. This implies that the resulting automaton is deterministic by construction.

Formally, given the distributed observation $OBS(\xi)$ of Equation (12.4), let

$$Space(obs_i(\xi)) = (\mathbb{S}_i, \mathbb{L}_i, \mathbb{T}_i, S_{0_i}, \mathbb{S}_{f_i}), i \in [1 .. v],$$

be the (deterministic, complex-observation) index space of $obs_i(\xi)$. The *spurious index space* $Sspace(OBS(\xi))$ is the deterministic automaton

$$Sspace(OBS(\xi)) = (\tilde{\mathbb{S}}, \tilde{\mathbb{L}}, \tilde{\mathbb{T}}, \tilde{S}_0, \tilde{\mathbb{S}}_f),$$

where

$\tilde{\mathbb{S}} \subseteq \mathbb{S}_1 \times \mathbb{S}_2 \times \cdots \times \mathbb{S}_v$ is the set of states;

$\tilde{\mathbb{L}} \subseteq \bigcup_{i=1}^{v} \mathbb{L}_i$ is the set of labels;

$\tilde{\mathbb{T}} : \tilde{\mathbb{S}} \times \tilde{\mathbb{L}} \mapsto \tilde{\mathbb{S}}$ is the transition function, defined as follows:

$$S \xrightarrow{(\ell,C)} S' \in \tilde{\mathbb{T}}, S = (S_1, \ldots, S_v), S' = (S'_1, \ldots, S'_v)$$

if and only if

$$\forall i \in [1 .. v] \begin{cases} (\ell, C) \in \Omega_i(\xi) & S_i \xrightarrow{(\ell,C)} S'_i \in \mathbb{T}_i \\ (\ell, C) \notin \Omega_i(\xi) & S_i = S'_i \end{cases}$$

$\tilde{S}_0 = (S_{0_1}, \ldots, S_{0_v})$ is the initial state;

$\tilde{\mathbb{S}}_f = \{S = (S_1, \ldots, S_v) \mid S_i \in \mathbb{S}_{f_i}, i \in [1 .. v]\}$ is the set of final states.

A state $S \in \tilde{\mathbb{S}}$ is *consistent* if and only if $S \in \tilde{\mathbb{S}}_f$ or there exists a transition $S \xrightarrow{(\ell,C)} S' \in \tilde{\mathbb{T}}$ such that S' is consistent. Transition $S \xrightarrow{(\ell,C)} S' \in \tilde{\mathbb{T}}$ is consistent if and only if S' is consistent.

The *index space* of $OBS(\xi)$ is the deterministic automaton

$$Space(OBS(\xi)) = (\mathbb{S}, \mathbb{L}, \mathbb{T}, S_0, \mathbb{S}_f)$$

obtained by selecting all the only the consistent states and transitions within $Sspace(OBS(\xi))$, that is,

$\mathbb{S} = \{S \mid S \in \tilde{\mathbb{S}}, S \text{ is consistent}\};$

$\mathbb{T} = \{T \mid T \in \tilde{\mathbb{T}}, T \text{ is consistent}\};$

$\mathbb{L} \subseteq \tilde{\mathbb{L}};$

$S_0 = \tilde{S}_0;$

$\mathbb{S}_f = \tilde{\mathbb{S}}_f.$

Example 12.5. Figure 12.5 shows the single index space of the distributed observation $OBS(\Theta) = \{W_1(\Theta), W_2(\Theta), W_3(\Theta)\}$ defined in Example 12.4. This space has been obtained by composing the three index spaces depicted in Figure 12.4.   ◇

5.3   Several Mutually Independent Index Spaces

Intuitively speaking, selecting this option means that the v views of the distributed observation of Equation (12.4) are partitioned into p sets such that the views within each set may be mutually overlapping while views belonging to different sets are not. Each set of views gives rise to its own index space, which is independent of that of the other sets in that each contextual label is relevant to one index space only.

More rigorously, if this option is selected, it is necessary to partition the set $\mathbb{C}$ of the components of the considered cluster ξ into p parts $\{\mathbf{C}_1, \ldots, \mathbf{C}_p\}$. Likewise, the whole set of views belonging to $OBS(\xi)$ is partitioned into p parts $\{\mathbf{W}_1, \ldots, \mathbf{W}_p\}$, such that, for each $j \in [1\mathinner{..}p]$, the following conditions of mutual independence hold:

(1) the set $\mathbb{C}_{\mathrm{obs}_j}$ of components that are observable within each part $\mathbf{W}_j$ is a subset of $\mathbf{C}_j$, that is $\mathbb{C}_{\mathrm{obs}_j} \subseteq \mathbf{C}_j$, where

$$\mathbb{C}_{\mathrm{obs}_j} = \bigcup_{\Omega_r(\xi) \mid (\Omega_r(\xi), obs_r(\xi)) \in \mathbf{W}_j} \mathbb{C}_{\mathrm{obs}}(\Omega_r(\xi));$$

(2) the set of components that are observable within distinct parts of the partition of views are disjoint, that is

$$\forall j \in [1\mathinner{..}p], \forall k \in [1\mathinner{..}p], j \neq k, \mathbb{C}_{\mathrm{obs}_j} \cap \mathbb{C}_{\mathrm{obs}_k} = \emptyset.$$

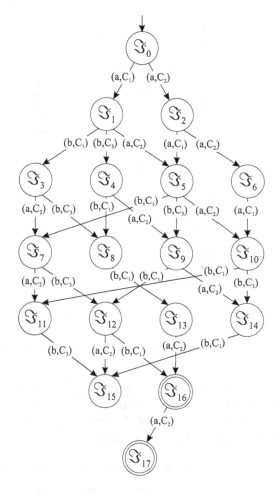

Figure 12.5. Single index space of a distributed observation.

Afterwards, for each part $\mathbf{W}_j$, $j \in [1..p]$, an index space, denoted by $Space(\mathbf{W}_j)$, is created, which is inherent to all of the views belonging to $\mathbf{W}_j$. Such an index space is produced by exploiting the method illustrated in Section 5.2 for generating a single index space inherent to all of the views of a distributed observation, where the distributed observation to be considered here is $OBS_j(\xi) = \mathbf{W}_j$.

During the (monolithic) reconstruction of the behavior of the system based on the distributed observation $OBS(\xi)$, each active space state will contain p indexes, one for each index space. Since $p \leq v$, the size of the search space is reduced. However, the major advantage of this option over that considering a separate index space for each view is that, during the reconstruction process, in order to check the consistency of

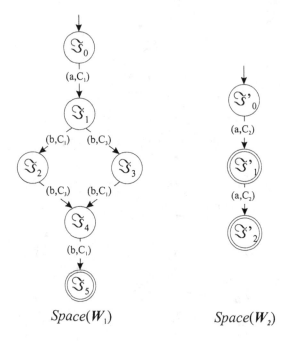

Figure 12.6.   Mutually independent index spaces of a distributed observation.

a transition of component $C \in \mathbf{C}_j$ with respect to the observation, it is necessary to take into account an index space only, i.e. $Space(\mathbf{W}_j)$, instead of all of the index spaces of the views whose observer can see a contextual label generated by that transition.

Example 12.6. Consider once again the distributed observation

$$OBS(\Theta) = \{W_1(\Theta), W_2(\Theta), W_3(\Theta)\}$$

of the previous two examples inherent to the system displayed in Figure 12.1. The set of components of Θ,

$$\mathbb{C}(\Theta) = \{C_1, C_2, C_3\},$$

has been partitioned into two parts, namely

$$\mathbf{C}_1 = \{C_1, C_2\}, \mathbf{C}_2 = \{C_3\}.$$

Two parts of the views belonging to $OBS(\Theta)$ have correspondingly been produced, namely

$$\mathbf{W}_1 = \{W_1(\Theta), W_2(\Theta)\}, \mathbf{W}_2 = \{W_3(\Theta)\},$$

which fulfill the two conditions of mutual independence stated above.

Finally, the two index spaces corresponding to the two view parts have been drawn, $Space(\mathbf{W}_1)$ and $Space(\mathbf{W}_2)$, which are shown in Figure 12.6. Note that $Space(\mathbf{W}_2)$ equals $Space(obs_3(\Theta))$, displayed on the right of Figure 12.4, since it is indeed an index space corresponding to the observation graph of a single view.

$Space(\mathbf{W}_1)$, instead, has been produced by exploring the search space depicted in Figure 12.7. Each search space state is the composition of two indexes, representing the current states of $Space(obs_1(\Theta))$ and $Space(obs_2(\Theta))$, respectively. So, the initial state of $Space(\mathbf{W}_1)$ is the composition of state $\Im_0$ of $Space(obs_1(\Theta))$ and $\Im_0'$ of $Space(obs_2(\Theta))$. In the figure, each search space state is denoted by the subscripts of the two composed states, e.g. (00) for $(\Im_0, \Im_0')$.

In order to perform the search that leads to $Space(\mathbf{W}_1)$, we have to remind that the observers relevant to the two composed index spaces, i.e. $\Omega_1(\xi)$ and $\Omega_2(\xi)$, share contextual labels (a, C_1) and (b, C_1). So, an edge marked by either of these contextual labels creates a successor state in the search space only if it updates both the current states of the two composed index spaces, i.e. $Space(obs_1(\Theta))$ and $Space(obs_2(\Theta))$. This is the case for the edge marked by contextual label (a, C_1) exiting from (00), as well as for the edges marked by contextual label (b, C_1) exiting from (11), (41), and (62).

Instead, the edge marked by contextual label (a, C_1) exiting from (11), which is a copy of the edge exiting from state $\Im_1$ of $Space(obs_1(\Theta))$, does not generate any successor state since there is no edge exiting from state $\Im_1'$ of $Space(obs_2(\Theta))$ marked by the same label. The edge is inconsistent since, being (a, C_1) a shared label, when it is detected by the former observer, it has to be detected also by the latter. Therefore, a backtracking has to be performed on this edge. The same applies to all of the others dangling edges. By removing from Figure 12.6 all of the dangling edges, the resulting automaton is the spurious index space $Sspace(\mathbf{W}_1)$. By removing from it all of its inconsistent states and edges, which are dashed, $Space(\mathbf{W}_1)$, displayed on the left of Figure 12.6, is obtained. $\diamond$

The case when a single index space of a distributed observation is generated, presented in Section 5.2, is indeed an instantiation of the case when several mutually independent index spaces are produced, presented in Section 5.3. For solving distributed diagnostic problems, in the following we will consider (one or more) mutually independent index spaces, while dropping the option of an index space for each view, presented in Section 5.1, since it leads to a cumbersome management of behavior reconstruction when distinct views are overlapping (while it becomes an instantiation of the mutually independent index spaces method in case no view is overlapping).

The index space $Space(OBS(\xi))$ of a distributed observation is then the composition of p mutually independent index spaces, that is,

$$Space(OBS(\xi)) = \{Space(\mathbf{W}_1), \ldots, Space(\mathbf{W}_p)\}.$$

This means that $Space(OBS(\xi))$ is a deterministic graph composed by p deterministic graphs. The initial state of $Space(OBS(\xi))$ is the composition of the p initial states of the p distinct graphs, that is,

$$S_0 = (S_{1_0}, \ldots, S_{p_0}).$$

Likewise, the set of final states of $Space(OBS(\xi))$ is the union of the p sets of final states of the p distinct graphs, that is,

$$\mathbb{S}_f = \bigcup_{i=1}^{p} \mathbb{S}_{i_f}.$$

The *extension* of $Space(OBS(\xi))$, denoted by $\|Space(OBS(\xi))\|$, is the (finite) set of sequences of contextual labels corresponding to the whole set of paths from S_0 to a state $S_f \in \mathbb{S}_f$. Given this definition of the extension of a distributed-observation index space, we claim that the property expressed by the following proposition holds.

Proposition 12.1. *Theorem 9.1 of Chapter 9 (for uncertain observations) is valid for distributed observations too.*

5.4   Monolithic Resolution

We will now consider how the check for the triggerability of a transition from the point of view of the message it emits has to be updated with respect to that given in Section 4.4 of Chapter 10 for the (single) index space of a complex observation in case several mutually independent index spaces of a distributed observation are taken into account. Such a check is exploited during the reconstruction of the behavior of a system/cluster.

Assume a distributed diagnostic problem $\wp(\xi) = (OBS(\xi), \xi_0)$. Let $\{\mathbf{C}_1, \ldots \mathbf{C}_p\}$ be the partition of system components corresponding to the partition $\{\mathbf{W}_1, \ldots \mathbf{W}_p\}$ of the views of the system, where such partitions fulfill the two conditions of mutual independence stated in Section 5.3. The central point of the reconstruction of the system behavior based on several mutually independent index spaces is that the composition of the states of all of the index spaces is used as the index field in the search

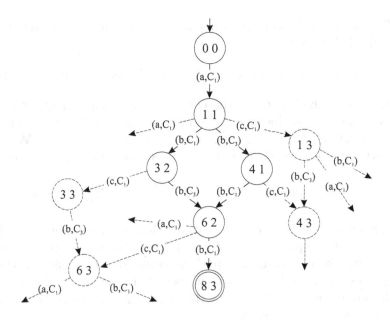

Figure 12.7. Search space for the construction of an index space corresponding to two overlapping views.

space. In the initial state of the search, the index field is the composition of the initial states of all of the index spaces.

Let

$$T = S \xrightarrow{\alpha|\beta} S'$$

be a transition of component $C \in \mathbf{C}_j$, $N = (\sigma, \Im, D)$ be the current node of the search space, where S belongs to σ and $\Im_j \in \Im$ is the current state of the index space $Space(\mathbf{W}_j)$ of the part $\mathbf{W}_j$ of the partition of views corresponding to part $\mathbf{C}_j$ of the partition of components, where

$$Space(\mathbf{W}_j) = (\mathbb{S}_j, \mathbb{L}_j, \mathbb{T}_j, S_{0_j}, \mathbb{S}_{f_j}).$$

From the point of view of the consistency with the observation, transition T is triggerable from N only if at least one of the following conditions hold:

(1) T is *not* observable, that is, $(m, Out) \notin \beta$;

(2) C is *not* visible, that is, $C \in Ukn(OBS(\xi))$;

(3) T is both visible and observable but it may generate a null message, that is,

$$(m, Out) \in \beta, C \notin Ukn(OBS(\xi)), \epsilon \in \|m\|;$$

(4) T is both visible and observable and it can generate a viable contextual label, that is,

$$(m, Out) \in \beta, C \notin Ukn(OBS(\xi)), ContExt(m,C) \cap Viable(\Im_j) \neq \emptyset,$$

where $ContExt(m,C)$ is the contextual extension of message m of component C, defined in Section 3.4 of Chapter 10, and $Viable(\Im_j)$ is the set of viable contextual labels of the considered part $\mathbf{W}_j$ of the distributed observation, that is (see Section 3.4 of Chapter 10),

$$Viable(\Im_j) = \{(\ell, C_j) \mid \Im_j \xrightarrow{(\ell,C)} \Im'_j \in \mathbb{T}_j, C_j \in \mathbb{C}\}.$$

In particular, the last condition states that an observable and visible transition of component $C \in \mathbf{C}_j$ is triggerable (as regards its consistency with the observation) when the contextual extension of the relevant message includes at least one out of the contextual labels whose sender component is C marking the edges leaving the current state of the index space $Space(\mathbf{W}_j)$.

If T is triggerable[1], a new state $N' = (\sigma', \Im', D')$ is generated within the reconstruction search space. Fields σ' and D' are determined as usual, while the new value $\Im' = (\Im'_1, \ldots, \Im'_p)$ of the index field is generated, starting from the old value $\Im = (\Im_1, \ldots, \Im_p)$, based on the following rules, corresponding to the four above conditions, respectively:

(1) and (2) $\Im' = \Im$ and the edge $N \xrightarrow{T} N'$ is created in the reconstruction space;

(3) $\Im' = \Im$ and the edge $N \xrightarrow{T(\epsilon)} N'$ is created in the reconstruction space;

(4) $\Im'$, where

$$\Im_j \xrightarrow{(\ell,C)} \Im'_j \in \mathbb{T}_j, (\ell, C) \in (ContExt(m,C) \cap Viable(\Im_j)), C \in \mathbb{C},$$

while for each $i \in [1 .. p]$, $i \neq j$, $\Im'_i = \Im_i$, and the edge $N \xrightarrow{T(\ell)} N'$ is created in the reconstruction space.

Thus, there exists a new value $\Im'$ for each different viable contextual label belonging to the contextual extension of message m.

---

[1] As usual, given the current node of the reconstruction search space, T is triggerable if it complies not only with the observation but also with the situation of links.

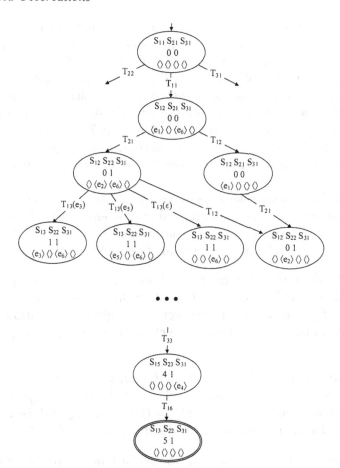

Figure 12.8. Monolithic reconstruction inherent to a distributed diagnostic problem.

Example 12.7. Let

$$\wp(\Theta) = (OBS(\Theta) = \{W_1(\Theta), W_2(\Theta), W_3(\Theta)\}, \Theta_0 = (S_{11}, S_{21}, S_{31}))$$

be a distributed diagnostic problem inherent to the system Θ displayed in Figure 12.1, where $OBS(\Theta)$ is the same as in Example 12.6. Suppose that, in order to solve this problem, a monolithic reconstruction is performed based on the two index spaces drawn in Example 12.6 and displayed in Figure 12.6. A fragment of the reconstruction space is shown in Figure 12.8, wherein the two observation indexes are represented within each state by the subscripts of the corresponding index space states of $Space(\mathbf{W}_1)$ and $Space(\mathbf{W}_2)$ of Figure 12.6. Recall that $Space(\mathbf{W}_1)$ is inherent to labels coming from components C_1 and C_3 while $Space(\mathbf{W}_2)$ is inherent to labels coming from C_2.

Only transition T_{11}, exiting from S_{11} of component C_1, generates a successor state starting from the initial state of the reconstruction space, while transition T_{22}, exiting from S_{21} of component C_2, and transition T_{31}, exiting from S_{31} of component C_3, do not. The reason why a backtracking is performed on T_{22} is that such a transition generates label d and there is no edge exiting from the initial state $\Im'_0$ of $Space(\mathbf{W}_2)$ that is marked by (C_2, d). Analogously, transition T_{31} is subdued to backtracking insofar as it can generate uncertain message $\{b, c\}$ but there is no edge exiting from $\Im_0$ of $Space(\mathbf{W}_1)$ marked by either (b, C_3) or (c, C_3).

Figure 12.8 encompasses a sample final state: each final state of the reconstruction space is characterized by the emptiness of the dangling sets and by observation indexes that are all final. In fact, $\Im_5$ is a final state of $Space(\mathbf{W}_1)$ and $\Im'_1$ is a final state of $Space(\mathbf{W}_2)$. $\diamond$

Given the definition of a distributed-observation extension provided in Section 3.4 and the reconstruction algorithm whose check for the consistency of a transition with respect to the given observation has been described in the present section, we claim that the following proposition holds.

Proposition 12.2. *Theorem 9.2 of Chapter 9 is valid also for distributed diagnostic problems, provided that the general expression of an uncertain diagnostic problem $\wp(\xi) = (\Omega(\xi), OBS(\xi), \xi_0)$ be replaced by that of a distributed diagnostic problem $\wp(\xi) = (OBS(\xi), \xi_0)$.*

Based on the above proposition, the reconstruction operator is able to generate the active space inherent to any given distributed diagnostic problem.

5.5   Modular Resolution

Assume a distributed diagnostic problem $\wp(\xi) = (OBS(\xi), \xi_0)$. Let $\{\mathbf{C}_1, \ldots \mathbf{C}_p\}$ be the partition of system components corresponding to the partition $\{\mathbf{W}_1, \ldots \mathbf{W}_p\}$ of the views of $OBS(\xi)$, where such partitions fulfill the two mutual independence conditions stated in Section 5.3. Let $Space(\mathbf{W}_j)$ be the index space of $\mathbf{W}_j$. Problem $\wp(\xi)$ can be recursively split into subproblems according to a reconstruction graph. Each node of the reconstruction graph is a subproblem inherent to a subcluster ξ' of ξ: such a subproblem is the projection $\wp_{\langle \xi' \rangle}(\xi)$ of the given problem on the relevant subcluster (see Section 4.1). In order to solve each distributed subproblem, an (adapted) process of monolithic reconstruction is exploited. Such a process requires that the index space relevant to the distributed observation of each considered subproblem is available.

5.5.1 Index Space Restriction

Let $Space(OBS(\xi)) = \{Space(\mathbf{W}_1), \ldots, Space(\mathbf{W}_p)\}$ be an index space of a distributed observation $OBS(\xi)$ of cluster $\xi = (\mathbb{C}, \mathbb{L}, \mathbb{D})$. The restriction of $Space(OBS(\xi))$ on subcluster $\xi' = (\mathbb{C}', \mathbb{L}', \mathbb{D}')$, $\mathbb{C}' \subset \mathbb{C}$, is the composition of the restrictions of its p mutually independent index spaces, that is,

$$Space_{\langle \xi' \rangle}(OBS(\xi)) = \{Space_{\langle \xi' \rangle}(\mathbf{W}_1), \ldots, Space_{\langle \xi' \rangle}(\mathbf{W}_p)\},$$

where each restriction $Space_{\langle \xi' \rangle}(\mathbf{W}_j)$ is determined by following the same steps as for determining the restriction of a complex-observation index space (see Section 3.5.1 of Chapter 10).

Given this definition of a distributed-observation index space restriction, the following proposition can be stated.

Proposition 12.3. *Theorem 9.3 of Chapter 9 (for uncertain observations) is valid for distributed observations too.*

5.5.2 Modular Reconstruction

Given a distributed diagnostic problem $\wp(\xi) = (OBS(\xi), \xi_0)$, based on Proposition 12.3, the index space inherent to each subproblem $\wp_{\langle \xi' \rangle}(\xi)$ can be generated either as $Space(OBS_{\langle \xi' \rangle}(\xi))$ or as $Space_{\langle \xi' \rangle}(OBS(\xi))$.

Embracing the first choice means computing the observation restriction on ξ' of each of the p parts $\mathbf{W}_j$ of the considered partition and then determining the index space inherent to each of such restrictions. All these activities have to be performed for each subproblem of the reconstruction graph.

Making the second choice means determining, once for all, p mutually independent index spaces of $OBS(\xi)$ and, then, for each subproblem of the reconstruction graph, projecting each of such spaces on the relevant cluster.

The technique for a modular reconstruction in order to solve distributed diagnostic problems is analogous to that already introduced for uncertain diagnostic problems. Moreover, also the equivalence between monolithic and modular reconstruction holds, as claimed by the following proposition.

Proposition 12.4. *Theorem 9.4 of Chapter 9 (for uncertain diagnostic problems) is valid for distributed diagnostic problems too.*

6. Summary

- *Observer*: entity capable of receiving and seeing a (possibly empty) set of labels coming from each component. The sets of labels inherent to distinct components may be different and/or overlapping.

- *Observer restriction*: projection of the capability of an observer to observe a cluster on one of its subclusters.

- *View*: pair formed by a complex observation and the observer that has gathered it.

- *Overlapping views*: views inherent to observers that can see one or more common labels of one or more components.

- *View restriction*: projection of a view inherent to a cluster on one of its subclusters.

- *Distributed observation*: composition of one or more distinct views, all inherent to the same cluster.

- *Distributed diagnostic problem*: diagnostic problem featuring a distributed observation.

- *Restriction of a distributed diagnostic problem*: projection of a diagnostic problem inherent to a cluster on one of its subclusters.

- *Distributed-observation index*: index to be used within each state of the reconstruction space, so as to represent the part of the distributed observation that has already been generated by all the paths leading from the initial state to the current state. Theoretically such an index is the composition of the indexes of the complex observations of all the views belonging to the distributed observation.

- *Distributed-observation restriction*: projection of a distributed observation inherent to a cluster on one of its subclusters.

- *Distributed-observation instance*: totally temporally ordered sequence of contextual labels that is compliant with a given distributed observation.

- *Distributed-observation extension*: the set of all the instances of a given distributed observation.

- *Single index space of a distributed-observation*: deterministic graph produced by processing the index spaces of the complex observations of all the views belonging to the distributed observation. In such a graph each node is an index of the considered distributed observation and each transition is marked by one or more contextual labels such that the reception of any of them causes the same index change.

- *Mutually independent index spaces of a distributed observation*: given two sets of views such that the views within one set do not overlap any view in the other, the single index spaces corresponding to the two sets are mutually independent.

- *Extension of the index space of a distributed observation*: set of all the sequences of contextual labels compliant with a given index space. It is claimed that it equals the extension of the observation itself.

- *Restriction of the index space of a distributed observation*: projection of the index space of the distributed observation inherent to a cluster on one of its subcluster. It is claimed that it equals the index space of the restriction of the observation itself.

Chapter 13

SAMPLE APPLICATION

Abstract

The research on diagnosis of active systems originated from a real problem: the localization of short circuits in power transmission networks. The occurrence of a short circuit on a transmission line can be automatically detected by monitoring the level of the line impedance. This task is carried out by specific protection devices that are expected to open relevant breakers in order to isolate the line from the rest of the network. However, the protection devices may be faulty when reacting, typically, by mistakenly localizing the short circuit or by failing to open or to close the breakers. The misbehavior of the protection apparatus is bound to cause several lines to be isolated, as protection devices may act as a backup to other faulty devices. This faulty reaction leaves the network in a state in which the localization of the short circuit is problematic to the operator who is in charge of reconnecting the part of the network that is not affected by the short circuit and to leave isolated the (unknown) shorted line only. Typically, the operator works under tight time constraints based on the state of the network after the reaction and a large set of messages sent by the protection devices during the reaction. The idea is to model the protection apparatus as a polymorphic system where components are protection devices and transmission lines. The set of messages generated during the reaction represents the system observation of a diagnostic problem. The reconstruction of the reaction makes it possible to find out possibly faulty devices and to eventually localize the short circuit.

1. Introduction

A power transmission network has the form of a meshed graph where nodes and edges correspond to the high voltage substations and to the high voltage three-phase lines, respectively. This structure aims to guarantee great robustness in case of accidents, mainly represented by short

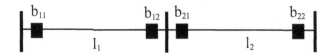

Figure 13.1.   Simple power transmission network.

circuits on transmission lines. The protection system is in charge of detecting dangerous conditions, of disconnecting a component (such as a line, a bus, a transformer, or a generation group) as soon as it starts operating in a dangerous way, and of keeping in operation non-faulty components as much as possible, in order to avoid a black-out. This is achieved by tripping the circuit breaker associated with each protection.

Several sorts of protective devices are used in transmission networks, depending on the electrical apparatus to be protected. Lines are guarded by distance protections, which recognize the presence of short circuits and command the opening of relevant breakers. Each protection is devoted to one component, but also works as a backup to the other protections nearby. Distance protections are located at both ends of a line and measure current and voltage in order to evaluate the impedance and, possibly, recognize the presence and the distance of a short circuit on the network. The protection is directional, as it only 'sees' the variations of impedance on the protected line and on the lines which are connected to that line in the same direction.

According to the measurements, the protection determines the timing of its intervention: the shorter the estimated distance of the short circuit, the faster the opening command sent to the breaker. This timing is discretized in steps defined by fixed impedance thresholds, which are set in order to discriminate the location of the short circuit: a tripping of the protection at the first step corresponds to the presence of a short circuit on the protected line; a tripping at the second step, excluding a few exceptions, to the presence of a short circuit on a line which is directly connected to the protected line, and so on.

When reacting to a short circuit, all the protective components send logical signals (messages) to a control room. These messages consist of a unique address of the event source, an event code, and possibly a timestamp. Operators have to decide within stringent time constraints where the short circuit is located and what recovery actions have to be applied. When the protection system reaction is faulty, such a localization is generally not straightforward at all. Therefore, due to severe time constraints and to the criticality of fault localization, an automated tool for fault diagnosis is quite advisable.

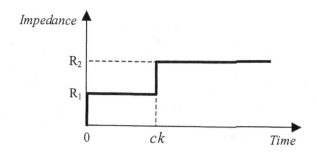

Figure 13.2. Simple impedance vs. time characteristics of breaker.

For the sake of simplicity, we consider a very simple network topology, as displayed in Figure 13.1. Such a network consists of two transmission lines, l_1 and l_2, that are connected to each other through a bus (depicted as a vertical bar). We assume that the network is closed, that is, no other lines are connected to the network. Each line is protected by two breakers, namely b_{11} and b_{12} for l_1 and b_{21} and b_{22} for l_2. As detailed below, such breakers are an abstraction of the distance protection device.

In normal conditions, the occurrence of a short circuit on a line causes the relevant breakers to open. For example, if the shorted line is l_1, both b_{11} and b_{12} are expected to open. In such a case, the localization of the short circuit is univocal and the operator is not in trouble.

However, if the behavior of b_{12} is faulty, for example, it fails to open, breaker b_{22} is expected to operate as a backup of b_{12} by opening at the second step. At the end of this faulty reaction, the opened breakers will be b_{11} and b_{22}. In such a case, the localization of the short circuit is not as straightforward as when the isolated line is just one: the shorted line might be either l_1 or l_2.

In a simplified view, we assume the breakers to open either immediately, when the short circuit is perceived on the protected line, or in a second step, when the short circuit is perceived on an adjacent line. This behavior is specified by the impedance vs. time characteristics displayed in Figure 13.2.

The origin of the time axis denotes the occurrence of the short circuit, when all the breakers 'starts' reacting. If the impedance is below R_1, the short circuit is perceived within the protected line, so that the relevant breakers open immediately. If the reaction of the two breakers is correct, the impedance on the lines is expected to increase beyond the threshold R_2, so that no backup actions are required for breakers of other lines.

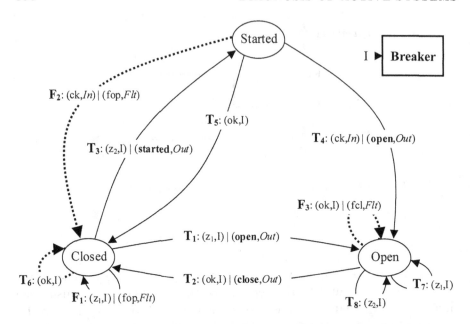

Figure 13.3. Model of breaker.

Instead, a steady value of the impedance between R_1 and R_2 means that the short circuit has not extinguished and, at time ck, backup behavior is required for relevant breakers in the adjacent lines.

The task of the operator is to reconnect the isolated part of the network that is not affected by the short circuit, specifically, all the lines but the shorted one. Thus, if there are good reasons for assuming the short on l_1, the operator will deliberately open b_{12} (if possible) and then close b_{22} so as to minimize the discomfort to users depending on l_2. Otherwise, assuming the short circuit on l_2, b_{21} will be opened while b_{11} will be closed, thereby keeping l_1 in operation.

Notice that the assumption of the operator may be wrong, especially when the isolated subnetwork encompasses several lines. If the short circuit is permanent and the shorted line is mistakenly reconnected, the protection system will react again to the new occurrence of the short circuit, thereby causing additional trouble to the operator (and to the users too).

2.   Modeling

Shown in Figure 13.3 is the graphical representation of our simplified model of the protecting breaker. Specifically, on the right is the topological model, namely, the set of input and output terminals, in our case,

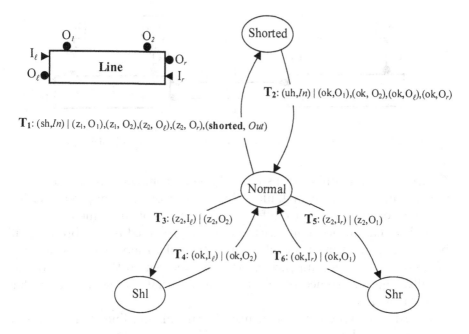

Figure 13.4. Model of line.

input terminal I. The automaton representing the behavioral model involves states *Closed*, *Started*, and *Open*, and transitions $T_1 \cdots T_8$ and $F_1 \cdots F_3$. Normal transitions are depicted as plain arrows, while faulty transitions are represented by dotted arrows.

For instance, transition T_1 is triggered by input event (z_1, I) and generates the (observable) output event $(open, Out)$. Such a transition brings the breaker to state *Open*. Notice that the same event (z_1, I) is associated with the faulty transition F_1, which does not change the state of the breaker. The faulty event (fop, Flt) represents such a fault, where *fop* is a shorthand for *failed to open*. Transitions T_2 and F_3 behave similarly to T_1 and F_1, respectively, when the state is *Open* (faulty event *fcl* stands for *failed to close*).

The peculiarity of such a breaker is that, as pointed out above and shown in Figure 13.2, it opens through a two-phase mechanism when the input event is z_2. First, it moves from *Closed* to *Started* through the observable transition T_3. Then, at the arrival of the clock event *ck* from the external world (standard input *In*), it moves from *Started* to *Open* by means of the observable transition T_4. Still in this case, the breaker may fail to open, which is modeled by the faulty transition F_2. On the other hand, once started, the breaker will not open if event (ok, I) occurs, which triggers transition T_5, that moves the breaker to state

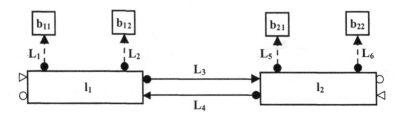

Figure 13.5.   Topology of system Σ relevant to the network of Figure 13.1.

Closed. The remaining transitions $T_6 \cdots T_8$ complete the model. The model supports transient faults, that is, the breaker may react either correctly or abnormally to different occurrences of the same event.

A short circuit occurring within the line protected by our breaker will be signaled by event z_1, while z_2 is intended to model the occurrence of a short circuit within the line that is adjacent to the protected one. Finally, the occurrence of event *ok* means that the short circuit has extinguished.

In order to represent a sequence of transmission lines (this is only a simplified topology for a power transmission network, which is in the general case a graph), we need to model a line as well, as shown in Figure 13.4. A line has two input terminals, namely I_ℓ and I_r, and four output terminals, namely O_1, O_2, O_ℓ, and O_r. Terminals I_ℓ and O_ℓ are meant to connect the line to another line on the left. The same applies to terminals I_r and O_r, which are for an adjacent line on the right. Instead, output terminals O_1 and O_2 are a means of connecting the line with the two protecting breakers, one on the left and the other on the right, respectively.

The behavioral model of the line involves states *Normal, Shorted, Shl*, and *Shr*. When the line is *Normal*, no short circuit has been detected, neither on such a line nor on any of the adjacent lines. Transition T_1 is triggered by a short circuit within the line (*sh* is a shorthand for *short circuit*, which is an event coming from th external world). Such a transition generates event z_1 on both terminals O_1 and O_2 (that is, directed towards the two protecting breakers). At the same time, it informs the two adjacent lines by means of the output event z_2 at terminals O_ℓ and O_r. For the sake of simplicity, we assume that transition T_1 is observable, generating message *shorted*. This allows us to focus on the diagnosis of the protection system rather than the localization of the short circuit.

When the short circuit is detected within an adjacent line, the line makes a transition from *Normal* to either *Shl* or *Shr*, that corresponds to transition T_3 or T_5, respectively (*Shl* and *Shr* stand for *short on the left* and *short on the right*, respectively). In such a case, only the breaker that faces the shorted line is informed by event z_2 at the relevant terminal. When *Shorted*, the line returns to *Normal* when the short circuit disappears (event *uh*, from the external world, stands for *unshort*). Specifically, transition T_2 informs both the protecting breakers, by generating the *ok* event at terminals O_1 and O_2, and the adjacent lines, by means of terminals O_ℓ and O_r. When in *Shl* or *Shr*, the line becomes *Normal* anew when the short circuit in the relevant adjacent line has extinguished, which is modeled by input event *ok* at the corresponding terminal: the *ok* event is generated for the breaker that faces the adjacent line.

The actual system is specified by instantiating the given models on actual components and by connecting such components by means of links. We assume that lines are connected to each other by asynchronous links, with capacity 1 and saturation policy *WAIT* (see Section 3 of Chapter 3), while the protecting breakers are connected with the line through synchronous links. Shown in Figure 13.5 is the topology of the polymorphic system Σ modeling the simple network depicted in Figure 13.1.

3. Simulation-Based Diagnosis

We are now ready to apply the diagnostic techniques devised for polymorphic systems to our reference system Σ. First of all, we define a diagnostic problem for Σ and solve it by means of simulation-based diagnosis. The diagnostic problem is defined as follows:

$$\wp(\Sigma) = (\Omega(\Sigma), OBS(\Sigma), \Sigma_0), \tag{13.1}$$

where all components are assumed to be visible, the observation is linear, namely $OBS(\Sigma)$ is identified by the unique item

$$\langle shorted(l_1), open(b_{11}), started(b_{22}), open(b_{22}), close(b_{11}), close(b_{22}) \rangle,$$

while the initial state is such that both lines are *Normal* and all breakers are *Closed*.

The reconstruction of the system behavior is represented by the active space depicted in Figure 13.6, whose nodes are detailed in Table 13.1. Notice that system transitions marking the edges of the graph are contextualized by relevant components. For example, the edge connecting node N_1 with node N_2 is marked by the system transition

$$\tau = \{T_1(b_{11}), F_1(b_{12})\},$$

meaning that breaker b_{11} performs the normal transition T_1 in parallel with the faulty transition F_1 of breaker b_{12}. In other words, b_{11} perceives the short circuit on line l_1 and reacts immediately by opening (messages

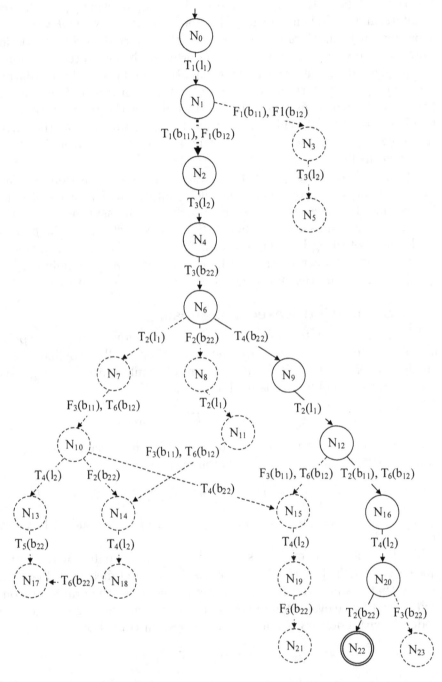

Figure 13.6. Active space relevant to the monolithic reconstruction of the diagnostic problem (13.1).

Table 13.1. Node details relevant to the active space displayed in Figure 13.6 (*Nor, Sho, Ope, Clo,* and *Sta* are shorthands for *Normal, Shorted, Open, Close,* and *Started,* respectively).

N	l_1	b_{11}	b_{12}	l_2	b_{21}	b_{22}	K	L_1	L_2	L_3	L_4	L_5	L_6
N_0	Nor	Clo	Clo	Nor	Clo	Clo	0	$\langle\rangle$	$\langle\rangle$	$\langle\rangle$	$\langle\rangle$	$\langle\rangle$	$\langle\rangle$
N_1	Sho	Clo	Clo	Nor	Clo	Clo	1	$\langle z_1\rangle$	$\langle z_1\rangle$	$\langle z_2\rangle$	$\langle\rangle$	$\langle\rangle$	$\langle\rangle$
N_2	Sho	Ope	Clo	Nor	Clo	Clo	2	$\langle\rangle$	$\langle\rangle$	$\langle z_2\rangle$	$\langle\rangle$	$\langle\rangle$	$\langle\rangle$
N_3	Sho	Clo	Clo	Nor	Clo	Clo	1	$\langle\rangle$	$\langle\rangle$	$\langle z_2\rangle$	$\langle\rangle$	$\langle\rangle$	$\langle\rangle$
N_4	Sho	Ope	Clo	Shl	Clo	Clo	2	$\langle\rangle$	$\langle\rangle$	$\langle\rangle$	$\langle\rangle$	$\langle\rangle$	$\langle z_2\rangle$
N_5	Sho	Clo	Clo	Shl	Clo	Clo	1	$\langle\rangle$	$\langle\rangle$	$\langle\rangle$	$\langle\rangle$	$\langle\rangle$	$\langle z_2\rangle$
N_6	Sho	Ope	Clo	Shl	Clo	Sta	3	$\langle\rangle$	$\langle\rangle$	$\langle\rangle$	$\langle\rangle$	$\langle\rangle$	$\langle\rangle$
N_7	Nor	Ope	Clo	Shl	Clo	Sta	3	ok	ok	ok	$\langle\rangle$	$\langle\rangle$	$\langle\rangle$
N_8	Sho	Ope	Clo	Shl	Clo	Clo	3	$\langle\rangle$	$\langle\rangle$	$\langle\rangle$	$\langle\rangle$	$\langle\rangle$	$\langle\rangle$
N_9	Sho	Ope	Clo	Shl	Clo	Ope	4	$\langle\rangle$	$\langle\rangle$	$\langle\rangle$	$\langle\rangle$	$\langle\rangle$	$\langle\rangle$
N_{10}	Nor	Ope	Clo	Shl	Clo	Sta	3	$\langle\rangle$	$\langle\rangle$	ok	$\langle\rangle$	$\langle\rangle$	$\langle\rangle$
N_{11}	Nor	Ope	Clo	Shl	Clo	Clo	3	ok	ok	ok	$\langle\rangle$	$\langle\rangle$	$\langle\rangle$
N_{12}	Nor	Ope	Clo	Shl	Clo	Ope	4	ok	ok	ok	$\langle\rangle$	$\langle\rangle$	$\langle\rangle$
N_{13}	Nor	Ope	Clo	Nor	Clo	Sta	3	$\langle\rangle$	$\langle\rangle$	$\langle\rangle$	$\langle\rangle$	$\langle\rangle$	ok
N_{14}	Nor	Ope	Clo	Shl	Clo	Clo	3	$\langle\rangle$	$\langle\rangle$	ok	$\langle\rangle$	$\langle\rangle$	$\langle\rangle$
N_{15}	Nor	Ope	Clo	Shl	Clo	Ope	4	$\langle\rangle$	$\langle\rangle$	ok	$\langle\rangle$	$\langle\rangle$	$\langle\rangle$
N_{16}	Nor	Clo	Clo	Shl	Clo	Ope	5	$\langle\rangle$	$\langle\rangle$	ok	$\langle\rangle$	$\langle\rangle$	$\langle\rangle$
N_{17}	Nor	Ope	Clo	Nor	Clo	Clo	3	$\langle\rangle$	$\langle\rangle$	$\langle\rangle$	$\langle\rangle$	$\langle\rangle$	$\langle\rangle$
N_{18}	Nor	Ope	Clo	Nor	Clo	Clo	3	$\langle\rangle$	$\langle\rangle$	$\langle\rangle$	$\langle\rangle$	$\langle\rangle$	ok
N_{19}	Nor	Ope	Clo	Nor	Clo	Ope	4	$\langle\rangle$	$\langle\rangle$	$\langle\rangle$	$\langle\rangle$	$\langle\rangle$	ok
N_{20}	Nor	Clo	Clo	Nor	Clo	Ope	5	$\langle\rangle$	$\langle\rangle$	$\langle\rangle$	$\langle\rangle$	$\langle\rangle$	ok
N_{21}	Nor	Ope	Clo	Nor	Clo	Ope	4	$\langle\rangle$	$\langle\rangle$	$\langle\rangle$	$\langle\rangle$	$\langle\rangle$	$\langle\rangle$
N_{22}	Nor	Clo	Clo	Nor	Clo	Clo	6	$\langle\rangle$	$\langle\rangle$	$\langle\rangle$	$\langle\rangle$	$\langle\rangle$	$\langle\rangle$
N_{23}	Nor	Clo	Clo	Nor	Clo	Ope	5	$\langle\rangle$	$\langle\rangle$	$\langle\rangle$	$\langle\rangle$	$\langle\rangle$	$\langle\rangle$

generated during the reaction are not explicitly reported on the active space) while the behavior of b_{12} is faulty insofar as, even though perceiving the short circuit on the same line, it fails to open.

The non-spurious part of the reconstruction consists of a single history, which involves one fault relevant to transition $F_1(b_{12})$, namely $fop(b_{12})$. This means that breaker b_{12} has failed to open at the occurrence of the short circuit on line l_1. The dynamics of the reaction is detailed in Table 13.2.

Now we solve the same diagnostic problem in a modular way. Our reconstruction plan is based on the decomposition of Σ as shown in Figure 13.7, where two clusters are considered, one relevant to each

Table 13.2.   Dynamics of the reaction relevant to the history displayed in Figure 13.6.

System transition	Explanation
$\{T_1(l_1)\}$	A short circuit occurs on line l_1
$\{T_1(b_{11}), F_1(b_{12})\}$	Breaker b_{11} opens, while b_{12} fails to open
$\{T_3(l_2)\}$	Line l_2 signals the short circuit on the left to its breakers
$\{T_3(b_{22})\}$	Breaker b_{22} starts reacting as a backup
$\{T_4(b_{22})\}$	Breaker b_{22} opens at second step
$\{T_2(l_1)\}$	Line l_1 returns to its normal state
$\{T_2(b_{11}), T_6(b_{12})\}$	Breaker b_{11} closes (b_{12} remains closed)
$\{T_4(l_2)\}$	Line l_2 returns to its normal state
$\{T_2(b_{22})\}$	Breaker b_{22} closes

protected line. Specifically, cluster ξ_1 involves l_1, b_{11}, and b_{12}, while cluster ξ_2 embodies l_2, b_{21}, and b_{22}. Following the notation introduced in Section 3 of Chapter 5, we have

$$\Xi(\Sigma) = \{\xi_1, \xi_2\}, \text{ where } Front(\Xi(\Sigma)) = \{L_3, L_4\}.$$

The diagnostic problems $\wp(\xi_1)$ and $\wp(\xi_2)$ are defined as the projections of $\wp(\Sigma)$ on ξ_1 and ξ_2, respectively. The reconstruction graph involves the reconstruction relevant to $\wp(\xi_1)$ and $\wp(\xi_2)$, and the eventual join of the obtained (partial) active spaces.

Shown in Figure 13.8 is the active space relevant to $\wp(\xi_1)$, whose (linear) observation is the projection of $OBS(\Sigma)$ on cluster ξ_1, namely

$$OBS(\xi_1) = OBS_{\langle \xi_1 \rangle}(\Sigma) = \langle shorted(l_1), open(b_{11}), close(b_{11}) \rangle.$$

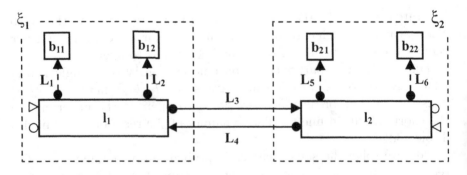

Figure 13.7.   Decomposition of system Σ.

Node details are given in Table 13.3. Even in this case, the active space embodies a single history that brings ξ_1 from the initial state N_0 to the final state N_7. As a matter of fact, such a history is the projection of the history of Σ, shown in Figure 13.6, on ξ_1.

The reconstruction of the behavior of cluster ξ_2 based on the relevant diagnostic problem is depicted in Figure 13.9 (see Table 13.4 for node details). Even in this case, the only history incorporated in the active space is the projection of the history of Σ (see Figure 13.6) on cluster ξ_2.

Once generated the partial active spaces relevant to $\wp(\xi_1)$ and $\wp(\xi_2)$, the active space corresponding to the original diagnostic problem $\wp(\Sigma)$ is obtained by merging them by means of the join operator introduced in Section 5.1 of Chapter 5, namely

$$Act(\wp(\Sigma)) = \mathcal{J}_{\wp(\Sigma)}(\{Act(\wp(\xi_1)), Act(\wp(\xi_2))\}).$$

The result of such a join is displayed in Figure 13.10, where each node of the graph embodies the pair of cluster states (left), the observation index (center), and the dangling set (right). As expected, the resulting (unique) history equals the history displayed in Figure 13.6, thereby

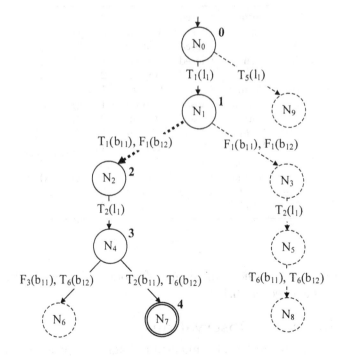

Figure 13.8. Active space for cluster ξ_1.

Table 13.3. Node details relevant to the active space for ξ_1 displayed in Figure 13.8.

N	l_1	b_{11}	b_{12}	K	L_1	L_2
N_0	Normal	Closed	Closed	0	$\langle\rangle$	$\langle\rangle$
N_1	Shorted	Closed	Closed	1	$\langle z_1\rangle$	$\langle z_1\rangle$
N_2	Shorted	Open	Closed	2	$\langle\rangle$	$\langle\rangle$
N_3	Shorted	Closed	Closed	1	$\langle\rangle$	$\langle\rangle$
N_4	Normal	Open	Closed	2	ok	ok
N_5	Normal	Closed	Closed	2	ok	ok
N_6	Normal	Open	Closed	2	$\langle\rangle$	$\langle\rangle$
N_7	Normal	Closed	Closed	3	$\langle\rangle$	$\langle\rangle$
N_8	Normal	Closed	Closed	2	$\langle\rangle$	$\langle\rangle$
N_9	Shl	Closed	Closed	0	$\langle z_2\rangle$	$\langle\rangle$

Table 13.4. Node details relevant to the active space for ξ_2 displayed in Figure 13.9.

N	l_2	b_{21}	b_{22}	K	L_5	L_6
N_0	Normal	Closed	Closed	0	$\langle\rangle$	$\langle\rangle$
N_1	Shl	Closed	Closed	0	$\langle\rangle$	$\langle z_2\rangle$
N_2	Shl	Closed	Started	1	$\langle\rangle$	$\langle\rangle$
N_3	Shl	Closed	Closed	1	$\langle\rangle$	$\langle\rangle$
N_4	Shl	Closed	Open	2	$\langle\rangle$	$\langle\rangle$
N_5	Normal	Closed	Started	1	$\langle\rangle$	ok
N_6	Normal	Closed	Closed	1	$\langle\rangle$	ok
N_7	Normal	Closed	Open	2	$\langle\rangle$	ok
N_8	Shl	Closed	Closed	1	$\langle\rangle$	$\langle z_2\rangle$
N_9	Normal	Closed	Closed	1	$\langle\rangle$	$\langle\rangle$
N_{10}	Normal	Closed	Closed	3	$\langle\rangle$	$\langle\rangle$
N_{11}	Normal	Closed	Open	2	$\langle\rangle$	$\langle\rangle$
N_{12}	Shl	Closed	Closed	3	$\langle\rangle$	$\langle z_2\rangle$
N_{13}	Shl	Closed	Open	3	$\langle\rangle$	$\langle z_2\rangle$

verifying once again the equivalence of monolithic and modular reconstruction (see Theorem 6.1).

4.   Uncertain Observations

Based on the techniques for uncertain diagnostic problems introduced in Chapter 9, we now consider a subsystem Σ of Σ' that is composed

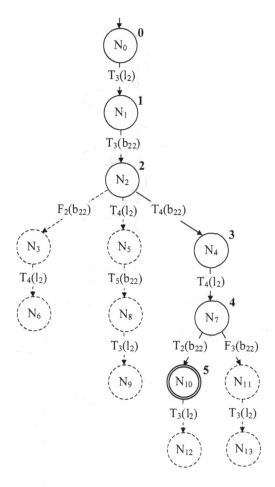

Figure 13.9. Active space for cluster ξ_2.

of the protected line l_1 along with relevant breakers. For the sake of conciseness, we rename l_1, b_{11}, and b_{12} as l, b_1, and b_2, respectively. System Σ' is depicted in Figure 13.11. Notice that dangling terminals on the right are supposed to be connected with the rest of the network, namely they belong to the set $\mathbb{D}_{on}$ of on-terminals of Σ' (see Section 4 of Chapter 3).

For system Σ' we consider the uncertain observation $OBS(\Sigma')$ depicted in Figure 13.12. The observation graph of $OBS(\Sigma')$ involves four nodes, $\omega_1 \cdots \omega_4$, each relevant to a (possibly uncertain) message. Specifically, the only actually uncertain message is in node ω_3, whose possible values are $close(b_1)$ and the null label. This means that the message relevant to breaker b_1 might not be generated.

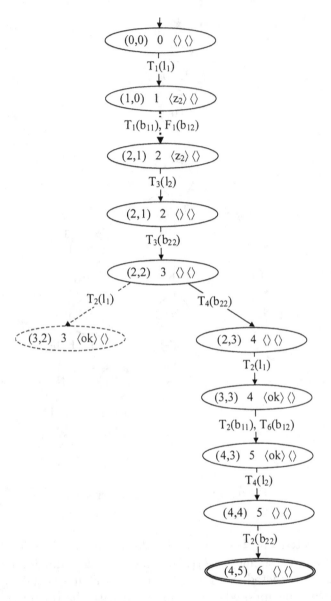

Figure 13.10. Join of the active spaces relevant to clusters ξ_1 and ξ_2 (see Figures 13.8 and 13.9).

Besides logical uncertainty (the imprecision of labels of nodes), temporal uncertainty causes uncertain messages to be accommodated within a DAG whose edges represent partial temporal ordering among such uncertain messages. In our example, ω_1 is precedent to ω_2 and ω_3, while ω_2 is precedent to ω_4. Since the ordering is only partial, some pairs of

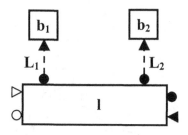

Figure 13.11. System Σ'.

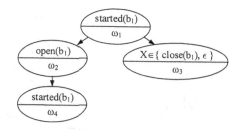

Figure 13.12. Uncertain observation for system Σ' displayed in Figure 13.11.

nodes may possibly be temporally unrelated. For instance, no temporal relationship is established between ω_2 and ω_3, nor is it defined between ω_3 and ω_4.

Assuming an observer for $OBS(\Sigma')$ with complete visibility, namely $Ukn(OBS(\Sigma')) = \emptyset$, we aim to solve the uncertain diagnostic problem $\wp(\Sigma')$ based on the uncertain observation of Figure 13.12. To this end, we need to generate the index space of $OBS(\Sigma')$, whose nodes will be used as index $\Im$ within nodes of the relevant reconstruction graph (active space of $\wp(\Sigma')$).

From left to right, Figure 13.13 outlines the rough index space, the nondeterministic index space, and the deterministic index space, the latter being the index space of $OBS(\Sigma')$), which consists of nodes $\Im_0 \cdots \Im_6$.

Based on such an index space, the reconstruction graph for the diagnostic problem $\wp(\Sigma')$ is made up as displayed in Figure 13.14 (node details are given in Table 13.5). Owing to the existence of a cycle, the number of histories in the active space is unbounded. However, only two possible (either shallow or deep) diagnoses are possible: one empty and the other involving the fault fcl (failed to close) relevant to breaker b_1.

5.   Monitoring-Based Diagnosis

Finally, we apply the monitoring-based diagnostic technique introduced in Chapter 8 to the system Σ' displayed in Figure 13.11. For the sake of clarity, we first generate the universal space $Usp(\Sigma', \Sigma'_0)$, assuming that, before the reaction, the breakers are closed and the line is in normal state. The resulting graph is displayed in Figure 13.15 (see Table 13.6 for node details).

Before coping with monitoring-based diagnosis of Σ', we introduce a slight extension to the technique presented in Chapter 8, where only shallow diagnoses were considered. As a matter of fact, the technique works as well considering deep diagnosis. Specifically, the element $\mathcal{F}$

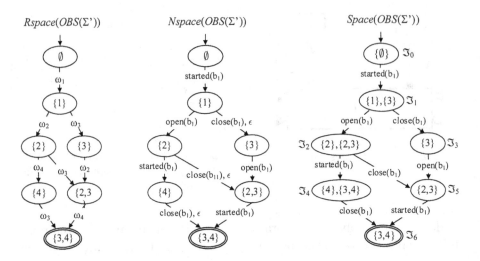

Figure 13.13. Rough index space (left), nondeterministic index space(center), and deterministic index space (right) relevant to the uncertain observation outlined in Figure 13.12.

marking a transition in the monitoring graph (see Section 5 of Chapter 8), namely

$$S \xrightarrow{(\mu, \mathcal{F})} S',$$

can be considered as the set of faults relevant to the system transition, rather than the set of faulty components. This way, a more accurate diagnostic information can be provided during monitoring.

The monitoring graph relevant to the universal space of Figure 13.15 is displayed in Figure 13.16, where labels A, B, C, and D stands for faults $fop(b_1)$, $fcl(b_1)$, $fop(b_2)$, $fcl(b_2)$, respectively.

Consider the evolution of system Σ' relevant to the following sequence of system messages:

$$\langle \{shorted(l)\}, \{open(b_1)\}, \{close(b_1)\} \rangle.$$

When no messages are considered, the initial set of candidate diagnoses (historic diagnostic set) coincides with the local diagnostic set of the root N_0, namely

$$\Delta(\langle \rangle) = \Delta^\ell(N_0) = \{\emptyset\}.$$

At the occurrence of message $shorted(l)$, a (non-faulty) transition from the root of N_0 to the root of N_1 occurs, which gives rise to a snapshot diagnosis that equals the local diagnostic set, namely

$$\mathfrak{D} = \{\emptyset, \{A, C\}\}.$$

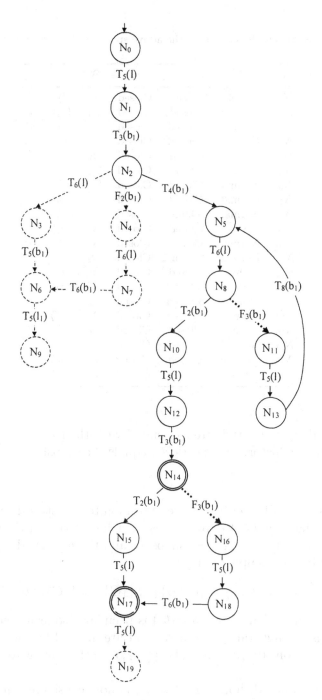

Figure 13.14. Active space relevant to the uncertain diagnostic problem $\wp(\Sigma')$.

Table 13.5.   Node details relevant to the active space displayed in Figure 13.14.

N	l	b_1	b_2	$\Im$	L_1	L_2
N_0	Normal	Closed	Closed	$\Im_0$	$\langle\rangle$	$\langle\rangle$
N_1	Shr	Closed	Closed	$\Im_0$	$\langle z_2\rangle$	$\langle\rangle$
N_2	Shr	Started	Closed	$\Im_1$	$\langle\rangle$	$\langle\rangle$
N_3	Normal	Started	Closed	$\Im_1$	$\langle ok\rangle$	$\langle\rangle$
N_4	Shr	Closed	Closed	$\Im_1$	$\langle\rangle$	$\langle\rangle$
N_5	Shr	Open	Closed	$\Im_2$	$\langle\rangle$	$\langle\rangle$
N_6	Normal	Closed	Closed	$\Im_1$	$\langle\rangle$	$\langle\rangle$
N_7	Normal	Closed	Closed	$\Im_1$	$\langle ok\rangle$	$\langle\rangle$
N_8	Normal	Open	Closed	$\Im_2$	$\langle ok\rangle$	$\langle\rangle$
N_9	Shr	Closed	Closed	$\Im_1$	$\langle z_2\rangle$	$\langle\rangle$
N_{10}	Normal	Closed	Closed	$\Im_5$	$\langle\rangle$	$\langle\rangle$
N_{11}	Normal	Open	Closed	$\Im_2$	$\langle\rangle$	$\langle\rangle$
N_{12}	Shr	Closed	Closed	$\Im_5$	$\langle z_2\rangle$	$\langle\rangle$
N_{13}	Shr	Open	Closed	$\Im_2$	$\langle z_2\rangle$	$\langle\rangle$
N_{14}	Shr	Started	Closed	$\Im_6$	$\langle\rangle$	$\langle\rangle$
N_{15}	Normal	Started	Closed	$\Im_6$	$\langle ok\rangle$	$\langle\rangle$
N_{16}	Shr	Closed	Closed	$\Im_6$	$\langle\rangle$	$\langle\rangle$
N_{17}	Normal	Closed	Closed	$\Im_6$	$\langle\rangle$	$\langle\rangle$
N_{18}	Normal	Closed	Closed	$\Im_6$	$\langle ok\rangle$	$\langle\rangle$
N_{19}	Shr	Closed	Closed	$\Im_6$	$\langle z_2\rangle$	$\langle\rangle$

Since the diagnostic attribute associated with the root of N_0 is the singleton $\{\emptyset\}$, the historic diagnostic set equals the snapshot diagnostic set, namely

$$\Delta = \mathfrak{D}.$$

The occurrence of $open(b_1)$ moves the monitoring state from the root of N_1 to the root of N_4, by means of a faulty transition marked by "$open(b_1), C$". Thus, both the snapshot and the historic diagnostic sets are given by the composition

$$\Delta^\ell(N_4) \otimes \{\{C\}\} = \{\emptyset, \{B\}\} \otimes \{\{C\}\} = \{\{C\}, \{B, C\}\}.$$

The occurrence of message $close(b_1)$ is peculiar insofar as there are two transitions leaving states within N_4 that are marked by such a message, specifically, from $(9, \{\emptyset\})$ and $(17, \{\{B\}\})$, both moving to the root of N_0.

It must be noted, however, that N_0 is now considered in a different context with respect to the initial monitoring node, thereby yielding different diagnostic sets. In fact, even though $\mathfrak{D}$ is still $\{\emptyset\}$, the historic

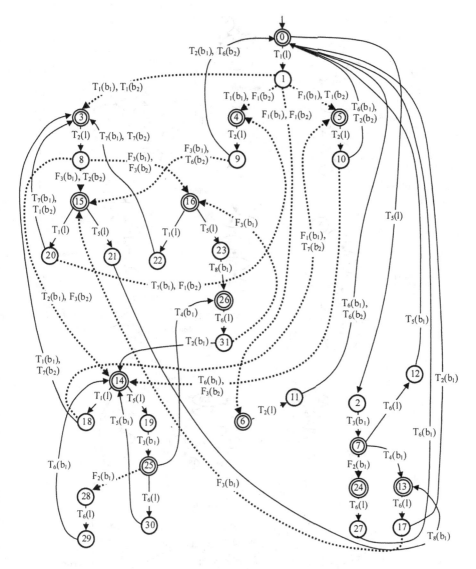

Figure 13.15. Universal space relevant to system Σ' outlined in Figure 13.11.

diagnostic set Δ is computed based on the relocated diagnostic attributes associated with the relevant (leaving) internal nodes.

Considering $(9, \{\emptyset\})$, such a relocation corresponds to the combination

$$\{\emptyset\} \otimes \{\{C\}\} = \{\{C\}\}.$$

Instead, for $(17, \{\{B\}\})$, the relocation is

$$\{\{B\}\} \otimes \{\{C\}\} = \{\{B, C\}\}.$$

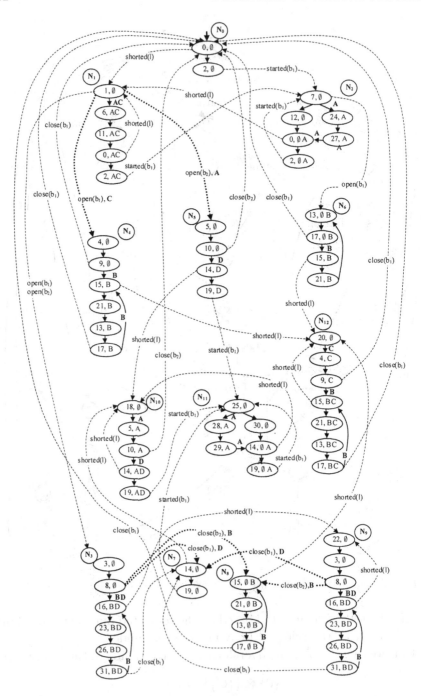

Figure 13.16. Monitoring graph relevant to the universal space displayed in Figure 13.15. Labels A, B, C, and D stands for faults $fop(b_1)$, $fcl(b_1)$, $fop(b_2)$, $fcl(b_2)$, respectively.

Table 13.6. Node details relevant to the universal space displayed in Fig. 13.15.

Node	b_1	l	b_2	L_1	L_2
0	Closed	Normal	Closed	$\langle\rangle$	$\langle\rangle$
1	Closed	Shorted	Closed	$\langle z_1 \rangle$	$\langle z_1 \rangle$
2	Closed	Shr	Closed	$\langle z_2 \rangle$	$\langle\rangle$
3	Open	Shorted	Open	$\langle\rangle$	$\langle\rangle$
4	Open	Shorted	Closed	$\langle\rangle$	$\langle\rangle$
5	Closed	Shorted	Open	$\langle\rangle$	$\langle\rangle$
6	Closed	Shorted	Closed	$\langle\rangle$	$\langle\rangle$
7	Started	Shr	Closed	$\langle\rangle$	$\langle\rangle$
8	Open	Normal	Open	$\langle ok \rangle$	$\langle ok \rangle$
9	Open	Normal	Closed	$\langle ok \rangle$	$\langle ok \rangle$
10	Closed	Normal	Open	$\langle ok \rangle$	$\langle ok \rangle$
11	Closed	Normal	Closed	$\langle ok \rangle$	$\langle ok \rangle$
12	Started	Normal	Closed	$\langle ok \rangle$	$\langle\rangle$
13	Open	Shr	Closed	$\langle\rangle$	$\langle\rangle$
14	Closed	Normal	Open	$\langle\rangle$	$\langle\rangle$
15	Open	Normal	Closed	$\langle\rangle$	$\langle\rangle$
16	Open	Normal	Open	$\langle\rangle$	$\langle\rangle$
17	Open	Normal	Closed	$\langle ok \rangle$	$\langle\rangle$
18	Closed	Shorted	Open	$\langle z_1 \rangle$	$\langle z_1 \rangle$
19	Closed	Shr	Open	$\langle z_2 \rangle$	$\langle\rangle$
20	Open	Shorted	Closed	$\langle z_1 \rangle$	$\langle z_1 \rangle$
21	Open	Shr	Closed	$\langle z_2 \rangle$	$\langle\rangle$
22	Open	Shorted	Open	$\langle z_1 \rangle$	$\langle z_1 \rangle$
23	Open	Shr	Open	$\langle z_2 \rangle$	$\langle\rangle$
24	Closed	Shr	Closed	$\langle\rangle$	$\langle\rangle$
25	Started	Shr	Open	$\langle\rangle$	$\langle\rangle$
26	Open	Shr	Open	$\langle\rangle$	$\langle\rangle$
27	Closed	Normal	Closed	$\langle ok \rangle$	$\langle\rangle$
28	Closed	Shr	Open	$\langle\rangle$	$\langle\rangle$
29	Closed	Normal	Open	$\langle ok \rangle$	$\langle\rangle$
30	Started	Normal	Open	$\langle ok \rangle$	$\langle\rangle$
31	Open	Normal	Open	$\langle ok \rangle$	$\langle\rangle$

Δ is then the combination of $\Delta^\ell(N_0) = \{\emptyset\}$ with the union of such relocations, namely

$$\Delta = \{\emptyset\} \otimes (\{\{C\}\} \cup \{\{B, C\}\}) = \{\{C\}, \{B, C\}\}.$$

Shown in Table 13.7 is the application of the monitoring algorithm (see Appendix A of Chapter 8) to the sequence of system messages listed

Table 13.7.   Generation of the diagnostic sequence.

μ	$\mathcal{M}'$	$\mathfrak{D}'$	Δ'
	$\{(N_0, \{\emptyset\}, \{\emptyset\})\}$	$\{\emptyset\}$	$\{\emptyset\}$
$shorted(l)$	$\{(N_1, \{\emptyset\}, \{\emptyset\})\}$	$\{\emptyset, \{A, C\}\}$	$\{\emptyset, \{A, C\}\}$
$open(b_2)$	$\{(N_5, \{\{A\}\}, \{\{A\}\})\}$	$\{\{A\}, \{A, D\}\}$	$\{\{A\}, \{A, D\}\}$
$close(b_2)$	$\{(N_0, \{\emptyset\}, \{\{A\}\})\}$	$\{\emptyset\}$	$\{\{A\}\}$
$started(b_1)$	$\{(N_2, \{\emptyset\}, \{\{A\}\})\}$	$\{\emptyset, \{A\}\}$	$\{\{A\}\}$
$open(b_1)$	$\{(N_6, \{\emptyset\}, \{\{A\}\})\}$	$\{\emptyset, \{B\}\}$	$\{\{A\}, \{A, B\}\}$
$shorted(l)$	$\{(N_{12}, \{\emptyset\}, \{\{A, B\}\})\}$	$\{\emptyset, \{C\}, \{B, C\}\}$	$\{\{A, B\}, \{A, B, C\}\}$
$close(b_1)$	$\{(N_0, \{\emptyset\}, \{\{A, B, C\}\})\}$	$\{\emptyset\}$	$\{\{A, B, C\}\}$

in the first column. For each message μ, the values of $\mathcal{M}'$, $\mathfrak{D}'$, and Δ' are outlined.

The historic diagnostic set associated with the last message $close(b_1)$ is the singleton

$$\Delta' = \{\{A, B, C\}\} = \{\{fop(b_1), , fcl(b_1), fop(b_2)\}\}.$$

This means that, during the operation of Σ', breaker b_1 both failed to open and close, while breaker b_2 failed to open only.

We expect such a historic diagnostic to coincide with the deep diagnostic set obtained by applying simulation-based diagnosis on the diagnostic problem for Σ', where the observation is in fact the list of messages outlined in Table 13.7. The verification of such an equivalence is left to the reader.

References

[Aghasaryan et al., 1998] Aghasaryan, A., Fabre, E., Benveniste, A., Boubour, B., and Jard, C. (1998). Fault detection and diagnosis in distributed systems: an approach by partially stochastic Petri nets. *Journal of Discrete Event Dynamical Systems: Theory and Application*, August:203–231.

[Aho et al., 1986] Aho, A., Sethi, R., and Ullman, J. (1986). *Compilers – Principles, Techniques, and Tools*. Addison-Wesley, Reading, MA.

[Atzeni et al., 1999] Atzeni, P., Ceri, S., Paraboschi, S., and Torlone, R. (1999). *Database Systems: Concepts, Languages and Architectures*. McGraw-Hill.

[Baroni et al., 1997a] Baroni, P., Canzi, U., and Guida, G. (1997). Fault diagnosis through history reconstruction: an application to power transmission networks. *Expert Systems with Applications*, 12(1):37–52.

[Baroni et al., 1997b] Baroni, P., Lamperti, G., Pogliano, P., Tornielli, G., and Zanella, M. (1997). Automata-based reasoning for short circuit diagnosis in power transmission networks. In *Twelfth International Conference on Applications of Artificial Intelligence in Engineering*, Capri, I.

[Baroni et al., 1998] Baroni, P., Lamperti, G., Pogliano, P., and Zanella, M. (1998). Diagnosis of active systems. In *Thirteenth European Conference on Artificial Intelligence – ECAI'98*, pages 274–278, Brighton, UK.

[Baroni et al., 1999] Baroni, P., Lamperti, G., Pogliano, P., and Zanella, M. (1999). Diagnosis of large active systems. *Artificial Intelligence*, 110(1):135–183.

[Baroni et al., 2000] Baroni, P., Lamperti, G., Pogliano, P., and Zanella, M. (2000). Diagnosis of a class of distributed discrete-event systems. *IEEE Transactions on Systems, Man, and Cybernetics – Part A: Systems and Humans*, 30(6):731–752.

[Barral et al., 2000] Barral, C., McIlraith, S., and Son, T. (2000). Formulating diagnostic problem solving using an action language with narratives and sensing. In *Seventh International Conference on Knowledge Representation and Reasoning – KR'2000*, pages 311–322, Breckenridge, Colorado.

[Barrow, 1984] Barrow, H. (1984). VERIFY: A program for proving correctness of digital hardware. *Artificial Intelligence*, 24(1).

[Brusoni et al., 1998] Brusoni, V., Console, L., Terenziani, P., and Dupré, D. T. (1998). A spectrum of definitions for temporal model-based diagnosis. *Artificial Intelligence*, 102(1):39–80.

[Cassandras and Lafortune, 1999] Cassandras, C. and Lafortune, S. (1999). *Introduction to Discrete Event Systems*. Mathematics and Its Applications. Kluwer Academic Publisher, Boston, MA.

[Chen and Provan, 1997] Chen, Y. and Provan, G. (1997). Modeling and diagnosis of timed discrete event systems - a factory automation example. In *American Control Conference*, pages 31–36, Albuquerque, NM.

[Chittaro et al., 1993] Chittaro, L., Guida, G., Tasso, C., and Toppano, E. (1993). Functional and teleological knowledge in the multi-modeling approach for reasoning about physical systems: a case study in diagnosis. *IEEE Transactions on Systems, Man, and Cybernetics*, 23(6):1718–1751.

[Codd, 1970] Codd, E. (1970). A relational model for large shared data banks. *Communications of the ACM*, 13(6):377–387.

[Console et al., 1989] Console, L., Dupré, D. T., and Torasso, P. (1989). A theory of diagnosis for incomplete causal models. In *Eleventh International Joint Conference on Artificial Intelligence - IJCAI'89*, pages 1311–1317, Detroit, MI.

[Console et al., 2000a] Console, L., Picardi, C., and Ribaudo, M. (2000). Diagnosis and diagnosability analysis using process algebras. In *Eleventh International Workshop on Principles of Diagnosis - DX'00*, pages 25–32, Morelia, MX.

[Console et al., 2000b] Console, L., Picardi, C., and Ribaudo, M. (2000). Diagnosis and diagnosability using PEPA. In *Fourteenth European Conference on Artificial Intelligence - ECAI'2000*, pages 131–135, Berlin, D.

[Console et al., 1992] Console, L., Portinale, L., Dupré, D. T., and Torasso, P. (1992). Diagnostic reasoning across different time points. In *Eleventh European Conference on Artificial Intelligence - ECAI'92*, pages 369–373, Vienna, A.

[Console and Torasso, 1991a] Console, L. and Torasso, P. (1991). On the cooperation between abductive and temporal reasoning in medical diagnosis. *Artificial Intelligence in Medicine*, 3(6):291–311.

[Console and Torasso, 1991b] Console, L. and Torasso, P. (1991). A spectrum of logical definitions of model-based diagnosis. *Computational Intelligence*, 7(3):133–141.

[Cordier and Largouët, 2001] Cordier, M. and Largouët, C. (2001). Using model-checking techniques for diagnosing discrete-event systems. In *Twelfth International Workshop on Principles of Diagnosis - DX'01*, pages 39–46, San Sicario, I.

[Cordier and Thiébaux, 1994] Cordier, M. and Thiébaux, S. (1994). Event-based diagnosis for evolutive systems. In *Fifth International Workshop on Principles of Diagnosis - DX'94*, pages 64–69, New Paltz, NY.

[de Kleer, 1976] de Kleer, J. (1976). Local methods for localizing faults in electronic circuits. Memo 394, MIT Artificial Intelligence Lab, Cambridge, MA.

[de Kleer, 1984] de Kleer, J. (1984). How circuits work. *Artificial Intelligence*, 24(1):205–280.

[de Kleer, 1990] de Kleer, J. (1990). Using crude probability estimates to guide diagnosis. *Artificial Intelligence*, 45:381–392.

[de Kleer and Brown, 1984] de Kleer, J. and Brown, J. (1984). A qualitative physics based on confluences. *Artificial Intelligence*, 24(1):7–83.

[de Kleer and Brown, 1986] de Kleer, J. and Brown, J. (1986). Theories of casual ordering. *Artificial Intelligence*, 29:33–61.

[de Kleer and Williams, 1987] de Kleer, J. and Williams, B. (1987). Diagnosing multiple faults. *Artificial Intelligence*, 32(1):97–130.

[de Kleer and Williams, 1989] de Kleer, J. and Williams, B. (1989). Diagnosis with behavioral modes. In *Eleventh International Joint Conference on Artificial Intelligence – IJCAI'89*, pages 1324–1330, Detroit, MI.

[Debouk et al., 2000a] Debouk, R., Lafortune, S., and Teneketzis, D. (2000). Coordinated decentralized protocols for failure diagnosis of discrete-event systems. *Journal of Discrete Event Dynamical Systems: Theory and Application*, 10:33–86.

[Debouk et al., 2000b] Debouk, R., Lafortune, S., and Teneketzis, D. (2000). A diagnostic protocol for discrete-event systems with decentralized information. In *Eleventh International Workshop on Principles of Diagnosis – DX'00*, pages 41–48, Morelia, MX.

[Dressler and Freitag, 1994] Dressler, O. and Freitag, H. (1994). Prediction sharing across time and contexts. In *Twelfth National Conference on Artificial Intelligence – AAAI'94*, pages 1136–1141, Seattle, WA.

[Dvorak, 1992] Dvorak, D. (1992). *Monitoring and diagnosis of continuous dynamic systems using semiquantitative simulation*. PhD thesis, University of Texas at Austin, TX.

[Dvorak and Kuipers, 1992] Dvorak, D. and Kuipers, B. (1992). Model-based monitoring of dynamic systems. In Hamscher, W., Console, L., and de Kleer, J., editors, *Readings in Model-Based Diagnosis*. Morgan Kaufmann.

[Elmasri and Navathe, 1994] Elmasri, R. and Navathe, S. (1994). *Fundamentals of Database Systems*. Benjamin-Cummings.

[Fattah and Provan, 1997] Fattah, Y. E. and Provan, G. (1997). Modeling temporal behavior in the model-based diagnosis of discrete-event systems (a preliminary note). In *Eighth International Workshop on Principles of Diagnosis – DX'97*, Mont St. Michel, F.

[Forbus, 1984] Forbus, K. (1984). Qualitative process theory. *Artificial Intelligence*, 24(1):85–168.

[Förstner and Lunze, 1999] Förstner, D. and Lunze, J. (1999). Qualitative modelling of a power stage for diagnosis. In *Thirteenth International Workshop on Qualitative Reasoning – QR'99*, pages 105–112, Lock Awe, UK.

[Förstner and Lunze, 2000] Förstner, D. and Lunze, J. (2000). Qualitative model-based fault detection of a fuel injection system. In *Fourth IFAC Symposium on Fault Detection, Supervision and Safety for Technical Processes – SAFEPRO-CESS'2000*, pages 88–93, Budapest, U.

[Friedrich and Lackinger, 1991] Friedrich, G. and Lackinger, F. (1991). Diagnosing temporal misbehavior. In *Twelfth International Joint Conference on Artificial Intelligence – IJCAI'91*, pages 1116–1122, Sydney, Australia.

[Garatti et al., 2002] Garatti, R., Lamperti, G., and Zanella, M. (2002). Diagnosis of discrete-event systems with model-based prospection knowledge. In van Harmelen, F., editor, *ECAI 2002. Proceedings of the 15th European Conference on Artificial Intelligence*. IOS Press, Amsterdam, NL.

[Genesereth, 1984] Genesereth, M. (1984). The use of design descriptions in auto-mated diagnosis. *Artificial Intelligence*, 24:411–436.

[Gilmore and Hillston, 1994] Gilmore, S. and Hillston, J. (1994). The PEPA work-bench: a tool to support a process algebra based approach to performance mod-elling. In *Seventh International Conference on Modeling Techniques and Tools for Computer Performance Evaluation*, LNCS 794.

[Guckenbiehl and Schäfer-Richter, 1990] Guckenbiehl and Schäfer-Richter (1990). Sidia: Extending prediction based diagnosis to dynamic models. In *First Inter-national Workshop on Principles of Diagnosis*, pages 74–82, Stanford, CA.

[Hamscher, 1991] Hamscher, W. (1991). Modeling digital circuits for troubleshooting. *Artificial Intelligence*, 51(1-3):223–271.

[Hamscher et al., 1992] Hamscher, W., Console, L., and de Kleer, J., editors (1992). *Readings in Model-Based Diagnosis*. Morgan Kaufmann, San Mateo, CA.

[Hamscher and Davis, 1984] Hamscher, W. and Davis, R. (1984). Diagnosing circuits with state: an inherently underconstrained problem. In *Fourth National Conference on Artificial Intelligence – AAAI'84*, pages 142–147, Austin, TX.

[Hillston, 1996] Hillston, J. (1996). *A compositional approach to performance mod-elling*. Cambridge University Press.

[Hopcroft and Ullman, 1979] Hopcroft, J. and Ullman, J. (1979). *Introduction to Automata Theory*. Addison-Wesley, Reading, MA.

[Iwasaki and Simon, 1986a] Iwasaki, Y. and Simon, H. (1986). Causality in device behavior. *Artificial Intelligence*, 29:3–32.

[Iwasaki and Simon, 1986b] Iwasaki, Y. and Simon, H. (1986). Theories of casual ordering: reply to de kleer and brown. *Artificial Intelligence*, 29:63–72.

[Kuipers, 1984] Kuipers, B. (1984). Commonsense reasoning about causality: Deriv-ing behavior from structure. *Artificial Intelligence*, 24(1–3):169–204.

[Kurien and Nayak, 2000] Kurien, J. and Nayak, P. (2000). Back to the future for consistency-based trajectory tracking. In *Eleventh International Workshop on Principles of Diagnosis - DX'00*, pages 92–100, Morelia, MX.

[Laborie and Krivine, 1997] Laborie, P. and Krivine, J. (1997). Automatic generation of chronicles and its application to alarm processing in power distribution systems. In *Eighth International Workshop on Principles of Diagnosis - DX'97*, Mont St. Michel, F.

[Lackinger and Nejdl, 1991] Lackinger, F. and Nejdl, W. (1991). Integrating model-based monitoring and diagnosis of complex dynamic systems. In *Twelfth International Joint Conference on Artificial Intelligence - IJCAI'91*, pages 2893–2898, Sydney, Australia.

[Lafortune and Chen, 1991] Lafortune, S. and Chen, E. (1991). A relational algebraic approach to the representation and analysis of discrete-event systems. In *American Control Conference*, Boston, MA.

[Lamperti and Pogliano, 1997] Lamperti, G. and Pogliano, P. (1997). Event-based reasoning for short circuit diagnosis in power transmission networks. In *Fifteenth International Joint Conference on Artificial Intelligence - IJCAI'97*, pages 446–451, Nagoya, J.

[Lamperti and Zanella, 1999] Lamperti, G. and Zanella, M. (1999). Diagnosis of discrete-event systems integrating synchronous and asynchronous behavior. In *Tenth International Workshop on Principles of Diagnosis - DX'99*, pages 129–139, Loch Awe, UK.

[Lamperti and Zanella, 2000a] Lamperti, G. and Zanella, M. (2000). Generation of diagnostic knowledge by discrete-event model compilation. In *Seventh International Conference on Knowledge Representation and Reasoning - KR'2000*, pages 333–344, Breckenridge, Colorado.

[Lamperti and Zanella, 2000b] Lamperti, G. and Zanella, M. (2000). Uncertain temporal observations in diagnosis. In *Fourteenth European Conference on Artificial Intelligence - ECAI'2000*, pages 151–155, Berlin, D.

[Lamperti and Zanella, 2001] Lamperti, G. and Zanella, M. (2001). Principles of distributed diagnosis of discrete-event systems. In *Twelfth International Workshop on Principles of Diagnosis - DX'01*, pages 95–102, San Sicario, I.

[Lamperti and Zanella, 2002] Lamperti, G. and Zanella, M. (2002). Diagnosis of discrete-event systems from uncertain temporal observations. *Artificial Intelligence*, 137(1–2):91–163.

[Lamperti and Zanella, 2003] Lamperti, G. and Zanella, M. (2003). EDEN: An intelligent software environment for diagnosis of discrete-event systems. *Applied Intelligence - The International Journal of Artificial Intelligence, Neural Networks, and Complex Problem-Solving Technologies*, 18:55–77.

[Lin, 1994] Lin, F. (1994). Diagnosability of discrete-event systems and its applications. *Discrete Event Dynamical Systems*, 4:197–212.

[Lunze, 1999] Lunze, J. (1999). Discrete-event modelling and diagnosis of quantised dynamical systems. In *Tenth International Workshop on Principles of Diagnosis – DX'99*, pages 274–281, Loch Awe, UK.

[Lunze, 2000a] Lunze, J. (2000). Diagnosis of quantised systems. In *Fourth IFAC Symposium on Fault Detection, Supervision and Safety for Technical Processes – SAFEPROCESS'2000*, pages 28–39, Budapest, U.

[Lunze, 2000b] Lunze, J. (2000). Diagnosis of quantized systems based on timed discrete-event model. *IEEE Transactions on Systems, Man, and Cybernetics – Part A: Systems and Humans*, 30(3):322–335.

[McIlraith, 1997] McIlraith, S. (1997). Explanatory diagnosis: conjecturing actions to explain observations. In *Eighth International Workshop on Principles of Diagnosis – DX'97*, Mont St. Michel, F.

[McIlraith, 1998] McIlraith, S. (1998). Explanatory diagnosis: conjecturing actions to explain observations. In *Sixth International Conference on Principles of Knowledge Representation and Reasoning – KR'98*, pages 167–177, Trento, I.

[Mozetič, 1992] Mozetič, I. (1992). Model-based diagnosis: An overview. In Marik, V., Stepankova, O., and Trappl, R., editors, *Advanced topics in Artificial Intelligence*, pages 419–430. Springer-Verlag.

[Nejdl and Gamper, 1994] Nejdl, W. and Gamper, J. (1994). Harnessing the power of temporal abstractions in modelbased diagnosis of dynamic systems. In *Eleventh European Conference on Artificial Intelligence – ECAI'94*, pages 667–671, Amsterdam, NL.

[Ng, 1991] Ng, H. (1991). Model-based, multiple-fault diagnosis of dynamic, continuous physical devices. *IEEE Expert*, 6(6):38–43.

[Pan, 1984] Pan, J. Y.-C. (1984). Qualitative reasoning with deep-level mechanism models for diagnosis of mechanism failures. In *First IEEE Conference on AI Applications*, pages 295–301, Denver, CO.

[Pencolé, 2000] Pencolé, Y. (2000). Decentralized diagnoser approach: application to telecommunication networks. In *Eleventh International Workshop on Principles of Diagnosis – DX'00*, pages 185–192, Morelia, MX.

[Pencolé et al., 2001] Pencolé, Y., Cordier, M., and Rozé, L. (2001). Incremental decentralized diagnosis approach for the supervision of a telecommunication network. In *Twelfth International Workshop on Principles of Diagnosis – DX'01*, pages 151–158, San Sicario, I.

[Poole, 1989] Poole, D. (1989). Explanation and prediction: An architecture for default and abductive reasoning. *Computational Intelligence*, 5(2):97–110.

[Poole, 1994] Poole, D. (1994). Representing diagnosis knowledge. *Annals of Mathematics and Artificial Intelligence*, 11:33–50.

[Raiman et al., 1991] Raiman, O., de Kleer, J., Saraswat, V., and Shirley, M. (1991). Characterizing non-intermittent faults. In *Ninth National Conference on Artificial Intelligence – AAAI'91*, pages 849–854, Anaheim, CA.

[Reiter, 1987] Reiter, R. (1987). A theory of diagnosis from first principles. *Artificial Intelligence*, 32(1):57–95.

[Rozé, 1997] Rozé, L. (1997). Supervision of telecommunication network: a diagnoser approach. In *Eighth International Workshop on Principles of Diagnosis – DX'97*, Mont St. Michel, F.

[Sampath et al., 1998] Sampath, M., Lafortune, S., and Teneketzis, D. (1998). Active diagnosis of discrete-event systems. *IEEE Transactions on Automatic Control*, 43(7):908–929.

[Sampath et al., 1995] Sampath, M., Sengupta, R., Lafortune, S., Sinnamohideen, K., and Teneketzis, D. (1995). Diagnosability of discrete-event systems. *IEEE Transactions on Automatic Control*, 40(9):1555–1575.

[Sampath et al., 1996] Sampath, M., Sengupta, R., Lafortune, S., Sinnamohideen, K., and Teneketzis, D. (1996). Failure diagnosis using discrete-event models. *IEEE Transactions on Control Systems Technology*, 4(2):105–124.

[Schiller et al., 2000] Schiller, F., Schröder, J., and Lunze, J. (2000). Diagnosis of transient faults in quantised systems. In *Fourth IFAC Symposium on Fault Detection, Supervision and Safety for Technical Processes – SAFEPROCESS'2000*, pages 1174–1179, Budapest, U.

[Schullerus and Krebs, 2001] Schullerus, G. and Krebs, V. (2001). Diagnosis of a class of discrete-event systems based on parameter estimation of a modular algebraic model. In *Twelfth International Workshop on Principles of Diagnosis – DX'01*, pages 189–196, San Sicario, I.

[Shannon, 1948] Shannon, C. E. (1948). A mathematical theory of communication. *Bell System Technical Journal*, 27:379–623.

[Shortliffe, 1976] Shortliffe, E. (1976). *Computer-based medical consultations: MYCIN*. American Elsevier, New York, NY.

[Srinivasan and Jafari, 1993] Srinivasan, V. and Jafari, M. (1993). Fault detection/monitoring using timed petri nets. *IEEE Transactions on Systems, Man, and Cybernetics*, 23.

[Struss, 1992] Struss, P. (1992). What's in sd? towards a theory of modeling in diagnosis. In Hamscher, W., Console, L., and de Kleer, J., editors, *Readings in Model-Based Diagnosis*. Morgan Kaufmann.

[Struss, 1997] Struss, P. (1997). Fundamentals of model-based diagnosis of dynamic systems. In *Fifteenth International Joint Conference on Artificial Intelligence – IJCAI'97*, pages 480–485, Nagoya, J.

[Struss and Dressler, 1989] Struss, P. and Dressler, O. (1989). Physical negation-integrating fault models into the general diagnostic engine. In *Eleventh International Joint Conference on Artificial Intelligence – IJCAI'89*, pages 1318–1323, Detroit, Michigan.

[Ullman and Widom, 1997] Ullman, J. and Widom, J. (1997). *A First Course in Database Systems*. Prentice Hall.

[Viswanadham and Johnson, 1988] Viswanadham, N. and Johnson, T. (1988). Fault detection and diagnosis of automated manufacturing systems. In *Twenty seventh IEEE Conference on Decision and Control – CDC'88*, pages 2301–2306.

[Williams, 1984] Williams, B. (1984). Qualitative analysis of mos circuits. *Artificial Intelligence*, 24(1):281–346.

Index